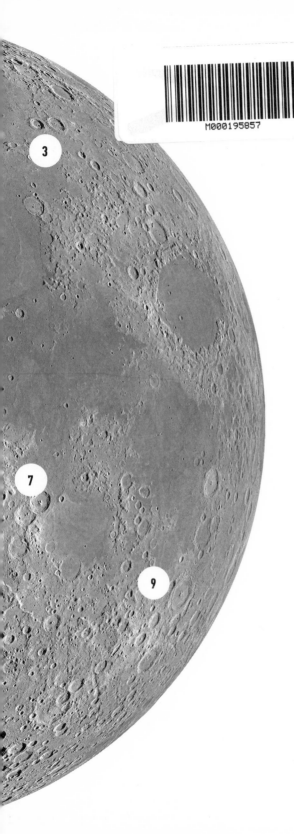

Still as Bright

Still as Bright

*An Illuminating History of the Moon
from Antiquity to Tomorrow*

CHRISTOPHER COKINOS

PEGASUS BOOKS

NEW YORK LONDON

STILL AS BRIGHT

Pegasus Books, Ltd.
148 West 37th Street, 13th Floor
New York, NY 10018

First Pegasus Books cloth edition April 2024

Interior design by Maria Fernandez

Library of Congress Cataloging-in-Publication Data is available.

ISBN: 978-1-63936-569-2

10 9 8 7 6 5 4 3 2 1

Printed in the United States of America
Distributed by Simon & Schuster
www.pegasusbooks.com

For my father, who gave me a telescope

CONTENTS

So, we'll go no more a roving
So late into the night,
Though the heart be still as loving,
And the moon be still as bright.
—Lord Byron, "So We'll Go No More a Roving"

The poet who walks by moonlight is conscious of a tide in his thought
which is to be referred to lunar influence. I will endeavor to separate the
tide in my thoughts from the current distractions of the day. I would
warn my hearers that they must not try my thoughts by a daylight
standard, but endeavor to realize that I speak out of the night.
—Henry David Thoreau, "Night and Moonlight"

To gaze upward at the Moon . . . is to look deeply into a well of history.
—Scott L. Montgomery, *The Moon and the Western Imagination*

PROLOGUE

E ach month, when the Moon returns at dusk, lit and waxing, the first "sea" I routinely observe is Mare Crisium—the Sea of Crises—and the first "lake" I look for is the foreshortened Lacus Temporis—the Lake of Time. That seems just so. Mare Crisium is a huge plain of solid lava close to the near side's eastern limb—it looks elliptical, you can see it with your own eyes—and in the telescope, Crisium is ringed by headlands, dotted with craters, and rippled with low ridges like cast-aside thoughts. Lacus Temporis, despite its weighty name, is a banal patch of gray rock, though shaped, appropriately, a bit like an hourglass.

The Moon is always there, at least in our lives, lit or dark, seen or not, a frigid or scalding globe of pits and rubble, of peaks and scars, our helpful companion in gravity and evolution, a symbol, a surface, a night-light during its days-long lunar morning and a silent siren for nocturnal predators, for mythic werewolves, lovers, geologists, and soon again, the astronauts. For me, until recently, as for other stargazers, the Moon was a bane that bleached out fainter gray-green galaxies and nebulae that one might otherwise see in the eyepieces of backyard telescopes or on the computer screens of professional observatories.

Not long ago, I found myself questioning, if not everything, then quite a bit—mortality, happiness, profession. I'd run my hand along the spines of philosophy titles in a used bookstore in Tucson, Arizona, tip one out, and purchase it to read at home in a certain silence. I'd lie down to meditate, trying to focus on breath instead of mistakes. I'd talk with my wife, Kathe, in a wide, gravelly yard behind our old adobe house beneath mesquites or the giant tamarisk. I'd fallen for that tree the first time we saw this place, a nineteenth-century ruin that Kathe would go on to restore and that would be our home for just two short years of our longer Tucson sojourn.

Too tired, too low, I couldn't bring myself to haul the telescope south of the city where skies darken and the Andromeda Galaxy blossoms enough to be seen with the naked eye. Ever since moving to the desert from four riverside acres in semirural northern Utah, I spent far less time under night skies than I used to—one of several touchstones I'd slowly lost in those clotted, frantic city years. I'd forgotten what mattered most, what Lord Byron called "the heart . . . still as loving."

One night, as Kathe and I sat beside a fire stoked in a battered woodstove—which was once the house's only source of heat, and which we had since moved outside—there it was: bright, unhurried, sliced by light and shadow, the Moon. I had never paid it much attention, despite a childhood smitten with Apollo flights, years when I would tape-record television reports of the last missions and those of Skylab and the Apollo–Soyuz rendezvous. I replayed the news anchors' voices—Walter Cronkite, Jules Bergman—on a cassette player at night in a dark bedroom in our trailer outside of Indianapolis. I was dreamy with imagined escape. Before then, I had sometimes used a three-inch Edmund Scientific reflector my father had given me and, with it, I would dash across the lunar landscape, and dash was right, because the telescope was rickety and hard to use. I caught the Moon in snatches. Too embarrassed to ask for help, I didn't say a word to my father, especially after he left my mother, sister, and me. Somewhere and sometime the telescope also disappeared.

That night in our Tucson backyard, I looked at the Moon with my own eyes and, touched by the shadows of delicate mesquite leaves that it cast, I felt something clear, something inevitable, something sweet without striving. Not a child's silver-spacesuit fantasy, not an adult's grasping for achievement and flattery, but a sober calm and, more, a grounded curiosity. *I could go there,* I thought. *I could watch and learn.* It felt like an epiphany, seeing that Moon—waxing past its "half-Moon" phase—brimming with potential, shining peacefully above ordinary darkness.

By then, I had learned that in a few weeks it would be the fiftieth anniversary of a forgotten, once-heralded photograph of the lunar crater named for Copernicus. The photo had been snapped by an orbiting probe sent to map the dangerous surface for safe landing sites before the astronauts descended. I was intrigued. I've always been interested in forgotten anniversaries. They tie me to history's chancy sweep, and they're a form of memento mori. They humble me and, somehow, make me feel loved. I decided to find that photograph, to learn why it had garnered so much attention. This small adventure in discovering something lost would be, like the moonlit night with Kathe, the beginning of my backyard journey to the Moon and its many histories. It would make my skin tingle. For the first time in a very long time, I would be part of something bigger than myself, and every month it would light my way.

On November 24, 1966, Lunar Orbiter 2 took the photograph with a dual-camera system, using a "velocity-height sensor" to adjust the film for orbital motion. (Some sources say the photograph was taken on November 23, another on November 28.) The craft then developed the image in its own darkroom bay, scanned the negative, and transmitted the data five days later to a world that would be so stunned by this first-ever otherworldly close-up, the image would be called the "picture of the century." Despite this, as the years passed, almost no one would remember the photo.

This was the decade of NASA's quest for the Moon, when, as a substitute for other competitions with the Communist world, including war itself, President John F. Kennedy declared in 1961 that we would, by the end of the decade, send Americans to the Moon and return them safely home. At the time of his challenge, America had all of fifteen minutes of suborbital space experience.

It's not that Kennedy particularly cared about space or for going to the Moon or for what the Moon could tell us about ourselves in the cosmos—which, it turns out, is still quite a lot; it's rather that he, like most Americans, had been stung by a string of Soviet space firsts, including the first satellite, Sputnik, in 1957, and the first-ever human flight by Yuri Gagarin four years later.

Crucial to the planned Moon landings were such automated efforts as Lunar Orbiter 2 and others in that series, along with Ranger's designed crash landings—photo after photo in dizzying closer then closer then closer then *bam!* sequences—as well as the more sophisticated Surveyor missions, which soft-landed and trenched lunar soil. These were, as a 1960s NASA publication put it, "blazing a trail for man to follow." Humanity aimed for the Moon and surely next, for Mars. (Or, at least, select parts of humanity were so aimed, namely Soviets and Americans—perhaps even just some Americans, for, as Gil Scott-Heron would later rap, "Whitey on the Moon.") To a child as to the engineer, all seemed possible. Amid nuclear brinkmanship, assassinations, and riots, space was aspiration, and, as a child, I sensed that too. I was only six when Apollo 11 landed, on the tail end of duck-and-cover drills and wondering about flashes of light. Lightning? Or an atomic blast targeting Eli Lilly and Company, where my father worked? I sensed the troubles of the wider world. There were troubles in the living room too.

Space was the inchoate hope of something better than an angry, broken family living in a postdivorce trailer with my shell-shocked mom and bratty sister, the cupboards nearly bare at the end of each month. When we moved into a decent apartment close enough to the junior high and high school that I could walk to them, I nurtured dreams of flight. I joined the Civil Air Patrol, which, on one memorable weekend, meant riding in a KC-135 Stratotanker with fellow cadets as we nestled next to the boom operator, watching a B-52 bomber refuel just yards away from the tanker, two mated metal insects thousands of feet above the fields and forests of the otherwise stupefying Midwest. Though we were not well-off, the school district was, and the high school had a flight simulator that allowed me to pass ground school, my first step to the sky. Model airplanes hung from my bedroom ceiling, and along with the requisite

Farrah Fawcett poster, I tacked up photos of astronauts and their needlelike T-38 Talons, still my favorite plane. After joining the Air Force, I would, of course, become an astronaut. And I looked at my View-Master frames of real and fictional spacecraft, those bright, backlit dioramas that tricked the eye into seeing three-worldly, colorful dimensions contained within a light box. This would be my future. I held it in my hands.

To this day, I remember Miss Hawk literally pulling me out of advanced algebra, though I don't know why—failure at a problem or screwing around or both?—and by the next class I was in remedial math, resigned, overnight, to never having wings pinned on a uniform. Back then, I gave up easily. I gave up not only my ambition to fly but also the sky itself, not returning to it till my late twenties when I sought starry vistas on my own, away from a fading first marriage. With a small spotting scope, I concentrated on birds, Jupiter, Saturn, and tiny, spindly galaxies, everything but the Moon, which was both too easy—it's not hard to find—and too confusing—there are, after all, a lot of holes in it. Even as I graduated to a much larger telescope, this neglect continued.

Now I know what I was missing: an entire world and its epic story and how some of us can travel there without ever leaving the Earth.

It was a grace note, the Copernicus photograph, taken between shots of possible Apollo landing sites in order to use up film and keep the system from jamming. Douglas Lloyd of Bellcomm, Inc., a NASA contractor, suggested the shot. Lloyd knew the crater photo would be spectacular because the crater is spectacular. A member of the probe's planning team would tell me that only one person objected to taking the photo as a "waste of film, but the rest of us . . . agreed that it would be useful." That was rare consensus. The attitude that only engineering mattered marked the early years of a space agency bent on just completing missions safely. It took a while for NASA to embrace the power of images to inspire the public. According to historian Roger Launius,

"Mercury project managers . . . worried that the small capsule [the first US crewed vehicle] could not accommodate a camera and its accouterments." But, encouraged by one NASA official, John Glenn took a modified drugstore camera with him on America's first orbital flight. Apollo 8 commander Frank Borman initially loathed the required photography and television portions of that first expedition to reach the Moon, the same mission that, ironically, captured the famous color *Earthrise* photo showing blue terra over stark luna. Slowly photography began to link terrestrials to the cosmic sublime.

Visible from Earth through binoculars, the result of a nearly billion-year-old impact, Copernicus is about sixty miles wide, big enough, as some point out, to swallow Rhode Island, and, if you were on the floor of the crater and wanted to get out, you'd have to scale terraced walls more than twelve thousand feet high, higher than most of the mountains in my adopted home state of Utah. Gullies and ridges radiate another sixty miles from the crater's edge, and rays of ejecta—bits and pieces of the Moon excavated and sprayed out by the impact—spread even farther, nearly four hundred miles. In the middle, on the crater floor, peaks were vaulted into place by rebounding, the mostly solid rock acting almost like a liquid to ultimately rise to nearly four thousand feet. From straight above and far away, the crater looks like a giant spotted flower.

Lunar Orbiter 2, however, didn't shoot straight down from a distance. It took its photo from about 125 to 150 miles south of the crater and from twenty-eight miles above—a glancing view. The angle revealed what humans had never seen before: the Moonscape as landscape. A real place. When I first saw the photo in different publications, in print and online, I gaped. I was also confused. There were different crops of the picture, and I set out one morning in Tucson to meet someone who could help me decipher what I was seeing.

The legendary planetary scientist William K. Hartmann—whose lunar research I'd encounter more thoroughly later—looked with me at the once-iconic image, along with maps and other photographs. We had to do a kind of flyover before I could finish a newspaper essay about the picture. I had climbed the steps to Bill's airy Tucson study, replete with many of his own fine paintings of planets and deserts, copies of his books, and canvases from the classic space artist Chesley Bonestell, whose dramatic lunar landscapes would challenge me to think about the relationship between desire and fact. Classical music CDs were neatly stacked, and the room had a bright sense of purpose and belonging. Gray-haired, smiling, altogether welcoming and energetic, Bill was the perfect guide for my small journey. Long before his scientific career, he had "grown up looking at the Moon through my telescope in the backyard," he remembers, "and making drawings of the craters." He'd long had what I was now beginning to seek: "A personal relationship with the Moon." Outside, I could hear wind clattering in the neighborhood palm trees, a sound I'd never quite grown accustomed to, but being in Bill's presence and, by extension, the Moon's, I felt at ease.

What confused us was that a version of the photograph from a NASA booklet had cut from the foreground the dark, near crater rim and the adjacent keyhole-shaped crater Fauth, features included in the image's initial publication in places like the *Los Angeles Times* and the *New York Times*, both of which put the photo on their front pages. I just couldn't tell what was the rim and what was the crater floor. I felt lost even as the chaos seemed like terrain I could, even should, enter.

Once we found a version online with the entire uncropped foreground, what appeared to be small rises were obviously the huge central peaks of the crater floor, those several mountains in shadow—"steep, jagged, irregular," science writer Walter Sullivan had called them. Beyond, the crater floor was flat on one side and hummocky elsewhere, blending into a jumble of slumping, daunting terraces and the distant rest of the rugged rim, a picture spanning some seventeen very real miles of the Moon from left to right and some 150 miles from the peaks to a promontory beyond Copernicus. One NASA source told the *Los Angeles Times* that "it's almost as though a photographer actually stood inside the crater to take these shots." Indeed, I felt like I was *in* the crater, the blue sky visible from Bill's windows replaced by relentless black.

The picture and others that followed were "important and striking," Hartmann told me, "because that was the first era in which we could begin to see famous lunar features at oblique angles, like looking out an airplane window."

Another world, seat 19A. "Awesome clarity," *Life* said of the photo. As Sullivan rhapsodized, it's like seeing "the western face of the Rocky Mountains in Utah." I smiled when I first read that. No junipers or Douglas firs on the Moon, but there was the same muscularity from two different orogenies. The Moon already had something to do with my own found ground.

At the time, NASA's Martin Swetnick called Lunar Orbiter 2's photo H-162 "one of the great pictures of the century." It was, as astronomy writer Peter Grego would later say, "perhaps the first truly spectacular image returned from space." It even competed for attention with the program's first robotic black-and-white photos of Earth rising behind the lunar limb, which Douglas Lloyd also helped to coordinate. "Inexpressibly desolate and inexpressibly grand," UPI said of the image. *Time* magazine invoked another earthly comparison: "Except for the black sky in the background, the photograph might have been mistaken for a composite of the scenic grandeur of the Grand Canyon and the barren desolation of the Badlands of South Dakota."

When it was beamed back to Earth, the photo's then-unique view made the Moon tangible, approachable in a way it had never been before. It was indeed a place, another world—stranger and more hostile—but one on which we could walk, discover, and maybe live.

Kodak engineer Arthur Cosgrove, a twentysomething stationed at a remote NASA satellite facility in Australia, was one of the first humans to see such images, when the Lunar Orbiters would transmit them oh-so slowly. He remembers cheap Foster's pints at each meal, social isolation, kangaroos, and dangerous outback roads. Mostly, however, he remembers the photographs. It was, he would tell me, "very emotional seeing something never before seen . . . The images are embedded in my mind." He compared the "elation, excitement, and a calm realization that you were equipped to appreciate what you were seeing" to a "philosophical discovery." Closer to home, literally, Kathy Lloyd, Douglas's daughter, recalls the stunning Copernicus photo hanging in their childhood house.

My twenty-first-century encounter with a photo nearly as old as I am, and with some of the personalities and stories behind it, began to bring other things to mind, including how, after much seeking and very little solace, and as I was approaching my later years, the Moon might be for me as the Lunar Orbiter pictures were for Art Cosgrove: a source of elation and calm appreciation and a kind of philosophical discovery, the Moon a wise sage made of basalt and anorthosite.

The Moon has proven to be a patient teacher.

The Moon is a provocation for culture, art, and science. As Scott Montgomery writes in *The Moon and the Western Imagination*, "To gaze upward at the Moon . . . is to look deeply into a well of history." The waters in that well run very deep. The Moon's constant inconstancy—its changeable but regular motions and phases—gave cultures around the world their first sense of time in between days and seasons; indeed, some consider the Moon's utility as a timepiece as not only practical but creative. In primitive hominids, time awareness derived from lunar changes may have sparked their first glimmers of self-consciousness. The mythology and folklore associated with the Moon is so vast that a complete survey would require several scholarly books. I've found creation stories including that of an Amazonian suicide whose head rolls around until it becomes the Moon and I've learned about the recurrent lunar binary of good and evil—when to look and what to say, when to cover windows against Moonlight on the bed. The Moon has reflected our hopes and fears, and for some ancient philosophers, was literally a mirror, sending the Earth's surface back to us.

Science showed otherwise. When Galileo turned his self-made telescope toward the Moon in 1609, he didn't find a perfect polished circle. He inferred terrain from caught light and cast shadow: circular valleys, high mountains, and wide plains that some would think were ocean. The Moon was a globe like the Earth, a world unto itself, and this revelation was part of the great revolution that science brought to understanding where we are in the universe. The reality of the Moon challenged the authority of the Church, as I would discover, and the first Space Race began not long after Galileo's observations, the race to affix names to the Moon's telescopic features for fame and putative immortality. More imaginatively, astronomers, writers, and theologians began to speculate about what creatures might live on the satellite above us, which, if with terrain, surely was teeming with life. From giant trees to slender bipeds, from resilient lichens to migrating insects, the Moon became a harbor for our unquenchable desire for cosmic company, even when this raised thorny questions of science (was there any water on the Moon?) and Western religion (would Jesus be crucified on other worlds?).

The latter is unanswerable, so far as I can tell, but formative lunar science did begin to constrain the Moon's environment; eventually, it was clear the Moon was airless, free of liquid water, and in turn very hot or very cold. Those facts, along with the countless craters of the near side, provoked speculations about the origin of the Moon (was it flung from the Earth where the Pacific now churns?) and the creation of its pockmarked surface (mighty volcanos or a barrage of meteorites?).

Yet even a dead Moon could not stay changeless. I would discover a controversial field called "transient lunar phenomena" (TLP), in which observers have claimed to see colors, clouds, and even lightning on the static Moon. The scientific study of TLP was pioneered by two women astronomers—including one of the first at NASA—who would fight many battles for acceptance and credibility in a largely male profession. Often derided as the province of cranks, TLP have been studied by professional astronomers and even seen by Apollo astronauts, yet its study and history remain polarizing. As the nineteenth-century French astronomer Camille Flammarion wrote, "This charming Moon has undergone in human opinion the vicissitudes of this opinion itself, as if it had been a political personage."

These controversies brought me to a surprising study of subjectivity, personified in the characters of painters and astronauts who committed themselves with various degrees of accuracy and kinship to the depiction of the Moon's actual environment. One of those painters, Alan Bean, was also an astronaut who walked on the Moon and came home to assiduously paint its terrain.

In the years ahead, while I gathered scientific papers, interviews, and books, while I struggled to remember the key features of lunar petrology (something called *KREEP* kept coming up), while I traced the rise and fall of lunar-formation theories, I did something just as important: I *looked*. I looked from one backyard then another and then another, as we changed houses to find our home. I looked from a mountain meadow, too, from a cabin that was a refuge. I became a lunar acolyte devoting long hours at the eyepiece, weather permitting, gazing at that white concretion in the black marl of night. I looked to learn—to experience—the features, the names, and the histories of the Moon's surface, traversing wrinkle ridges as I crossed hardened lava seas, plunging from the thin-lit rims of craters to shadows of black depth then ascending sun-fired peaks taller than anything on Earth. To write the Moon without seeing the Moon—distant exploration magnified, up close, startle-sharp—would be like writing about birds without watching them fly, perch, and sing. It is, for example, one thing to read about the violence of impacts on the surface of the Moon, asteroids vaporizing in their eons of hit-and-run. It is another to bear witness at the telescope, the hum of a highway or the cries of night birds nearby, as the vacuum-silent aftermath reveals itself as an ancient mountain range that was *uplifted in seconds*, like knuckles punching from a grave.

The mountains of the Moon helped create an appetite for mountains on Earth. The sublime—that feeling of wordless grandeur and dissolution and connection and estrangement all at once and brought on by the very powerful and the very large—is in part a legacy of lunar observation. It is to the sublime there and here that I would return, again and again, a wild and joyful breath. It was to the Moon's sublime surface that astronauts descended during my childhood, and the Apollo program's science, long overshadowed by the sheer marvel and achievement of visiting another world, has helped answer questions that stretch back to those cold winter nights when Galileo first spied the epiphanic craters.

Also overshadowed, at least in many of the popular accounts of Apollo, have been the origins of the rocket that first sent humans to the Moon, the

Saturn V, a direct legacy of a weapon built by slave laborers in the secret tunnels of a mountain. More than a dark chapter in the history of space exploration, the complicity and amnesia that has accompanied it raises troubling questions about how we can craft a long tomorrow in space, a human future beginning with the Moon—for that world does, it turns out, have water, lots of it, locked up as ice in permanently shadowed craters at the lunar poles. It's this water that can give us a Moon we might live on. Can we build a humane future in space if we forget the inhumanity at the origin of the Space Age?

As I'd near the end of my pilgrimage, all this would be on my mind, as I traveled to a gorge in New Mexico where astronauts simulated exploration as if they were on the Moon, as I looked at an original lunar atlas from the seventeenth century, and as I held a chunk of anorthosite—an important mineral on the Moon and one as bright as gratitude—right by a road cut in California. That exposure was where Apollo astronauts also trained, and soon after I saw it, I'd walk into a cathedral: the dome of the venerable sixty-inch reflector at Mount Wilson Observatory, where I'd experience the Moon as I never had before. I'd search hard for one of my grails: the sliver-thin channel of Rima Alpes, where lava flowed through a wide graben, the valley of which I can spy from my backyard. It was that delicate rille I wanted to see, an epilogue to a journey, though I know: I'll never stop looking.

Yet in all those scenes, all those adventures and their revelations to come, I could not know of the death of my father. Nor how I'd restore a version of the same telescope he had given me decades ago. In the dark surrounding the fire glow that night, I decided to set out on my lunar journey, the Moonlight, shining between mesquite leaves, seemed like a gesture of forgiveness. I got up. It was time to have my first-ever good, long look at the Moon three decades after becoming a stargazer. I set up the telescope, a long and bulky tube that one lifts and lowers to nestle in a simple cradle. I brushed off a bit of dust from the black metal.

Kathe added a branch of mesquite to the fire. It spat in protest. Inside, our three cats—Zinc, named for the metal; Burchfield, for the artist; and Shackleton, for the explorer—sprawled in their drowsy ways. Electronic screens were off. A mile away, cars and trucks whooshed along I-10. Our usually noisy neighbors were unaccountably quiet. It was a cool November evening, the air still as glass. "Steady seeing" it's called, when the agitations of air, its bubbles of burbling heat, its invisible currents, are smoothed out like sheets in the morning.

I knew enough to know where Copernicus would be. Through the smaller finderscope, the Moon was in the crosshairs, easy, and there was the blot that was Copernicus. I shifted to the main telescope's ten-inch aperture, dropped in a zoom eyepiece set at 7.5 millimeters, giving me 166-power magnification, equivalent to orbiting about fifteen hundred miles from the lunar surface. I took off my glasses, setting them on a green metal table by my study door, put down my wine, and looked up then, bent to the eyepiece.

Bright blur! I ratcheted the focus with my left hand, my right resting on then hovering above the optical tube. Then—a sharpness I'd not seen before. Preternatural clarity. Edges and shadows. I swept myself above the craters of the lunar highlands, over solid lunar seas, and moved deliberately along a white ray of ejecta I knew led back to Copernicus, points on a compass rose. I pushed the scope with my hands and nudged it with my body—no computer here, here the body is an instrument of direction, a living sextant—and I let go of the scope for vibrations to settle then looked again.

From 250,000 miles away, magnified, intensified, Copernicus was shockingly exact: low-angled hummocks, like a solidified froth of child-crafted mud, mounting from the plains to the jagged circular rim, one side of the rim in shadow, the other of white rock in sun, and a gray and black and arctic-white cascade of amphitheater walls and landslides having fallen in other ages, all dropping to the crater floor nubbly with hills and the shadows of hills and, in the middle, the sun-bright central mountain peaks shining like a row of spotlights. Astronomers once proclaimed the crater "the Queen of the Moon." That evening the air was clear of haze—"transparent"—and my telescope, though it had been little used of late, had its well-fashioned mirrors aligned. The view was adamantine, still one of the best I've ever had.

The cold metal telescope, wine, my wife singing some tune I liked but could not place, the smell of woodsmoke, and a light in the kitchen spilling into the backyard. I've had more backyards with telescopes than without. It is from such private places that we can best make the voyage from the Earth to the Moon. In that backyard, with its lovely tamarisk, began the rest of my life, such evenings there and elsewhere, lunar happy hours, studious staring. I'd even enroll in observation-training programs for amateur Moongazers so as to know that surface like the Utah trails I know by heart. I needed uplift, I needed solace, and I found both in the fugal discord of the Moon's terrain. It spread bright beneath my eye, shattering in its calm. I haven't stopped looking or learning since. Even on cloudy nights, I explore—traversing the surface with images from Lunar Orbiter, Apollo, and the Lunar Reconnaissance Orbiter, whose high-definition photographs and maps allow viewers to rocket across this other world. A friend who is a scientist studying the Moon says those images let her "get . . . lost on the fractal nature of the surface craters, finding the little rocks and huge boulders thrown out by craters, and literally seeing the footprints of robots and humans."

There are few moments we are given when we know things have changed. That night was one of them. The Moon would become as familiar and as powerful as the other gifts: a morning cup of coffee with Kathe, a cat on my lap, a bike ride, a book, a conversation in which shortcoming becomes kindness, the smell of cilantro chopped in a green fresh mound on a wooden countertop, a hard backpack up to a cold alpine lake.

The British astronomer Richard Baum, writing in the introduction to *A Portfolio of Lunar Drawings* by Harold Hill—one of the most obsessive and talented amateur lunar observers in history—says,

> For, in the shadows and highlights, we perceive something more
> than harsh reality. An indefinable something, caught on the edge
> of things, that is as revelatory of the man as the moonworld itself.

Intangible it may be, but it cloaks the scars of a violent past with a mystique as telling as any physical discovery, reminding us of the excitement which awaits those who embark on a personal journey of discovery and exploration.

The Sun shining on that surface throws not just crater rims into relief but the deepest questions about purpose, time, and acceptance, such questions that tend to rise from the plains in the second halves of our lives and that encourage exploration of why humans quest at all. Although astronomer Fred Price claims in *The Moon Observer's Handbook* that "personal cares and problems are best not brought to the telescope," those sent me to the Moon, where I could fly with my eyepieces and walk with my maps, where I would get lost less often.

Still as Bright is about the Moon, about its sublimity and science and history, its meanings, personal and cultural, about how it can be for each of us a country of curiosity, discovery, and contented adventure, as it is for me. A few days after Mare Crisium and Lacus Temporis are first visible, the line between light and dark passes over other lunar landforms: the seas named for Serenity, Tranquility, and Fecundity.

CHAPTER ONE

The Invention of Time

I magine the hand that held the bone and the hand that held the stone, a sharp edge pressed into the palm-sized piece of antler. Perhaps it's morning beneath a rough rock overhang, a shelter now called Abri Blanchard in what is now the Dordogne in southern France. Mist hangs over the Castel Merle Valley, though, if it has a name, it isn't that one. Ghostlike aurochs feed, the tips of bulls' horns piercing the white, wet veil like gray stars. Owls sleep on birch branches. What sound does the incision make?

It is twenty-eight thousand years ago, and the strong hand with the antler and the tool belongs to a Cro-Magnon. The tip, the edge, whatever the sharper surface is, scritches and scratches and presses deeper until tiny marks in the ovaloid bone become curvilinear, then—over how many mornings?— the marks resemble petite fattened blobs stippled inside their boundaries. Some look like small black ponds. But what these hands were making was not, it seems, symbolic of the Earth.

In an illustration, you can see plainly what no one had seen before until the 1960s and 1970s when the writer and independent scholar Alexander Marshack studied the bone. He proposed, in academic and popular work—and to no small controversy—that the figures on the bone were neither random nor practice nor some terrestrial image. They were the Moon or, rather, the

Moon waxing and waning across the sky of bone, so many iterations of the Moon, nearly seventy marks—I'm looking at a drawing of it right now—and to Marshack they represented two lunar months or "lunations," with the waning and Moonless nights set to the right, the Full Moons to the left, and the middle phases in between.

When he began his quest to understand prehistoric, preagricultural notations on relics like that bone, Marshack had been working on two other book projects—one about the Space Age and one, it happens, about the Moon. The former he finished, the latter, he did not. Marshack had become obsessed with how similar humans are, whether they're astronauts or, in his dated phrasing, "extremely primitive natives" or "starving farmers I had seen in India." We rightly cringe at these colonial descriptions, but Marshack was at least trying to find some skeins of inheritance and kinship across what he called our "time-factored" thought. He was looking for the trajectories his Western Space Age interviewees could not fathom: from the Moon on bone to humans on the Moon.

"Not one person I met," he wrote, "knew clearly why we were going into space or how it had come about." Early in his study *The Roots of Civilization*, Marshack suggested that the path from the prehistoric to modernity began at least as far back as cultures that had lunar calendars, from the Hindi to the Chinese, among others. Temporal awareness and timely organization had originated, however, far earlier than had been suspected, he thought. For Marshack, how we noticed the Moon is how we became ourselves.

He examined museum bones with a toy microscope and, over coffee, stared at reproductions of marks on a 6,500-year-old bone found near the headwaters of the Nile, a significantly later object than the Abri Blanchard bone. The marks appeared to be notations of the lunar cycle, in which the Moon rises, grows brighter, reaches full, lessens, and disappears for a short time (the "New Moon"), then returns like a lit whisper to dusk.

The words *Moon* and *month* share, of course, the same root. Language itself embodies the closeness. The interval between New Moons is twenty-nine-and-a-half days, the so-called synodic month. The sidereal month is the amount of time it takes for the Moon to complete an orbit around the Earth as seen against background stars; this is nearly twenty-eight days long. Marking phases and noting where the Moon was in the sky—farther to the south in this season?—would have introduced not only regularity but change: ordered change.

Marshack visited the National Museum of Antiquities near Paris—today called the Museum of National Archaeology—which he described as "the ugly, heavy-stoned repository of France's pre-Christian past," an old-school archive with "drab, poorly-lit side rooms . . . badly displayed cabinets without electricity . . . colorless, almost meaningless accumulations of Upper Paleolithic materials, crowded under glass with their aged yellowing labels." Arriving there gave him "the sudden chill feeling of an intruder in an abandoned graveyard. There was a huge silence in the musty air of the high-ceilinged stone chamber."

In such unpromising circumstances, Marshack found the Abri Blanchard bone, which, after study, he asserted was a Cro-Magnon lunar calendar. Other peoples around the world were then making similar timepieces, he argued, and there is speculation that such observational knowledge of the Moon's phases and sky-movements might even have been used by Neanderthals.

Yet this was more than a calendar, this antler with notches—more than a way, as we say, to "keep" time. It was the beginning of time—the beginning of our awareness of beginnings, middles, ends. Watching the Moon and recording its passage was, for Marshack, the fountainhead of complex time cognition and, so, the origin of hominid self-consciousness. The very act of observational recording—bringing together synecdoche (marks on a bone representing a changing, moving, eventually predictable light in the sky) and measurement (the marks are a sequence on bone that signify this passage in units)—was the, or at least one of the, acts that gave us ourselves: to be aware of time is to be aware of being a self in a larger reality. Jungian scholar Jules Cashford speculates that it was "women, calculating the timing of their menstrual cycles from Moon to Moon, [who] made the first reckoning of time." Such forms of self-in-world awareness helped lead our ancestors to grasp the arc of life, from birth to death.

The longer iterations of the Moon, and daily, the Sun, extended from animal sense to rational capture. If we could measure on bone and predict in our minds and communicate this, then we were inventing something other than the moment. That these pasts and futures looked largely the same still gave time a kind of cyclical texture for most of our history. Indeed, until Christianity; until the lockstep of time with economics; and until, especially, the Industrial Revolution, the temporal zone that humans, animals, gods, and the world occupied regarded the past, present, and future not as an arrow but a circle. That cyclical, everlasting sense of experience was one way we squared physical death with time. The Moon died each month. It came back. Why not us?

This originating temporal awareness and the beliefs engendered by our first attempts to understand the forces around us were epic and truly admirable. They were Moonlit. We come from there, and our first complex timekeeping—lunar timekeeping—began thousands and thousands of years ago as an empirical act, as a kind of science.

☾

So my Cro-Magnon ancestor was a scientist. And here I am, standing in a dusty Tucson yard, a sightseer. As I think about the Abri Blanchard lunar timepiece, I feel the sweep of time and am easily distracted, wowed by ceaseless views in my telescope. Yet the kinship I feel with my prehistoric Moonwatcher is leavened with chagrin. Moonwatcher studied. I glance. At the telescope, I cannot keep my eyes still. I try to see it all and so see almost nothing. While Moonwatcher *watched the Moon*, I just find it up there and take in what I can, my sightings a paraphrase of John Milton: *Chaos, the vast immeasurable abyss, / Outrageous as a sea, bright, wasteful, wild*. Despite the weaving of the Moon's motions in the sky and the reasons for its changing phases, or, perhaps because of them, for a while I avoid their study. It is the world I seek. When the Moon is up, my telescope is out, and I look at its surface, rather too effortlessly. For me, the Moon is inevitably a place, and I look at it as such: telescope world,

Apollo world, future world. But whatever Moonwatcher thought it was, its rhythms of light and passage forced consideration.

The light arrives and it changes. The light grows and it dies. The light moves across the sky in many ways. Moonwatcher would not have named it as we do, but saw the same thin "young" crescent low in the west at dusk—often so low in the sun-haze it can be hard to spy—then appearing higher and brighter as days and nights progress—waxing, we say—till the Moon is half-lit, half the size it would become in a few more days, what today we call First Quarter, meaning the first quarter of the Moon's phases for the month. It thickens, growing gibbous, growing older, like a fat bull, until one evening it rises as a circle of light, seeming larger than it will higher in the sky, like magic, and this Full Moon presages its waning, another two weeks of lesser light, rising about fifty minutes later each evening until the fading light leaves another crescent, facing the other direction, and then one night it is gone: the dark of the Moon, the New Moon, a time, I will come to learn, that many feared.

Whatever all that was, it was not a day. It was not a season. It was the bridging time, the in-between, repeated seemingly without end. This Moonwatcher notched on the Abri Blanchard bone, not knowing the reasons why the Moon's shape varied. We do. Phases result from the position of the Moon around the Earth at angles relative to the distant Sun. For example, at Full Moon, the much larger Sun is directly "behind" the smaller Earth, illuminating the Moon; the Moon also orbits the Earth at a five-degree inclination from the plane in which the Earth orbits the Sun.

Did Moonwatcher notice that the blue spots on the Moon—the darker basalt "seas" whose names I would chant for days and weeks to remember—never change position, though the Moon tilts its face at different times? You can see this with your own eyes. It's the same "face" we always see, not because the Moon doesn't rotate on its axis but because that rotation matches the rate at which the Moon it orbits the Earth. There is a near side and a far side, but both are sunstruck each month. There is no dark side, Pink Floyd notwithstanding.

This would have been beyond Moonwatcher's ken—especially Pink Floyd—much as the changes in the Moon's altitude and its position of rising and setting were his mastery. Here, Moonwatcher bests me. I make no academic study of this, skipping any explanatory pages and charts in my growing lunar library, working instead on the Moon as map. Slowly, on my own, stepping in one backyard then others to come, I'll begin to register the monthly and seasonal variations of the Moon's height and location, driven largely by guesstimating its visibility on any given evening in a city with roofs and trees, a matter of leisure for me. For Moonwatcher such variations had real significance—life-or-death consequences regarding spring, summer, autumn, and winter, the ebb and flow of game and fruits and nuts and blossoms, sustenance for the clan. To watch the Moon and understand its journey was to feed the belly and to plan for privation.

I look at a chart that shows a kind of triple helix. Just as the Sun's position and apparent trajectory vary by season, so too do the Moon's. In the northern hemisphere, for example, the First Quarter Moon is highest in March—a beautiful time to observe it, I find—then descends in its monthly appearances to an ebb in late summer and early fall. Then it climbs again. The Last Quarter Moon traces the opposite path, while the Full Moon takes the middle course. The Full Moon is highest in winter—lending sparkle to snow—just when the Sun (and New Moon) are lowest on the horizon.

The Moon's motions—there are even more that I'll encounter—have a name: lunar theory. And, despite my primer, it is so complicated that even as early as 1835 the astronomer Hugh Godfray complained, "Of all the celestial bodies whose motions have formed the subject of the investigations of astronomers, the Moon has always been regarded as that which presents the greatest difficulties, on account of the number of inequalities to which it is subjected." Those "inequalities" are dizzying. The Sun exerts eight perturbations on the position of the Moon, including such things as "revolution of the line of apsides" and "regression of the nodes." There are one hundred more such effects emanating even from as far away as Venus and Jupiter. Tackling the subject have been figures as brilliant as Sir Isaac Newton and the forgotten, diligent Ernest Brown, whose 1896 work on lunar theory was so good that NASA used it for navigation during the Apollo missions. Brown and a coworker spent some nine thousand hours on lunar theory, with upward of five million digits involved, according to lunar expert Robert Garfinkle in his three-volume *Luna Cognita* reference work.

I'm not up to that. Neither was Moonwatcher. I begin to rely on websites that tell me when and where the Moon will rise and set and what main features will be visible on any given evening or morning. Even so, I'm often befuddled, looking in the wrong place or being confused by the orientation of the Moon's face.

And I have trouble remembering the names of things. Over the years, I'll need less help, but for now, I touch the names of the maria on a foldout, laminated map and say them aloud, concentrating on the ten most important. It takes a while before I stop confusing Mare Nubium with Mare Nectaris. The maria, all these lava flats, are the most prominent landmarks on the visible face of the Moon, and with their help I begin to locate and name several major craters. One of the most spectacular is part of a four-crater sequence called the Great Eastern Chain, best visible on the third day after the New Moon. Petavius is unmistakable. That one I learn quickly. It's a huge walled crater—115 miles across—with a convex floor a thousand feet higher at center than at its edges, and it virtually snarls with a complex of mountain peaks, more than a mile high. The crater's many cracks and fissures are called Rimae Petavius (*rima*, singular, means rille). The most visible rille, a graben, is a huge swath that looks like a dangling long hand from an old analog watch. It's an antique hanging in a shop window. Petavius is a broken clock. Appropriately, it's named for a French theologian obsessed with historical chronology. One of his books was called *The History of the World: Or, an Account of Time.*

Water clocks, sundials, sand-filled hourglasses, pendula, clock towers, pocket watches, pocket sundials, wristwatches, cell phones, those old time-and-weather recordings you could call in whatever strange city you'd landed in, bank-sign digital clocks rotating like wind vanes over strip malls, windup alarm clocks with those big bell ringers on top, unfathomable atomic clocks: What if it all began with the bone of a dead animal in the hairy hands of a living human silvered with Moonlight then lit at dawn?

On mountain trails, I'll sometimes pick up a cow skull or find the ribs of a deer. I try to fathom the originating impulse to slice bone with marks that *mean something*. To organize passage. I might scratch a bit, careful with my blade, for I am a clumsy biped. I'll just feel, just let register, the weight of the bone—so light—and the force of an edge—carefully muscled. Then, to avoid slicing my fingers—I once nicked off part of a thumb making a model airplane as a boy—I'll set the bone back among vines of Virgin's Bower or a hillside of Mule's Ear. Once, as a guest at an Idaho physics camp for high-school teachers, I threw an atlatl. It flew like a toilet-paper roll batted by a cat. The Stone Age would not have humored me.

Ernest Becker, in his monumental book *The Denial of Death*, says that "each thing, in order to deliciously expand, is forever gobbling up others." We are predator and prey. We hunt, even if only online or at the aisles of the grocery, and we flinch at shadows that cross us quick, duck before the loud crash ceases, grip armrests of the airplane seat, watch seven thousand deaths in eight years on a TV show, see the oncoming car intercepting our path while we say *no no no*. We recoil at the dark twitch of action at the spooky edge of sight. We eat and sleep and wake and most of us live afraid of dying.

Write psychologists Sheldon Solomon, Jeff Greenberg, and Tom Pyszc-zynski: "Only we humans are, as far as anyone knows, *aware of ourselves*, as existing in a particular time and place." These and other such researchers form a school of social psychology called Terror Management Theory. It proposes that if we "take all the cultural trappings away . . . we are all just generic creatures barraged by a continuous stream of sensations, emotions, and events, buffeted by occasional waves of existential dread, until those experiences abruptly end." Our death-awareness—arising from our ability to figure time—energizes us into individual and cultural meaning-making.

The Moon gave birth to death, and death gave birth to everything else—everything human, that is, from self-awareness to religion, from tribal and political affiliation to art. Without these things, our lives would be meaningless, and, some might say, they still are. Death-awareness is usually quiet, a barely registered breeze in your yard. Yet we are so unsettled by subconscious reminders of mortality, experiments have found, that they reinforce our belief systems, increase effort and performance, and can make us distinctly less charitable, even hostile, to those who don't share our worldviews. The temporal Moon gave birth to our fears, and, in the heaven we conjecture, we'll have neither enemies nor clocks.

"Terror of death," the psychologists write in their book *The Worm at the Core: On the Role of Death in Life*, "is thus at the heart of human estrangement from our animal nature. It isolates us from our own bodies, from each other, and from other creatures with whom we share noses, lips, eyes, teeth, and limbs, everywhere on the planet." They quote Albert Camus: "Come to terms with death. Thereafter, anything is possible."

My mother's journals were full of terror at death, brimming with hope for reincarnation into something better. Sometime after she died, my father told me she'd suffered a nightmare going under anesthesia. For the rest of her life, she believed she had encountered Satan, and she shrunk from life as though it were that very monster. Once, as a boy, I had tried to get her to leave the trailer before sunrise to see a comet, but she refused. After a stroke, she finally admitted her broken decades of depression, anxiety, obsessiveness, and paranoia. Prescribed the appropriate medication, she smiled and laughed again, happy and secure for just one more year before she died, a few months freed of chain-smoking, Valium addiction, and more. I saw that transformation and tried, in phases, to think ahead: How should *I* live?

Decades later, while living in Munich, I walked into a store and found a T-shirt with a pixelated blue-black portrait of a Full Moon, beneath which occurred the phrase "Only passing through." I bought it immediately. This T-shirt is my manifesto. The phrase captures some of my sense of mortality and morality, a kind of compassionate nihilism in which we learn from mistakes, try to do better, and recognize that, in the end, everything goes away. It's my way of naming what Lao Tzu counseled: "Do your work, then step back. The only path to serenity."

I step back. I look up from the telescope. Just above the green wisps of the tamarisk the waxing Moon glides like a bird of prey. It's dark enough that the Moon lights the gravel yard behind the white adobe, and shadows pattern the ground like tapestry. The black metal telescope sits in its rocker box like a rocket without a nose cone. Beside it is a gray metal typing stand on which I place a lunar map and an atlas. I pull up a rusty metal folding chair, sit down, and pull up my spine to reach the eyepiece. I begin to focus.

Richard Baum speaks of "the aesthetic, philosophical and spiritual aspects of observing. After all, these may have the greatest value to the individual. Moreover, to observe the moon in all its phases is to be made aware of its place in the human psyche. But, to furnish the mind with the refined pleasure thus obtained, it is necessary to look for yourself."

I adjust the knurled knobs of the focus, raising and lowering the eyepiece by degrees, until I can look for myself. Adventure has never been quieter.

I am north of the first mare I learned—Mare Crisium—that seemingly circular gray expanse that looms large in the Moon's early phases—and I'm looking for the crater Endymion, which, I'll soon realize, is actually closer to the eastern terminus of another mare, Mare Frigoris, the Cold Sea. Endymion presents none of the jagged drama of Petavius. Surrounded by an impact-worn rim, the interior is covered in solidified dark lava, like the placid pupil of some eye. Years later, I'll search for what I read about: challenging craterlets within and a small hill in the middle and a lazy ridge. For now, I'm happy to have found the crater entire and realize that the subtle places on the Moon have their value too. The crater is as sleepy as its namesake. Its dark circle is calming.

I lift up from the eyepiece, step back. The evening is ongoing, and I realize I don't know what time it is. I've left my smartphone inside and I'm not wearing a watch. Air Force jets rumble overhead, night flights heading home to the nearby base. The ever-present highway thrums not far from us. Two windows in our house are golden squares. No bird calls. Pushing the telescope on its rotating base and adjusting the vertical position, I head south and relocate my first lunar lake, the Lake of Time, Lacus Temporis. Honestly, I don't see much. It's foreshortened, small, and relatively featureless as so much is on the limb of the Moon. That gray sort-of-figure-eight implies, though, the symbol for infinity. A lake of time. A startling image, not a flow of time, not the river Heraclitus invoked. No, it's a lake of eons, a pond of minutes, a puddle of seconds. Is there any ancient symbol for time as a lake? This time-lake is more like the view from modern physics, what Albert Einstein called the "block universe," whereby in the standard model there is no mathematical difference between past, present, and future. It's all the same. A lake holding all time in itself. Calm, inert, acted upon, acting upon, ravishingly simultaneous.

Strangeness overcomes me. What is this? Where am I? What is this passage? It's as though I am on a mountaintop, dissolving into matter, into contact, here in my city backyard, with my telescope, with a wavering gray surface magnified in a small, expansive view. Time? A human construct? An inch not the object that is three inches long? Arrow, river, lake, mathematical stasis, wheel? Something we waste, spend, fill? Later, I'll agree with what Russell West-Pavlov writes in his book *Temporalities*: "There is no 'time' outside of the multiple ongoing processes of material becoming, the constant transformations, often invisible, that make up the life of apparently inert things." He urges us to "relinquish the notion of stable entities," and so, if we do, "the notion of clear-cut ends and beginnings." That accomplished, we are free to pay attention, for "every moment of final entropic collapse is also a moment of autopoiesis, of self-creative activity, within another neighbouring system."

That's what Moonwatcher saw: collapse and return, collapse and return. Not just death and rebirth, no, maybe more like homecoming. And I thought I was just going to identify craters.

We have so many versions of time—diurnal, sidereal, lunar, clock, biological, market, and now 24/7, in which we are entrained, time that demands constant attention so that life has become a kind of digital prison of expected consumption and rigorous, exhausting productivity. Jonathan Crary says sleep is our only assured escape from 24/7. I decide never to answer work e-mails after the workday is done. And there's another escape from 24/7: watching the Moon. The lunar timepiece is my escape from time.

Scholars such as Judith Robinson have disputed Alexander Marshack's assertions about the Paleolithic lunar calendars. His detractors argue the evidence is too subjective, using, they say, any number or shapes of marks to make his arguments. Indeed, the claim that any kind of notational system precedes the blossoming of civilizations in the Middle East is largely conjectural. (As an aside, Marshack, while trained as a writer, pioneered techniques still used by archeologists.) There are historians of astronomy interpreting cave art in Europe—from nearly forty thousand years ago, older than the Abri Blanchard artifact—as depicting the lunar month. A 41,500-year-old ivory pendant from Stajnia Cave in Poland may be a lunar analemma (an illustration of the Moon moving across the sky). The evidence for this piece is not definitive and may never be. Nonetheless, the Moon's role in forming a conceptual bridge between the twenty-four-hour day and the seasons is undisputed. The only question is when this bridge was crossed.

The master horologist William J. H. Andrewes writes in *Scientific American* that "according to archaeological evidence, the Babylonians, Egyptians and other early civilizations began to measure time at least 5,000 years ago, introducing calendars to organize and coordinate communal activities and public events, to schedule the shipment of goods and, in particular, to regulate the cycles of planting and harvesting." The day, the solar year, and the lunar month all contributed to this way of looking at the world, but "before the invention of artificial light," Andrewes writes, "the moon had greater social impact. And for those living near the equator in particular, its waxing and waning was more conspicuous than the passing of the seasons. Hence, the calendars developed at lower latitudes were influenced more by the lunar cycle than the solar year." Farther north, in regions with marked seasonal changes, the opposite was true.

Practically, this might mean, as in Assyria, some seven hundred years before the Common Era, that when the first crescent Moon was seen by priests and proclaimed as returned, the month was underway again. Hawaiians and other Pacific Islanders have used separate names for each day of the lunar month and for each prominent lunar phase. Depending on the time of the year, the Full Moon would become an instrument of knowing and naming what was happening seasonally. For the Hopi of the American Southwest, May is the Moon of Waiting and June the Moon of Planting. Famously, in the northern hemisphere, the Full Moon closest to the start of fall is the Harvest Moon, a time traditionally used to reap crops under the bright evening light of a golden-white orb. In Mesopotamia, the calendar was based on twelve lunations, and ancient Vedic astronomy was also highly complex: life was a Moonlit concern.

Eventually, months as we more or less know them today were invented and adjusted—early Rome figures into that development—and even the post-pagan world of Judaism and Christianity remains tied to the Moon. Passover, for example, occurs on the first Full Moon on or immediately after the vernal equinox, and Easter occurs on the first Sunday after the same. (From the Midrash: "The Moon has been created for the counting of the days.") In Islam, the lunar calendar not only sets the days for when months begin and end but also for the timing of religious observations. Spotting the naked-eye crescent is one reason why the Arabic world developed such a sophisticated astronomy, one that would be foundational to post–Middle Ages European science. The Islamic lunar calendar or Hijri is many centuries old, and the young crescent Moon or Hilal features on some national flags as a potent symbol of the faith's living lunar heritage. Hindi and Chinese cultures also utilize lunations for setting the dates of celebrations.

Spotting a young crescent Moon was one of my early aspirations. It can be something of a bragging point—to see a Moon only a few hours old. On May 24, 1990, the astronomer and writer Stephen James O'Meara set the record of seeing a Moon without aid at only fifteen-and-a-half-hours after the Sun had dawned on its surface. From Iran, Mohsen Mirsaeed used binoculars to see a Moon almost four hours younger, another record. The average spotting time is about twenty-four hours after the dark New Moon.

My most memorable crescent was the one I saw through binoculars, the Moon as thin as intuition, in a February sky orange-pink with windblown sandstone dust in Utah's Valley of the Gods. I'm embarrassed that I didn't record how many hours old and now don't remember. Regardless, take the magnification away and I couldn't see that slender light. (The next night my eyes found it true.) I was camping with my friend Michael Sowder, a poet, and when I helped him navigate the binocular view to the correct spot, he exclaimed. We were standing in the gusting wind, evening coming on, cold, red rocks growing darker, red cliffs growing beetled and, while we had seen slender young Moons before, of course, it was another thing to seek one out with the knowledge I'd been gleaning, the Moon, the evening, facts making us kin to how many fellow humans, alive and dead?

That night we sat in firelight.

The first giant leap was from recording the shapes and movements of the Moon to predicting both. Not long ago, scientists from the University of Birmingham found what they claim to be the world's oldest timepiece, a ten-thousand-year-old series of pits laid out in an arc: a lunar calendar in what is now Scotland. On the Swedish island of Gotland, nearly four thousand grooves are carved into rocks, having both solar and lunar import; some of the clawlike grooves appear to mark when the Full Moon appears near bright stars. Ancient timekeepers could be amazingly accurate; the Maya, for example, calculated close to modern standards the number of Full Moons over a period of decades. Some of Stonehenge's monumental rocks are aligned to lunar positions. There are ancient observatories and rock imagery with lunar connections all over the world.

At Parowan Gap in southern Utah, I've had my own encounter with ancient lunar observing, if some archaeoastronomers are correct. A short drive off I-15, Parowan Gap is a well-preserved, right-by-the-road gallery of impressive rock imagery left by the ancient Fremont culture—called the Nungwu by today's Paiutes—dating back some fifteen hundred years. A community of farmers, they lived in pit-house villages nearby. The low, jumbled cliffs they wrote on—incising imagery into stone—once bounded a now-vanished river. Teeming with dots, lines, curves, circles, animal shapes (snakes, ducks, more), Parowan Gap is a veritable library of beautiful shapes and narratives, often mysterious, chipped into the dark desert varnish that coats the Navajo sandstone. One shape represents a womb and a woman's spread legs. Others tell of leaders in dangerous times. Another, a drowning. In the 1990s, researchers said they found evidence that much of the Parowan Gap related to the sky. The famous Zipper Glyph looks like a V with a bulbous bottom and might be a solar calendar. As well, on a concave line, three stylized figures "stand" with no legs or arms, rounded shoulders, long stem necks, and circular heads, each head with a dark part of rock in the center, like a single eyeball or a globe. Researchers propose that the figures represent "The Three Moons of Winter." Contemporary Paiutes say the Three Moons of Winter mean "Many have nothing to eat within a strong place, they brought low." The bands crossing the figured bodies represent, in this view, famine. Perhaps there is something true in each view—winter Moons light hard times, after all. The Parowan Valley is hot, dusty, open country, graced by the ephemeral Little Salt Lake, which, on my recent visit, was glittering in the high, spring Sun.

While staring at Parowan Gap, I thought of other places across the American West where the Moon has been figured, literally and metaphorically. At Chaco Canyon, I've stood beneath the famous star pictograph—an ancestral Puebloan sprayed ochre on the pale stone—showing, most believe, the daytime supernova of 1054 (which persists as the Crab Nebula), right beside a crescent Moon and a human hand. It's the oldest depiction of the Moon's shape I have encountered. At Chimney Rock, in Colorado, two natural pillars on a mesa mark when the Moon will rise at the same point on the horizon nearly every nineteen years. An ancient Chacoan Great House was likely built to take advantage of this observation; in fact, it seems the house was built when the Moon was rising between the two rocks. The Moon's orbit is wobbly, and it rises across different points on the eastern horizon, from north to south; for three years, every 18.6 years, it rises from essentially the same spot—called the Major Lunar Standstill, at its farthest north or south—for a night or two each month. That's what can be seen at Chimney Rock: Moonrise between stone pillars. Someday I hope to see it. For the ancestors of those who lived and worshiped at Chimney Rock, the event is still sacred. I cannot bridge that gap; it won't be sacred for me. But it will be a humbling reminder of the scale of the universe and its forces and of the precision and beauty of empirical observations that long predate Western science.

The astrophysicist Arlin Crotts, author of several controversial lunar studies, once said this of the Moon: "To pin it down we appeal to ancient connotations of goddesses and harvests, not the huge magma and rock spheroid orbiting through the vacuum. We are so accustomed to this alien planet hanging close overhead, and would be so horrified if we thought daily of its true magnitude, that we do not easily focus on it. It is too big and too reliable."

Despite the empirical triumph of observing, recording, and predicting the shape and position of the Moon in the sky, this reckoning could not make sense of what the Moon actually was, why it was there, or what effect, if any, it had on the Earth, apart from its illumination for safe travel at night and its clues about seasonal transitions. The Moon as passage was known but the Moon itself was not.

So we conjectured, sometimes based on the slenderest of clues, often out of whimsy or, most purely, out of gaping human need, out of desire and fear. The stories told about the Moon make up an almost unbelievably large storehouse of explanation and symbolism. These have largely faded, but in seeking their meanings we are seeking to understand aspects of ourselves. These tales—one is tempted to call them timeless—have a glow like Earthshine. That's when the young crescent Moon is bright and white with direct sunlight while the rest of the lunar disk is visible as an ashen blue-gray, like a cloud. You can see this for a few evenings after the Moon is back in the sky. That blue-gray is sunlight hitting the Earth, bouncing back to the Moon, then reflecting back to us as Earthshine. Our planet's own reflection in the Moon.

CHAPTER TWO

"There be wonderfull vertues of places . . ."

A ncestors trail us like leaves stirred on an autumn walk.

At the Population Reference Bureau, researchers make hominid living-to-dead estimates, deploying informed calculations about birth rates, mortality, and population growth for the past several thousand years. Roughly speaking, in this first fifth of the twenty-first century, there are about eight billion people alive today. There are more than 117 billion dead. They outnumber us, nearly fifteen to one.

I first encountered this perspective on the magnitude of passage as a 1970s middle schooler reading Arthur C. Clarke. In our trailer, one lot over from the wooded creek that harbored a terrifying owl—it resented boys playing war or talking about girls or furtively looking at tattered copies of *Playboy*—I read *2001: A Space Odyssey*. This was the novelization of the Stanley Kubrick movie my father had taken me to see one custody weekend during its second run. The film bewildered me and made me quiet and thrilled with visions of a technological future I didn't know was already in retreat—Howard Johnson's in orbit! A sprawling Moonbase where I'd work!—and, as the lonely spaceship *Discovery* with its lonely astronauts crawled toward Jupiter accompanied by a strings-only adagio of Aram Khachaturian, I cried in the theater, secretly. The book, considerably less artful, begins with its most memorable passage, a claim in the foreword that for each person alive there "stand thirty ghosts," a figure accurate for 1968 when both the novel and film were released.

I knew nothing of death or dying but the word *ghosts* and the image of them standing by us, by me, felt uncanny and important. Things were much bigger than the school bus. There were mysteries among the needle-nosed rockets. Other paperbacks had told me: There were lines on Earth carved by ancient aliens. Edgar Cayce had seen the future. If one tried hard enough, the mind could bend spoons just like Uri Geller did on talk shows. Things were bigger than Coach Bertalon screaming, "You pussy! I never want to see your face again." He'd arrived at football practice with a mouthful of bloody dental gauze, red spittle dripping from his chin as though he'd just gnawed a gazelle. Things were bigger than my mother saying I had a sick mind because I'd joked about Brooke Shields claiming, in her infamous ad, nothing came between her and her Calvin Kleins. Things were bigger than the silences of any given car ride.

They, too, would be dead someday, the tormentors adult and preteen, then I would be too, and we'd all join those corpses behind the next-living, who, I felt, would also be afraid for themselves—though perhaps a little less so if they had jet packs and videophones.

Recently I remembered that Clarke's first protagonist in *2001* is also named Moon-Watcher, though he looks at the sky from Africa, not Abri Blanchard. The Moon of most of the dead is not mine. Though poetical and nostalgic in its way, my Moon is largely scientific, empirical, materialist, and much bigger than that school bus and its backpack-tattered oracles of pseudoscience. (And yet the Moon is still, to borrow from Charles Darwin, full of grandeur.) However, for most people for most of time, the Moon has been a totem of superstition, of charms and curses, a deity with powers for good and ill, for life and death.

Death is still with me as I look through my telescope on this particular night. There is an old pot full of blood and wine and a dead frog boiling in the firepit in the meadow by our Utah cabin, our escape from Tucson summers. For this, I blame Henry Cornelius Agrippa. It's his recipe—technically a "suffumigation"—taken from a sixteenth-century book of his. The alchemist, suspected demon, putative vampire, and defender of at least one accused witch suggested the ritual as a means to whatever good ends the owner of blood, wine, a dead frog, and more might hope for.

It would be a lark, I thought, this ritual. Then, in reading more about lunar mythology, folklore, and religion, I realized I wasn't just compiling a catalog of prescientific beliefs—dazzling as they are—but contemplating what has motivated both spirit and data: the need to understand and to gain some handhold in a universe in which we die. Even more surprising was finding the truth in the words of the poet Robert Phillips, who asks, in his anthology *Moonstruck*, a compendium of lunar-themed poetry from around the world, "Can the old superstitious practices regarding the moon still be efficacious in charming the believer's unconscious, containing hints concerning our own psychological changes?" Although the Moon of the long-dead could never be mine—despite this boiling pot of grotesqueries—I'd find the answer is yes.

"It was lunar symbolism that enabled [people] to relate and connect such heterogeneous things," writes comparative mythologist and anthropologist Mircea Eliade, "as birth, becoming, death, and resurrection; the waters, plants, women, fecundity, and immorality; the cosmic darkness, prenatal existence, and life after death." The notions of "cycle, dualism, polarity, opposition, conflict, but also of reconciliation . . . were either discovered or clarified by virtue of lunar symbolism." And these ideas and symbols were not, as I would learn, merely passive. The Moon was a force.

From ancient poets to Sylvia Plath, Jules Cashford also argues for the foundational and enduring importance of the mythic Moon: "The evidence of stories and images from around the world suggests that most, if not all, ancient cultures passed through a stage of interpreting certain aspects of their reality through the Moon." This reality was often female, divine, and mutable.

A powerful example is the four-thousand-year-old Sumerian poem *The Descent of Inanna*, which is perhaps one of the oldest texts about the Moon. In it the cycle of life, death, and rebirth of the monthly Moon is dramatized. The goddess Inanna, as the Moon, disappears and becomes "a piece of rotting meat . . . hung from a hook on the wall." Other gods "sprinkle the food of life on it" and "sprinkle the water of life" on the corpse so "Inanna will arise." Death is not final, and there is evidence: the Moon always returns.

The Moon became "second to nothing in its influence on world-wide folk belief and practices," my old *Funk and Wagnall's Folklore Dictionary* declares. The Moon—its light, its phases, its rumored properties—became especially a totem for good: to win desired lovers, to reap bountiful harvests, to fill coffers full of coin. The Moon meant growth. This lunar association with fertility and life is profound. Whether to many African cultures or to the ancient Greeks, among so many others, the Moon was a deity who presided over crops and harvests. ("In planting season we begin to watch the moon closely, just as we watch the months before giving birth to the child. Without the moon, there would be no life," once said the Nigerian Ngas poet John A. Kwashi.) It was always best to plant seeds during the New Moon so the growth of its light in increasing phases would benefit the plants with leafy fullness. Harvests were best at Full Moon, that fat orb an emblem of food filling bellies during winter.

There is perhaps no more famous a Moon goddess than Diana or Luna, she of ancient Greece and Rome. A huntress (in the daylight form of Artemis), she also oversaw childbirth. A beautiful, athletic virgin, in her knee-length hunting dress, she carried a quiver and bow and was, as Lord Byron called her in her lunar form, "Queen of the sky, whose beams are seen afar!" Diana loved a sleeping shepherd, who, each night, she'd kiss with Moonbeams, causing him to stir. This was her gift to him, if gift it be, of sleeping forever on a mountain, touched without his knowing: Endymion, whose dark crater is not far from Lacus Temporis. Sometimes when I look at Endymion through my telescope, a crater tranquil as the shepherd's cave, I think of the mythic goddess moving across the night in her chariot pulled by a white horse. She passed over columned temples devoted to her, one of the most popular goddesses in the ancient world. Catullus had a fetching hymn to the magnanimous Moon, which goes like this:

> Latona, your mother, gave birth to you beneath Delos's sacred olive tree [and] made Diana (praise Diana) hill-goddess, forest-goddess, valley-and-river goddess, distant wending of rivers that becomes music, praise you, who women invoke in breath-caught labor, the mistress of night, of dark, of false dawns, praise you, Diana, who moves all seasons across all years so that autumn harvest comes in ripe with grain and good . . .

The Moon could yield other harvests. In many cultures, casting a spell under a Full Moon or jingling some coins or displaying new money in the waxing Moonlight, all would increase one's wealth. Simply making a wish upon spotting the new crescent Moon means the wish will come true—so long as you don't tell anyone what it is. If you're truly desperate for good luck, procure the left hind foot of a rabbit in a graveyard during the New Moon. Easier is a friend's tactic. Jennifer told me that one April during the Pink Moon—so-called for the colors of spring blossoms—she "put water in my ritual . . . container, filled with rose quartz, and left it outside in the moonlight overnight. Drank it in the morning. Now the pink moon is inside me."

The Moon has long been associated with water—it didn't take long to notice the Moon's pull on the tides—but to some pre-Columbians the Moon was also a divinity for intoxication. They would not be alone in thinking about Moonshine. A nineteenth-century witches' handbook includes this spell for vinification: "If drinking from this horn I drink the blood—/ The blood of great Diana—by her aid—/ If I do kiss my hand to the New Moon, / Praying the Queen that she will guard my grapes . . . / So may good fortune come into my vines."

The link between the Moon and blood would have been reinforced by the copper-red appearance of the Moon during frequent lunar eclipses, when sunlight passes through the atmosphere of the intervening Earth and the Moon transforms from white to various hues, dark red to dusky orange. "Blood on a dead Moon," said some. Many have linked the Moon to menstruation, saying, as some Amazonians do, that periods result from the Moon having intercourse with a girl or woman. (It is, however, a coincidence that menstrual cycles nearly match that of the lunar month.)

In many places, weddings traditionally are held on the New Moon to promote the couple's reproductive success; as the Moon grows, so too babies. Men are not always left out: the Moon can lead, some cultures have thought, to lengthening the penis, though precisely how I cannot say. The sexual nature of the Moon is global. The Moon and Sun are lovers, family members, or both. The Maya associated the Moon with weaving, the sea, and birth, among other things. For the Incas, Mama Quilla was the lunar goddess of menstruation and marriage and she cried silver tears.

In some New Guinean and African cultures, the Moon is a chief. In Uganda's Bantu kingdom of Buganda, royals have used lunar cycles to symbolize the rise and fall of kings. Among the Shambaa of Tanzania, before the government ended the tribe's kingships, a new leader approached from the east, where the Moon rises.

In many places, more practically, the Moon and weather were linked. For example, if a New Moon occurs on a Saturday, the weather will be bad. When the Moon rains down jaguars and boars, which we don't witness, the result is lightning and thunder, which we do, according to Bolivian lore.

Humans have agency with the Moon too. Almanacs once informed their readers of how the Moon influenced their affairs even to specific days. All sorts of advice exist about what to do under lunar light. Wood should be cut during the waning Moon because sap is weaker; the wood will season more quickly. According to one compendium, a "Swiss cure for illness is to cut the fingernails and toenails on a Friday night when the Moon is waning, then force the parings under the shell of a crab, which will take the fever away with him." Contemporary pagans are still advised to set out a mirror to Moonlight and use its reflections as they ask for Diana's help. The Moon card in a tarot deck advises, according to one 1910 guide, "Peace, be still; and it may be that there shall come a calm upon the animal nature, while the abyss beneath shall cease from giving up a form."

While these beliefs may, to some perspectives, seem as whimsical as Edward Lear's line "the cow jumped over the Moon," there is no denying that songs, poems, and festivals across the world celebrate "Moonstruck" lovers. In China, the Mid-Autumn Festival celebrates the Moon. A writer for *China Daily* claims that "the Moon Festival would have to be about the most romantic of holidays. An evening devoted to staring up at the sky, soaking in the silver light from the full moon while nibbling (sometimes) sweet cake and sipping red wine." Rulers of old used the celebration to make offerings to the beneficent Moon. The Moon and heavens, more generally, were deeply important to Chinese astrology, thus helping to account for the recordkeeping and predicative power of Chinese astronomy.

The deeper origins of this festival concern the beautiful Chang'e and the day the hero Hou Yi shot arrows at ten suns that were burning up the Earth. Destroying nine of them, he left one for us today. In thanks, the Empress of Heaven gave him a potion for eternal life, which he loaned to his wife, Chang'e, for safekeeping. Threatened by a thief who wanted the elixir, she drank it to keep the potion from getting into the wrong hands (or throat, in this instance). She ascended to the Moon, where she remains, hearing perhaps the sighs of her Earthbound lover while she's kept company only by a rabbit, whose shape on the face of the Moon we can still discern. But not Chang'e's. If we are motivated to various consolations by our fear of death, this tale is a caution. Immortality can be lonely, even with a pet.

The lore of the Moon—with its protective and fecund qualities—was in part bound up with the Moon's practical use as a night-light in the long era before artificial lighting. In *At Day's Close: Night in Times Past*, A. Roger Ekirch writes about how larger cities systematically began to use candle and oil lanterns in the late seventeenth century, but even then they were sometimes not lit continuously and might be used only on certain nights. When the Moon was not bright, crime was a major concern, and there was a sense among police and judges in the Western world that night crimes were especially awful because, in one prosecutor's words, "Then people are most unguarded." Lanterns fixed and portable could be put out by thieves or by weather, and the darkest nights—Moonless and cloudy, called "pitmirk"—were feared, especially for travel. Samuel Pepys spoke of "brave moonshine" while others called the Moon the "parish lantern." The dark was deadly. The writer Paul Bogard relates a shocking historical account. On one night in 1830 it was so dark in New York City that a watchman ran headlong into a post and killed himself.

Eventually, localities turned to candles and whale-oil lamps then finally electrical lighting. Some early arc lights were even called, as if in tribute, "moonlight towers."

🌚

"What good can come from a poor, dead frog?" Kathe asks, disgusted and a bit plaintive.

I've just adjusted the old pot on a bed of embers in our firepit by the cabin. A giant bullfrog stretches like an Olympian swimmer in a red broth of wine and other ingredients I've gathered with the aid of a bygone text.

As closely as I can, I've followed the recipe from Henry Cornelius Agrippa's *Three Books of Occult Philosophy*, from the 1500s, while trying to absorb his witchy thinking and rudimentary ecology concerning the Moon. Agrippa writes that the Moon has the dual property of earthiness and moisture—ancients around the world believed the Moon to be wet and cold—and that the orb "attracts to itself the Celestiall waters which being imbibed . . . doth by reason of its neerness to us power [pour] out, and communicate to us." The Moon is a friend of Saturn and is most like Venus. It can influence "the brain, lungs, marrow of the back bone, the stomack, the menstrues, and all other excrements, and the left eye, as also the power of increasing."

As it happens, I use my left eye at the telescope and had whacked it recently on the viewfinder, giving me a bloodshot mien. "Moongazing injury," Kathe remarked.

Many things on the Earth are "lunary," Agrippa says, including soil, water, oceans, and rivers, "and all moist things, as the moisture of Trees, and Animals, especially they which are White." Sweat, fat, egg whites, mucus, the taste of salt, white and green stones, silver, pearls, all lunary, things also that shine will "daily increaseth, or decreaseth as doth the Moon." Each time the Moon rises, Agrippa reports, palm trees grow a new bough. This is a contention readily discounted, however, by observing palm trees in Tucson.

Did Agrippa have a dry, nonlunary sense of humor? He claims, with neither change of tone nor a supply of anecdotes, that there is "a great enemy to the Moon," which is, and I would not have guessed this, quail. The wee, brown, ineffectual quail. It is true that they can startle underfoot, but how quail array their tiny, tasty, peckish bodies against the power of the mighty Moon is a question as yet unanswered by the pagan arts.

Happily, all felines are lunary because their eyes change size, so our three cats are also of the Moon, though they prefer patches of sun. Other lunary animals—those "that observe, and imitate the motion of the Moon"—include goats and pigs. "Bees of the putrefaction of Cows, small Flies of putrefied Wine" are lunary, but most especially dung beetles who, my conjurer says, burrow into cow patties for a month then burst forth when the Moon returns. This has a basis in fact: dung beetles can detect polarized Moonlight and navigate by it.

Agrippa is of his time, and his own sources are ancient and diverse, leading back at least as far as Pliny the Elder. Still, I wonder if the alchemist—or anyone—really believed that menstruating women would "walk naked about the standing corn" provoking pestilence and cankers. Why *exactly* would naked menstruating women walk about cornfields? Hatred of kernels? Had they grown weary of corn chowder? Were they in league with quail? The association between diseases and sickness—whether of plants or creatures—was also part of Agrippa's lore, despite the oft-sought link between the Moon and fertility.

There are other treasures from this alchemist that I prefer. On the canyon overlook a few yards in front of our cabin, near my ritual pot, and set as high as treetops of Douglas fir and willow, grow some peonies planted by the original builders of the cabin, the Hales, the husband of which, Heber, was, we've been told, "the singing postman of Cache Valley." The Hales planted lilacs, eastern red cedars, roses, irises, and more plants that came, as perhaps they did, from back east. Wistfully, Kathe remembers peonies from her aunt's house in Green Bay. I like them all the more because Agrippa claims the peony is the Moon's flower on Earth.

In the chapter "Of the vertue of places," the alchemist delineates the relationships among the planets and their associations with locales on Earth. The places on Earth that relate to the Moon include "wildernesses, woods, rocks, hils, mountains, forrests, fountains, waters, rivers, seas, sea-shores, ships, groves, high-waies, and granaries." I have nine of those fifteen here at our hillside mountain outpost in Utah, so I'm in the right place for my "suffumigation"—a Rocky Mountain forest with wild things (a mountain lion last winter), peaks and canyon rocks, groves of mountain ninebark and, by the river, of paper-birch and willow—the latter a tree that one contemporary chronicle of pagan beliefs says "can bring down the Moon's blessings." I'm to carry a switch of it if I want to "quell . . . fears of death." Places do have wonderful virtues. Of this, Agrippa was right.

I've been given no chant or charm to deploy while breathing in the smoke from my boiling pot of several things. As it burbles and bubbles, I find myself just looking at the canyon wall, built up over millions of years from the dead shells of small sea creatures. The Logan River down below took eons to cut through the canyon, creating a dramatic entrenched meander, the water staying with the course it cut, not veering. Such dedication is not unlike my habit of looking at the world through the fact of deep time. Trying on magic has not come easily to me. Agrippa is right about this much: places have their virtues, and this place is my real home.

Back in Tucson, to get started on Agrippa's instructions, a month before the trip north, I'd driven a few minutes from our old adobe house to a New Age shop called Metaphysics' World on 22nd Street. I'd imagined smiling at the counter and saying, "I need some supplies for a lunar ritual: camphor, frankincense, seeds from a white poppy. Also, a dead frog. And a cow's eye.

Bull, actually. And I need some goose blood, though I don't know how much, and the advice really is to use menstrual blood, but I'm guessing, you know, goose blood?" I can be a nervous talker, and, in all cases, menstrual blood was a bridge too far.

The proprietor and her daughter at the register seemed nice, and I'm not one to consciously surprise or mock in conversation, and, besides, things weren't going well at the store, what with the angry, tall, sweating, plaid-shirted guy with a ponytail berating the woman about opening plastic bags—how many times had he told her?—and for not turning on the AC earlier (her husband, I presumed, since even though he called her "mother," he looked to be about her age or older). It was hot in Metaphysics' World, a typically scalding Sonoran April afternoon battering the large windows facing the parking lot. Beyond, the mountains near the freeway rose like coprolites out of the desert, whose expanses, the color of cat litter, surround the city like a fever. In Tucson, even the heat has thorns, and the rivers are made of sand.

I was also uncomfortable because, before I'd left, I had fessed up to Kathe about the ingredients Agrippa required. I was pretty certain that Metaphysics' World wouldn't have the animal fluids and critter bits the ritual demanded.

"So, the problem," I'd said quickly, "is that the dissection supply company won't ship to our PO Box, so if I order, it's going to have to come to the house." As I would be leaving a few days before Kathe to open the summer cabin, that meant she'd be traveling with a bag full of cow's eye—and they are not small—as well as a preserved frog. I expected a firm no.

"It is a testament to how much I love you that I am willing to do this," she said, only after I assured her there would be no smell, though, to be clear, I didn't know that.

Tentatively, I wandered the aisles, ignoring the argument and another woman on her phone pleading for a ride. So this was magic's Walmart, with self-published novels about terrorists, a Tom Clancy thriller, various spiritual guides, meditation books, meteorites, statues of the Buddha and Hindu deities and, of all things, samples of Trinitite, a glassy melt that only formed at the site of aboveground nuclear tests in the American West. What can one do with

Trinitite apart from look at it and ponder our Faustian power? (A colleague responds, "You can put it in your pocket and increase your radiation dose.") Equally unsettling was the array of tall crystals big enough and sharp enough to cut open a face, the bayonets of a pissed-off shaman. I scooted to the incense section, grabbed camphor sticks and bag-your-own frankincense, and paid up.

At home, I put in my order for the cow's eye and dead frog, then paid for expedited shipping, which meant they would arrive in time for me to take them to Utah.

Once there, after an easy procurement of poppy seeds at the downtown Logan garden shop, I faced the prospect of replacing menstrual blood with that of some fowl. I pulled into a parking stall beside a tumble of buildings off Main Street that houses Horlacher's, a local butcher. Overly cheerful and concocting a story about a visiting friend "on an extreme paleo diet" and "wanting to make something special from an old Polish soup recipe with blood," I finally got to, "Can I get a pint or a half-pint of goose blood?"

The woman at Horlacher's had short, disheveled gray hair and a well-stained apron. She looked tired, and I suddenly felt like a jerk. "No, no, I don't think so," she said, her eyes growing large as my request sunk in. Still, I was surprised. This was, after all, a butcher's shop.

"No goose blood at all or any kind of blood, like duck blood?" I kept emphasizing the word *blood*.

Firmly, she said, "No," gesturing to a half-pint container by the cash register. She lifted a giant rack of ribs over the counter.

"Now," she explained, "you could squeeze some blood out of this." She held the rack high enough she could have struck me down. "But even to get a half-pint, you're talking, oh, $250 or $300." That was out of the question. So, for her trouble, I bought a frozen steak.

Therefore, I had resorted at the cabin to dribbling some blood from the package of ground bison we'd used for tonight's chili dinner. It had provided my suffumigation with perhaps a teaspoon. So little blood! Thing is, about magic, you can never really say. Go ahead and throw some shade at the Middle Ages—no internet, bad teeth—but they must have been awash in goose blood.

Then, with a macabre Bashō plop, I added the cow's eye, which sits atop the frog like a forlorn planet in a sci-fi pulp novel. Poppy seeds roil in the murk. So far the only scent is quite pleasant: sticks of frankincense and camphor burning slender and smoky in the coals.

"Perfumings . . . are of great force for the opportune receiving of Celes-tiall gifts," Agrippa writes, because heavenly bodies "do strongly work upon the Aire, and breath. For our breath is very much changed by such kind of vapours." Arabesques of smoke rise from the incense.

"Well," I say to Kathe, who is right to wonder what good can come from a dead frog, "that's the question."

She doesn't wait for my answer and heads to bed. Me, I'm surrounded by the detritus of my ritual: scissors used like calipers to haul the specimens out of plastic packaging, folded instructions on how to dissect the frog and the eye, an old cloth to put over my mouth if the smell becomes a stench.

The Sun sets, a chill comes on. I don't have the heart to hold my nose directly over the pot, so I breathe in the incense on the fringe of ancient vapors. The scent reminds me of my days as an altar boy, and I decide to wish that a friend not be upset (he wasn't, it turned out) and ask the Moon for a healthy early retirement (as yet unfulfilled).

Yet I cannot see the Moon, likely another ritual error. It's up though not vis-ible behind the ridges off toward Right Hand Fork. Is Diana waning tonight? Waxing? I can't even remember, which is some pretty slack magic.

As ever, though, I look at the canyon walls with an almost painful gratitude. I am rather disbelieving that circumstances have accrued to create such moments, and my pagan immersion has reminded me of its roots: powerlessness. The resort to magic has long been the domain of the downtrodden. In Tucson, beneath dusty, wind-whipped palms, the shopping carts of the unhoused gather and disperse in clanky armadas. I see it every day. Part of my discomfort I must literally own: I'm reminded of my privilege—what we used to call "good for-tune"—even as I think back to my mother teetering on bankruptcy. We had come close to losing the owl-haunted trailer in which I dreamed of space, and, as an adult, I have felt less-threatening kinds of powerlessness, often of my own origin, but I can still hear the taunts and alarms from childhood, the echoes of learned helplessness. I recall that for the Buddhists the Moon is an emblem of enlightenment. Perhaps it is to me as well.

Rather than magic, I have resorted to therapy, conversations, philosophy, and self-help books by neuroscientists. And my telescope. Here, by the fire, I thought I could charm something from the Moon by *pretending* to charm something from the Moon, and instead I've confronted history and my own psyche. Robert Phillips's hard question was, "Can the old superstitious practices regarding the moon still be efficacious in charming the believer's unconscious, containing hints concerning our own psychological changes?" That cow's eye is staring at the sky, and, though I am staring at that eye, I have come to see other things lit by a healing Moon.

Hermit thrushes sing, and I step back from the smoke. So . . . that's it? Agrippa doesn't say how long the perfuming should take, nor what, if any sign, announces its success or failure. I'm on my own. I've smelled the incense, felt no supernatural jam, pondered of necessity a variety of debts, made two wishes, am cold, and that cow's eye is haunting me.

"Namaste, old pot of gross things," I say and trudge to the door.

In the cabin, Kathe says, "That was better in theory than in practice." I have to agree.

In the middle of the night, the nearside Moon is in day, bathed in particles of solar radiation, patiently breaking down the lunar topside. Micrometeorites pepper the dust. Along the line between light and dark, the terminator, electrostatic charges raise the dust above the surface like a gown trailed along a temple floor. Perhaps deep inside some mineral shiver occurs. In the middle of the night, I stand by a second-story window. The pot I'll throw out is cooling. The lilacs by the door below are mounds of blossoms made of silver. "Once in a blue Moon," we say. "Over the Moon," we say. We see Moonlight and speak of lunacy and lunatics. What is metaphor but secular alchemy? Tomorrow is Monday, Moon day, and yesterdays outnumber us too.

It's been said that souls go to the Moon forever. Or that they wait there among the rocks till they find passage to a new realm. Most poignant, I think, is the claim that all of our lost things are heaped up there. From socks vanished by a dryer to a former lover left adrift, from a parent to people you've disappointed or hurt to everything you had as a child that you misplaced or set aside, everything you have ever lost, trivial and tragic, it's up there on the Moon: the solar system's giant lost-and-never-found. I thought about this while cleaning out closets in Tucson.

As if to italicize loss—or, at least, change—Kathe and I were going to move. The old adobe house she'd restored with a fierce kind of love had piqued the interest of an unexpected buyer, and, as is said, we could not pass up the offer. When we bought "the mud house," tatters of sheet metal served as a desultory roof. There were dirt floors and crumbling walls. Nesting pigeons and graffiti. A low-budget horror film had been shot there. Kathe had saved this historic adobe and made it our home. It went from ruin to refuge. It was like living in a Spanish farmhouse and we embraced the neighborhood and we knew it couldn't last. The sprawling, rough-walled, wood-planked house was far too big for two writers, three cats, and one telescope. If we'd each been shedding unhelpful ambitions and perspectives, as we had, this was now a practical matter of shedding objects. We were downsizing in more ways than one. I gave away half my clothes, five hundred books, lots of movies and music, and culled childhood belongings, though I kept the fifth grade diary that included my detailed description of the starship *Enterprise* and the poem in which I claimed, "Happiness is beating my sister in an argument." Chairs, pillows, plates, silverware, more, gone.

Though we would lose the mud house itself—never to step again among its dusty slants of light—it was good to lighten the load. I thought often of the Elizabeth Bishop poem, "One Art," in which she famously asserts, "The art of losing isn't hard to master." She cites among her losses the places she called home. I was going to miss the mud house. I still do. By way of staying calm, I looked, of course, at the Moon.

One evening during the preparations to leave, over that giant, 150-year-old tamarisk I loved and that the new owner would cut down for a pool, I glassed the Moon in the telescope. I looked at a now-easy and attractive signpost: three large craters, the first of which is the youngest and most complicated, Theophilus, which lines up with the craters Cyrillus and Catharina, each older than its neighbor. Theophilus is a large crater and it's complex—meaning terraced in stair-step fashion from rim to floor and with a central peak complex, like Copernicus. The crater spans sixty-two miles, plunging down more than two, and I really paid attention—long, quiet minutes, gentle nudges on the scope to keep the place in view—and I saw that the mile-high central mountain isn't one peak but is broken into notches of shadow and tips of light, a mountain range unto itself. I saw how the crater rim cut into that of Cyrillus, indicating its younger age. I followed the undulations of the crater's edge. To the north, the land was rife with furrows, the nearby spray of ejecta forming an apron or "nimbus" about the crater. After a bit, I looked again at the massive central mountain complex. There were three impressive peaks, their heights cut by valleys, still in darkness, then a fourth massif to the south separated from the others by a triangular abyss that pointed like an arrow.

What I could not see was any token of my various losses, what could have been on the Moon: The creek that seemed so far below what must have been a modest embankment in the woods behind the trailer. My G.I. Joe (or was it Matt Mason?) in his silver spacesuit. The tall grass prairie of the Flint Hills where Kathe and I fell in love. Friendships lost to inertia or harm. Blue ribbons and plaques from speech tournaments and journalism competitions. My mother's rare laugh and my father's early temper. My homeless nephew. Autumn days with my sister and cousins. A child my girlfriend and I did not have decades ago. Episodes of *Space: 1999* and books on the Bermuda Triangle. I didn't see the homes I've left or that glow-in-the-dark UFO model that calmed me when my parents argued downstairs. The only human things left behind on the Moon are old spacecraft, science instruments, trash, and Soviet hammer-and-sickle medallions. It's not the art of losing one must master. It's the art of letting go.

To a smaller, recently built house, modeled in the Sonoran style, we moved, just down the street. So early in my lunar pilgrimage I'd already looked from three different places—from the mud house, from the cabin, and from what we'd dub Casa Luna. Casa Luna's backyard is not expansive, but there's a decent view toward the Moon's arc across the southern sky. From there, I would make the most headway on the lunar-observing programs that sharpened my skills and knowledge. Despite the narrowed sky, my horizons expanded.

That the Moon and its places compass me in these ways made it even harder to imagine the fear it has provoked as an agent of evil. I've had moments of real terror, spiky crescendos rising from the receding background of my childhood's white noise of unease. Years ago, Kathe and I topped out on a ridge, just under ten thousand feet, when the placid overcast churned to black. Metal-framed packs on our backs, we crouched and ran among boulders as hail and lightning and thunder crashed around us. I feared for our lives as we ran, boots on wet rock, in descent. I have driven a hundred miles per hour in Nebraska outrunning a tornado. In junior high, I was hit by a car and sent into a short suborbital flight before smashing into the pavement of 10th Street in Indianapolis. My ex-wife and I were so brittle that in a half-awake slumber I once heard something in the bedroom and jumped up yelling to whomever had entered our house, screaming until I realized it was just a blanket that had slipped off a closet shelf.

If I were of a certain cast of mind, I'd seek reasons beyond psychology and stochasticity. If I were old-school lunary, I'd blame the Moon. There have been bad Moons rising over the homes and huts, villages and veldts of this human Earth for a very long time. It might be that some of the darker folkloric associations with the Moon arose when rituals like mine, doubtless better orchestrated, failed to achieve their ends. It did not help that the Moon itself went dark each month. Of course it comes back. Doesn't it?

This interval between lunar months, during the dark of the New Moon, was when witches gathered, especially at crossroads of three or more, where they cast their dangerous curses. Witches would draw down the Moon, using spells to squeeze poison from its light. In the dark, rulers weren't to travel by carriage, doctors were not to touch the ill. The boundary between the practical—it *was* more dangerous to travel without Moonlight—and the magical—witches made this worse—was blurred.

The widespread association of Moonlight with moisture led to blaming that silvery glow to all manner of maladies. As late as the eighteenth century, many Europeans and Americans avoided sleeping outside under the Moon or having Moonlight in their bedrooms. The fear continued in rural pockets of superstitious Europe into the twentieth century; apparently Benito Mussolini feared Moonlight. A Full Moon could drive a woman mad or even kill her. In 1842, Britain passed a Lunacy Act, attributing forms of madness to the Full Moon. No such link has been proven, though scientists have found that people go to sleep later and sleep less during a Full Moon. Perhaps that biological restlessness was part of the worry about Moon-borne illness.

Trouble comes by looking at the Moon over your shoulder. Trouble comes when you point toward a new Moon. Trouble comes when a star is near the Moon. From different lunar phases come drought or death. In a New Year, be careful what you are doing under the first Full Moon; you are destined to repeat it for a month. Do not sleep with your face under the Full Moon on summer nights; you'll become a werewolf. On the Moon live the Fates, who spin the threads of our lives, cutting them at will; perhaps the Fates made house in Lacus Mortis—the Lake of Death—a lava plain in which the crater Bürg looms like a tumor.

Diana, that goddess of fertility and wisdom and the hunt, with her owl and arrows, was also Hecate, the queen of Hades, deity of witches, goddess of graveyards and crossroads.

They developed in Italy their own secret gospel, and Diana became a kind of consort to Lucifer. The witches worked against the Catholic Church, which they saw as a tool of the rich and powerful, the Church that noticed Adam's first wife, Lilith, was associated with the Moon, with urges primal and sexual. Perhaps it was witches such as these who walked naked about the corn of feudal lords to dispense disease upon their yields? Or perhaps they went after their oppressors directly with spell and curse? Shakespeare writes that "the moon, the governess of floods, / Pale in her anger, washes all the air, / That rheumatic diseases do abound." The transgender figure of Baphomet is also devilish and sometimes associated with the Moon.

Against these dark arts, against the troubles of any life, including mine, as small as they have been, we can invoke a Sufi story involving the wise trickster Nasrudin, who proclaimed to guests in a teahouse that "the Moon is more useful than the sun." When asked why, he replied, "We need the light more during the night than during the day."

☾

Yet unanswered in our survey is how Nasrudin's useful light got there in the first place. As with beliefs about the Moon's powers, its origin stories are diverse. The first woman in Wichita creation stories becomes the Moon. The North Star's grandmother, for the Crow, is the Moon. The Inuit Moon god and supreme power is the hunter named Igaluk. According to the Navajo, two elderly men gave their souls to the Moon and Sun so they would move across the sky. Lest one think the Moon for the Diné is a matter only of their past, the tribe was offended when NASA deposited the ashes of lunar geologist Eugene Shoemaker on the Moon's surface without consulting a people for whom the Moon remains a sacred presence. The Moon's cultural importance is manifold and ongoing.

In Africa, the Bushongo, a Bantu tribe from the central part of the continent, tell how the supreme being Bumba was lonely and ill. His stomach ached, and he vomited up the Sun. Bumba was still sad and sick. He vomited up the stars. Then the Moon. Out of distress comes the world.

Astrology writer Sarah Bartlett relates an Aboriginal tale in which the Moon is a cheerful, dumpy, awkward flirt seeking a wife or, at least, some companionship. When two sisters heard him sing, they thought he must be handsome; but when they caught sight of the Moon, they jumped in a canoe to escape. The Moon begged them to give him a lift—he couldn't row and he needed a ride across the water. They relented only to be angered when he tickled their arms. They dumped the Moon in the river to drown, which he did, growing thinner and thinner till he vanished. Each month the Moon rises, growing bolder again, and each month he is disappointed, fading in the water to end another endless cycle of desire.

This yearning that permeates so much lunar mythology—even the gods are weak—is strikingly present in an Amazonian lunar-creation tale, one that proceeds along grotesque lines until the startling conclusion. It is the story of the Moon that most holds me. Some men, it's told, had retrieved the head of a suicide, a man who had cut his own throat; they wanted to show it to fellow villagers, though why is not said. The head kept falling out of their sack and rolling around on its own until finally they gave up and left it behind. But the head had other ideas. It followed them. It asked for fruit, which they shook from a tree, pitching the fruit far away so the head would leave them. But it followed them once more. Eventually, the head rolled into the village, where people huddled in their homes, afraid. Head wondered what it might become, and several possibilities presented themselves but all ended poorly: "If it turned into wheat they would eat it . . . If it turned into a bean they would cook it." He'd considered becoming the Sun, becoming rain, becoming milk.

"I will turn into the Moon," the head declared. Outside his hut, he yelled, "Open the doors, I want to get my things." No one obeyed. Head pleaded for just "two balls of twine," with which, and a stick, he could climb the sky. The villagers complied this time and warily went outside to watch it ascend. Head somehow unspooled the twine, somehow managed the stick.

"You going to the sky, Head?" they yelled.

He did not answer.

The story ends: "As soon as it got to the Sun, Head turned into Moon. Toward evening, Moon was white and beautiful."

Here is beauty that originates in tragedy: Who was that man who cut his own throat? What grief had filled him? Before he lost the will to live, what had he lost? What amends had he failed to make to himself and others? He died to become the Moon. His belongings, his cargo for life on Earth, stayed under wet and rotting thatch, dark and dank below, white and beautiful above, his Moonbeams shafting through the jungle canopy.

CHAPTER THREE

Mirror

At the campsite we found a knee-high fire-ring of thick stones, comfort made dignified by soot. There were even three small stone benches.

"A palace!" Michael had announced upon arrival.

Around this rustic royal retreat, home for the next four days, at nearly eleven thousand feet, grew night-dark clematis, regal larkspur, drooping bluebells, many small sunflowers, and understory fleabane, interrupted here and there by paintbrush, pale or ruby-red. Like a reminder, a stream tumbled past granite boulders to a lake by whose nearest shore water lilies blossomed. They quivered in the alpine wind. All around, firs, green or dead-brown from bark beetles, stood by the lake, stood up the rocky slopes, pierced skyward beside fell-fields, and onward down a valley and up the next run of mountains. My friend and I were in the High Uintas with its crenulated cliffs and long, sloping, rounded peaks where trees give way to krummholz, to the austerities of summer ultraviolet and winter entropy. They are Utah's wildest range.

For decades as I have looked up, first to stars and galaxies and nebulae, now to the Moon, I have also witnessed mute rock upon mute rock and have heard the calls of night creatures on Earth. I left hermit thrushes singing by the cabin to come here. A scientist who studies Moon rocks once told me, a bit sheepishly, that he didn't know his way around the place his samples came from. That surprised me. Now I understand. We can grow so intent that the source behind intent becomes obscured.

Michael and I had backpacked, so I had no telescope as I watched the nearly Full Moon rise above a brown-white columned cliff. No telescope, but I did haul up *The Cambridge Companion to Arthur Schopenhauer*, at which Kathe shook her head. (Well, I thought, I could always wield it against a bear. And it would help me sleep.) The naked-eye Moon. Already I had grown more used to the Moon as a magnified presence. So I was a little uncertain. Could I savor the Moon only with my eyes, as our Moonwatcher ancestor did? Could I bear witness to the Moon of the ancient philosophers, those Arabic and Greek astronomers and thinkers who considered the orb not a god or goddess (or not only) but as a thing to explain, an actual object in the blue and the black?

At the campsite, I'd slung a lightweight hammock between two firs so I could see the Moon and I sipped red wine that evening, reaching down to fetch the metal cup from duff, out of which grew, in this shade, a single white columbine. The white lunar highlands shone bright as those petals. The darker maria were as blue as the daylight sky itself, as though they were holes in the Moon letting through the light. In another Elizabeth Bishop poem, the entire Moon is a hole in the sky for the ill-fated "man-moth" who tries to pass through, to escape, and each time fails, thus "proving the sky quite useless for protection."

Like a bemused syllogism, a world hung over the mountains.

To see the Moon as given to our eyes is often to impose pattern: pareidolia, seeing something recognizable in random shapes. Anyone who still takes the time to look at passing cumuli knows the feeling—there's an elephant!— though the tendency evolved not out of leisure but survival. Pareidolia helped our ancestors identify predators in the grasses, landmarks in the distance, prey in the forest.

On the Full Moon we have recognized much: a man carrying sticks, a woman, a four-eyed cat. One Asian view is of a rabbit pounding rice. In Africa, Mexico, Tibet, China, and elsewhere, people see a frog or rabbit. In Peru, a fox has clawed marks on the Moon. In Malaysia, the figure on the left side of the Full Moon is, according to my folklore encyclopedia, "a hunchback making a fishing line" beneath a banyan tree, which would be Mare Frigoris, the long rooflike maria at the top of the Moon. There's a rat consuming the line as it's made, and we should be grateful for that, for if the hunchback finished, he'd use his line to fish out every sea on our planet. Siberians see "a girl with a pail near a shrub . . . rescued by the sun and moon from a pursuing wolf." Cook Islanders see a girl making tapa. Scandinavians, a boy, a girl, and their water bucket, the archetypal Jack and Jill.

There's the Man in the Moon, of course, but I could never quite figure what parts of the Full Moon's surface were supposed to be what parts of the face. Then I saw a drawing that darkened the features: Mare Imbrium is the left eye, the combined maria of Serenity and Tranquility the right eye, extending a bit so that Mare Vaporum becomes the nose. Apparently Mare Nubium is the mouth (which seems surprised). From the hammock I still couldn't see it. (In the southern hemisphere, by the way, the Moon's top and bottom are reversed from what we see in the north.)

All these figures eluded me until I read in an old encyclopedia that some Native Americans see a woman—a goddess-creator—stooped over a kettle with a dog beside her. The woman is made up of Mare Imbrium and Oceanus Procellarum on the Moon's left side; the dog consists of the interconnected maria on the right. I see the kettle as the intrusion of the white southern highlands in between. This I can witness from the mountains: a figure in the Moon, the only deity I have ever seen.

A telescope is no guarantee against pareidolia. I'm coming to rely on it, in fact. When I see the white Montes Recti—a narrow mountain chain called in English the Straight Range—I see a severed spinal column. When I see the swept-back white stripes fanning out from the crater Proclus, I know they were created by ejecta from the impact that formed the crater, but I see a falcon's bright wings. Some people have described the magnified shapes of mountains and other formations as an arrowhead, a baby, a skull, and a valentine. As I'd come to learn, systematizing these pattern recognitions into nomenclature would involve not only terrestrial analogues but also vanity and sheer imagination.

As to the Moon's phases, which reveal or obscure the face of the Moon, they were explained by the Hindus as their night god rotating a lantern. Others, that when the Moon goes hunting we see it wane or even that the Moon wanders between two wives. Our very sight of the Moon is heartfelt.

In Stanislaw Lem's novel *Solaris*, a character says, "We have no need of other worlds. We need mirrors." In the story, astronauts encounter physical manifestations of their deepest fears and shames. The protagonist, a psychologist named Kris Kelvin, is dispatched to the space station orbiting Solaris to investigate mysterious troubles. While there he encounters a "copy" of his ex-wife, Rhea, who killed herself years before. Somehow, and seemingly without self-awareness, the ocean world of Solaris creates "phi-creatures," avatars of one's own guilty conscience. Long after humans were decentered as the measure and gravitas of the cosmos—a process the Moon helped enable—the fictional planet of Solaris returns humans to the center by mocking them: the god-gloried Earth is not the center, it's our worst selves.

So we see ourselves. Or can. But we can also see what's really there, myopia becoming vista.

There were mirrors in the ancient world with stems fashioned as the bodies of goddesses. The mirrors themselves were usually burnished bronze. The Etruscans used them, so too Greeks and Romans. Pliny the Elder tells us there were even crude glass mirrors, with a silver or tin coating, but for some reason they did not become popular. An illustration of an Egyptian bronze mirror from 1500 B.C.E. features a Hathor-goddess stem, a woman's face, and the crescent horns of a bull, emblems of the Moon. Did the stories of Hathor affect the image seen in reflection?

Stories can tell us truths. (Perhaps all the insights we need into our fears and ambitions are there in the first epic in the West, the story of Gilgamesh, which is thousands of years old.) It's when, however, we elevate certain world-building stories that are not self-correcting—when they presume to fix reality—that there's trouble. So I don't take this next one literally, but it is a conceptual leap toward an empirical Moon. One version of the Moon's creation credits Phanes, a son of Chronos. Phanes was a mythic cyborg: "Golden wings, four eyes, the heads of different animals, the voices of a bull and a lion, male and female genitalia, and many names . . . the very icon of natural fecundity and power, an image of excess and superabundance," writes Scott Montgomery in his short but rich book *The Moon and the Western Imagination*. Perhaps appropriate to a god made of many real things, the Moon that Phanes built was a *place*—quite real—with "many mountains, many cities, many mansions," according to the poet Orpheus.

It is not a long journey from this kind of construction to thinking critically about the Moon's physical reality. Adds Montgomery:

> In addition to its size, changing aspect, and brightness, the Moon is
> unique in being the only celestial body with surface features visible
> to the unaided eye. In drier climates, such as those that have gov-
> erned the Mediterranean region during the past several thousand
> years, these features present all the more starkly and magnificently.
> It is impossible that the Greeks, in particular, did not perceive the
> lunar maria on a regular basis. Indeed, there is much evidence in
> the textual remains we possess that they did and that they applied
> to these "spots" a range of competing interpretations.

Philip J. Stooke of the University of Western Ontario writes that the idea that the Moon literally mirrored the surface features of the Earth dates to Clearchus, "a pupil of Aristotle, and survived in European and Middle Eastern folklore until the nineteenth or twentieth century." Even before Clearchus, around 500 B.C.E., disciples of Pythagoras also thought the Moon a mirror, believing that the bright and dark variegations reflected the Earth's landforms and oceans. Pythagoras and his followers also thought the Moon a more blessed version of Earth, replete with larger, more attractive flora and fauna, the latter being so much better than creatures of this planet that they neither urinated nor had bowel movements. (Pythagoras apparently had a thing about that, advising followers, most sources say, to avoid eating beans because they housed dead spirits, which gives a rather macabre twist to flatulence.)

A few centuries after Pythagoras, Plutarch wrote in "On the Face in the Orb of the Moon" that the Moon could *not* be a mirror because everyone knew the Earth's ocean was a single entity and not broken up into various seas. He claimed that the Moon was a world like ours with "open regions of marvelous beauty and mountains flaming bright and . . . zones of royal purple with gold and silver . . . bursting forth in abundance on the plains or openly visible on the smooth heights." Plutarch was right about the mirror part but wrong about most of the rest. His phrase of "mountains flaming bright" or "mountains of eternal light" struck a chord; it is used even today by lunar observers and scientists to describe peaks in the polar regions where the Sun never sets. This is because the Moon's axis is barely tilted, so light grazes the high polar summits relentlessly. (Plutarch did apply the first names to the Moon— Hecate's Recess for Oceanus Procellarum and the Gates for the three connected eastern maria of Serenitatis, Tranquillitatis, and Fecundiatatis.)

Centuries after Plutarch, Leonardo da Vinci raised his own objections to the Moon-as-mirror, writing in his notebooks that the Moon showed always the same aspect. If it were a mirror, the view would change: "Since the shape of the lands is irregular, when the Moon is in the east it would reflect different spots from those it would show when it is above us or in the west; [but] the spots . . . never vary in the course of its motion." The Moon had to be its own place—and its brightness was not reflective of our world. Still, da Vinci also thought the Moon was a version of our world, with mountains and valleys and seas and rivers and prairies.

Could that be? What *was* the Moon? Its own world? But why did it shine like a little sun in the night? These questions had vexed philosophers for a long time.

Three men in Miletus had some things to say about these matters. The first was Thales, who had predicted a solar eclipse in 585 B.C.E., and for whom water was the primary cosmological substance; he asserted that the Moon was similar to the Earth, passed before the Sun to cause eclipses, and was bright because it reflected sunlight. A. C. Grayling writes in *The History of Philosophy* that Thales "is the first person we definitely know of who wondered about the nature and origins of the universe [. . . and] put forward ideas about them which are distinctively philosophical rather than religious or mythological in nature."

His student Anaximander disagreed, believing the Moon shone of its own accord; he reportedly thought the Sun, Moon, stars, and planets were fires that are seen through holes in the air—like those of a flute—and that lunar phases and eclipses resulted from the holes getting filled up then emptied. Born in the year of the 585 eclipse, a student of Anaximander, Anaximenes, also turned to fire as an explanation. The fire grew out of vapors rising from our planet and moved at different speeds; the Sun-fire was fast and bright, the Moon-fire was slow and dull. All three explanations are notable for being nonmythological, relying instead on observation, analogue, and reasoning.

Subsequent conjectures about the Moon's movements, its origins, its light, and its substance are largely variations on these themes and images: flat or spherical, fiery or watery in different degrees, creating its own light or borrowing it from another source, hotter or colder, and so on. For example, in 500 B.C.E., Parmenides in his *On Nature* suggested that the Moon, which was cold and black, originated, like the hot and white Sun, from the Milky Way. Parmenides thought the Moon's colder mien accounted for the spots. He thus argued for the Moon shining by reflected sunlight while orbiting the Earth. The first-century Indian mathematician Aryabhata also argued that the Moon shone by reflected sunlight. At about the same time, Empedocles also suggested the Moon was cold, saying it had congealed from frozen air "just like hail."

Other notable philosophers—including Democritus, Diogenes, and Anaxagoras—argued for a Moon with some degrees of similarity to the Earth. Born around 460 B.C.E., Democritus would live a long life—to one hundred—and be the prime exponent of atomism, the belief that all things were made of tiny, indivisible particles with hooks and rings so they could infinitely combine. Because atoms were everywhere, Democritus reasoned that there was an infinity of worlds, and on one of them, the Moon, there were hills whose shadows caused the dark maria. Diogenes made a connection between volcanic rock and a meteorite he had once seen, reasoning that stuff in the sky must also be volcanic, including the Moon. (This contention would be revived with post-Galileo observations of craters, which suggested volcanic calderas to many astronomers.) Anaxagoras also thought the Moon and other heavenly bodies were admixtures of stone and heat and light and that the Moon had a topography like Earth's. He said too much for the Athenians, unfortunately, when he proposed the Moon hosted living beings. For this he was exiled.

Plato, despiser of the decrepitude of all Earthly things, would have liked that last touch. For him, nothing Earthly could infect his perfect vision of a perfect cosmos. All bodies in the sky were perfect round souls in a perfect round universe created by a perfect (presumably round) God. They were all there for our imperfect contemplation. Certainly the perfect round ones were not composed of such things as dirt and water and rocks and hail and fire and all that decaying garbage Plato thought our lives were and were cast upon. All heavenly objects were indeed heavenly and made of some kind of perfect glass. The line from Plato to Christian perfectionism is indeed a straight one. (Plato also despised Democritus and wished that all his books had been burned.) As to Aristotle, who, like Plato, would exert a huge influence on human affairs far past his day, his Moon was lit by friction, possibly the domain of living creatures (including men), and was a kind of ecotone between the Earth and the rest of the heavens.

It was another pupil of Plato's, Eudoxus, who was more interested in approaching the sky with math and geometry rather than speculation. He solved the problems of eclipses, lunar phases, and planets' retrograde motion (when they appear to move "backward") by developing a cosmos of spheres centered on the Earth. Variants of this system—from Callippus to Aristotle to Hipparchus to Ptolemy to Brahe—took hold up till the time when Copernicus revived the heliocentric system in the sixteenth century and Galileo in the seventeenth century provided more evidence for it.

Heliocentrism's first advocate was the astronomer Aristarchus, who worked at the great library in Alexandria, Egypt, and died in 230 B.C.E. Not only did he put the Sun at the center of things, but he also argued that Earth rotated on its axis. He was not heeded, though he now has named for him a giant lunar plateau, riven with a valley, the alleged source of lights from outgassing events, if some are to be believed. Named for him on this plateau is the brightest large crater on the Moon. By contrast, the one named for Plato appears as featureless as a humorless idea. It is beautiful, I hasten to add, but bears an informal name that seems appropriate to Plato's dour ways—the Black Lake.

If the Moon were a mirror, a mirror, say, of place across time—not a static reflection of the present, but a kind of temporal glass, a time machine—what continents, what seaways would return to my sight? What would waver into shape from the surface of some other Earth? The Moon has hung there for eons. A mirror-Moon would've shown change. From primal rock I stood on, the song of the solitaire pinging the cliff by our Uintas camp, the nodding flowers closed, I wondered.

There, the long-vanished Western Interior Seaway was reflected from the western portion of lunar glass. Beneath the real waters swam dodgy paddlefish, timeless stingrays, fishes not yet fossil-caught. In the rivers that filled the sea, mooneye foraged for insects; a version still does, in northeast Wyoming. Overhead passed a frigate bird, now extinct, and sandhill cranes, still alive. A kingfisher, like the ones that yet rattle up and down rivers, plunged into water. They reside in stone.

The light that reaches our eyes is always from the past. The Moon we see is more than a second old, the Sun eight minutes past. What our eyes perceive, from world or mirror, are sights of what no longer is. We can't help it. We're tricked. We must take the present as given—we have to, in order to live—and there, reflected on my imaginary lunar surface, bordering the old sea, are the high ranges I hiked to reach this near-summit, but I called them something else, the Telescope Mountains, from which I looked at the Moon with my naked eye.

The confusing, bright southern highlands of the Moon? Terrains of far tomorrow, when the Sun has become a red giant, blasting from the surface of our home all air, all waters, all creatures, all green, all songs, like a bored god sweeping a table of its food.

Then I let my vision go.

There was the actual matter in the sky, bright lunar swathes, dark lunar spots—the seas, once thought and so-called, which we know now are plains of solidified lava. Already from memory I could identify the major maria: On the eastern limb, Mare Crisium. The three rough circles of Mare Serenitatis, Mare Tranquillitatis, Mare Fecunditatis, from north to south. Above them, across the north, Mare Frigoris, shaped like a strait (or, as one nineteenth-century observer, Edmund Neison, said, it could be mistaken for a thin, passing cloud). And below them, to the south, Mare Nectaris, though its small size made it hard to truly discern it as different from Fecundity and Tranquility, into which it seems to blend. But, squinting, I could see its dark blotch at the bottom of this sequence of seas.

Rising from the south another shape—a white bird's head, wings on either side: the chaos of the cratered highlands, clawing their way into the maria. Its impact sites I could hardly yet name, so many craters overlaid on each other it's like trying to hold in mind the shape of each leaf in a windy forest. Easier, for now, the maria. The wide expanses of Mare Imbrium and Oceanus Procellarum in the west, below which were the two smaller seas of Nubium and Humorum. Where they began and ended, I thought I could make out. The Full Moon was no longer more nameless variations of blue and white. This pleased me. We are homeless without names.

Eccentric lunar astronomer W. H. Pickering once created a list of features on the Moon one can spy with the naked eye. The features are increasingly difficult, a test not only of one's way around the surface but of one's eyesight. At this point in my apprenticeship, my knowledge was hardly thorough, and at this point in my life, my eyesight was aging in that Plato-doesn't-like-it sort of way. So, from the Uintas, what could I see from his list of twelve?

The maria Nectaris and Humorum, even if their edges weren't always crisp. They are second and third on Pickering's challenge. The first was easy, almost an afterthought: the bright region around Copernicus, the white rain of ejecta, and white rays of rock snaking across the surface like a Chihuly sculpture. Could I see the bright region around Kepler? Well, there are lots of bright regions, and I could see some of them. The Full Moon shows crater rays to great effect. They detonate in streaked mayhem without motion or sound. Where Kepler lay, I could not then recall. Later, I'd see it, a kind of white island west of Copernicus, in the midst of Oceanus Procellarum, refuge for sailors in the Ocean of Storms, and I would never forget its midocean coordinates. From the hammock, then, three of twelve—not particularly impressive, but a start.

I took a moment to swing in the hammock, wondering if that lightness felt at all like the Moon's lesser gravity, a sixth of the Earth's. I'd weigh twenty-eight or twenty-nine pounds there. Some of the twelve Moonwalkers slept in hammocks in their landers. Then I shivered in the dark, roused up, and picked my way across rocks and fallen branches toward my tent and that thick book I'd read for about two minutes.

Though I could imagine the Moon a mirror shining over the Uintas, and I often think of myself as a time traveler, the truth is the globe up there more nearly resembled Empedocles's hailstone, one of the many that drop, it seems, almost daily in these summer mountains. The Moon looked colder in the night, though that sunlit surface was hotter than the water we boiled with our camp stove. My neck ached. Smoke wafted from our palace-camp. Back there, Michael tended a fire that would warm me after my naked-eye sojourn while I remembered a gray-green Moon laid over an illustration of North America in my *Golden Guide to the Stars* from childhood. That world spanned North America, and it struck me suddenly that the childhood vision of a Moon brought close to Earth was the one I was in fact living now. I could be in two places at once, and both together constituted home.

At the cabin and back in Tucson, I would study maps, charts, and books at lunch, at bedtime, with coffee, to emplace myself on the Moon. I made up rhymes and mnemonics for maria and, slowly, tricks to remember the more intimidating spill of dropped books that makes up the library of lunar crater names. I whispered back the names of the Moon.

From early Christendom into the Middle Ages, there appears a new kind of doubleness in the Moon. The Church adopts it as a varied symbol, while scholars continue to speculate about its appearance and what that might say about its nature.

When the Moon is portrayed in churches or in monastic manuscripts, the effect is both decorative and metaphoric, with the Sun and Moon positioned above Christ at his crucifixion. The Moon embodies Christ's human side with all the implications of the corruptible and corrupting Earth. Oddly, the Moon of the Church becomes both stylized as a man while it becomes symbolic of the Virgin Mary. Perhaps the shift to Mary was a semi-aware Christian appropriation of the pagan powers of fertility associated with the Moon. The movement from the female pagan Moon goddess to a man's face in the Moon is part and parcel of the movement toward a male-dominated sky-god religion. And when the Moon was "female," it was necessarily virginal, a way of preserving the secondary role of Mary in relation to God and Christ.

Meanwhile, various scholars immersed in Greek, Latin, and Islamic traditions carried on the philosophical speculations about the Moon's appearance, in particular the bluish maria. For some, the Moon's spots were the marks God put there to remind us of sin. Albert the Great thought instead that the Moon's material was thicker in some parts, which is why those areas were darker. Others continued the theme of mixed materiality on the Moon—the dark spots were Earthly, the lighter areas more like the Sun. It might not be a mirror, but it was smooth and round—echoes of Plato. A major figure of the Islamic Golden Age and a pioneer in vision and optics, Ibn al-Haytham opted for the explanation that the Moon receives and reflects sunlight, but that it also becomes partially self-illuminating based on its substances. Tellingly, he admits that if the Moon shines by reflection alone, then surface topography *must be* the reason for the dark spots. This is a profound—indeed, revolutionary—suggestion of an actual world beyond the Earth.

This quiet suggestion of Earthliness was echoed in the fourteenth century by Nicholas Oresme, who continued with the naturalistic dogma of a smooth, perfect, glassy Moon with variations of density. But, to help illustrate his argument, he invokes a metaphor, as Scott Montgomery notes. The Moon's variegations are like those in alabaster. Montgomery says of this seemingly minor comparison that "for the first time since antiquity perhaps, literal pieces of the Earth are proposed as the key to the nature of the Moon." Oresme thus "introduces a new world of immediate, visual evidence that one can physically hold before the eye." This secularization seems of a piece with the fact that early manuscripts devoted to astronomy and time included lunar representations and phases in their charts. The Moon grew closer.

It was a Christian theme that gave us our first well-known and undisputed representational portrayal of our companion. It was not until Jan van Eyck painted a masterpiece called *The Crucifixion* sometime between 1420 and 1425 that—amazingly—the Moon was actually portrayed more or less how it looked to the naked eye. One of a handful of van Eyck's works showing the Moon, this one gives us a realistic view of the lunar surface for the first time in history, though its placement in the sky in the afternoon is fictional. Near the horizon, the fat Moon shines in a sunny, partly cloudy sky, and the largest maria are visible. I find it breathtaking that so much time had passed before anyone thought the Moon was worthy of such realism. The story is complicated. Art had begun to shift from its didactic religious function to something more ample, and, with the invention of perspective and the increasing felicity with naturalism, artists found the world more interesting. Shadows, which had not appeared in art for hundreds of years, also returned. Still, van Eyck couldn't help himself. He darkened the maria just enough so this very real Moon also became a skull, the death's head of memento mori. The realistic Moon was also symbolic.

Are there older, "public" surface representations of the maria? Philip J. Stooke considers five-thousand-year-old clusters and lines of swooping, curvilinear carvings at an Irish tomb to be renderings of the Moon, perhaps even a crude naked-eye map of the maria thus predating van Eyck by quite a bit. The evidence, however, is not entirely clear and rests on the assumption that Neolithic people were producing at least some representational art. In North America, around 1000 C.E., the Mimbres people were making pottery in which a rabbit represented the lunar face. But, as Stooke says, "It is clear the depiction is of a rabbit representing the moon, not the moon itself." This too is a metaphorical Moon. Despite the notable realism of his painting, van Eyck participated, it seems, in a tradition that once again crosses time and cultures: the lunar mixing of the thing itself with mythmaking and symbols of desire.

One desert evening after my Uintas trip, windy and warm and therefore unsuited for the telescope, I stared hard at the gibbous Moon, so bright above a neighbor's house, and saw where Mare Serenitatis blends into Tranquillitatis; there's a little curving sweep of bright "terra" or rocky highlands on the left. I couldn't see the crater Plinius, but I saw where it would be. And this pleased me. I had gotten another feature—the Plinius region—off the Pickering list. Then, just to the west of that white upland was the small dot of Mare Vaporum, distinctly smaller than Mare Crisium. Another target.

I wasn't sure I made out Sinus Medii to the south. I'd have to look again. I'd also have to wait until I had a big Full Moon before I'd try for the region where the crater Gassendi should be, as well as the Lubiniezky region in the south. The faintly shaded area near Sacrobosco crater, the dark spot at the foot of Mons Huygens and the isolated Montes Riphaeus or Riphean Mountains—well, I doubted I could spot any of them with my eyes alone. (I never have.)

Perhaps I'd have my best chance with a Full Moon low against the horizon, that time when it appears bigger than it actually is. Surprisingly, the explanation for "the Moon illusion" still eludes us. Hold your thumb at arm's length whenever the Moon is in the sky. It's basically the same size, give or take depending on the Moon's slightly changeable distance from Earth. Some thought the atmosphere produced a magnifying effect, but you can disprove this with the aforementioned thumb experiment. Having the Moon near some Earthly object or point of reference—trees or buildings, which appear smaller—may make the Moon look bigger. This is called the Ebbinghaus illusion. Yet the Moon still looks larger when lower in the sky as viewed from the sea, where there are no trees or houses. The Ponzo illusion suggests that our brain makes things farther away look bigger than they seem to be. Yet—and who thought of doing this?—the effect vanishes, it is said, if you bend over and look at the Moon between your legs. I have never done this nor do I plan to. Convergence micropsia explains the illusion by noting that our eyes are less "converged" when looking at the Moon close to terrestrial references. It makes the Moon seem bigger.

In any case, tonight's Full Moon does seem big, hovering like a golden eyeball over metal roofs and power lines of the barrio. Even palm trees look lovely silhouetted by the Full Moon.

The most dramatic Full Moonrise I've ever seen was one cold April at the cabin. All around me stood junipers, Douglas firs, Woods' rose, maples, and barely budded aspens. I sat on my haunches on the rocky slope of grasses slightly brighter and brighter beside black trunks and limestone outcrops. The sky was mostly clear and a gray-blue light fell upon the canyon wall across the river. Thin clouds just arrived thickened into scudding clumps, looking like so much liquid marble, two large clouds casting two giant shadows, pillars in the sky, and with the clouds taking on a rose cast. Soon, the branches, even the needles, of the ridgeline firs became defined lines. In a moment coldly molten, the Moon's sphere cleared the ridge and moved with astounding speed. The clouds split open as though on cue, and the canyon was blue-lit on one ridge, black on the other.

The Moon averages an orbital distance of 239,000 miles. It was eighty thousand miles away four billion years ago—three times as large from Earth. Can you imagine it between clouds then? I love the sight of a Full Moon but every time I see it now, I think of how much telescope time I've had during its waxing. And I wonder whether I can keep myself up even later or rise earlier to see its waning. Sleep is one of my passions, but I am hungry for prospect.

It would fall to Leonardo da Vinci to first draw the Moon as just the Moon, forgoing embellishment for the sake of accuracy. He did so five times that we know of, between 1505 and 1514, including a view of the crescent Moon with Earthshine. Like others, da Vinci got the essentials of Earthshine correct but, like most who attributed terrestrial characteristics to the Moon, he believed the dark spots to be oceans and the bright sections to be land.

If Jan van Eyck recast the Moon as a more realistic object in an historical sky, though one deeply suffused with religious meaning, then da Vinci's Moon—drawn from the naked eye nearly a century before Galileo's first telescope views—is one whose only context is simply itself: an object whose depiction served as explanation and vice versa. In that sense, it is da Vinci's Moon, not van Eyck's, that first truly renders that world as we had literally seen it for millennia. Da Vinci's Moon is strictly an observation, though it was kept private in his notebooks. But it is the Moon freed of dogma, liberated from the obfuscation of desire. It is the first Moon of science and it is beautiful.

Leonardo da Vinci was a lunar pioneer in another way. He suggested that someone "construct glasses to see the moon magnified." It was a good idea.

CHAPTER FOUR

Collimation

Old telescopes never die. They are just laid away.
— Leslie Peltier, *Starlight Nights*

Why should our special hatred of death render fragility so odious?
— Galileo, *Letters on Sunspots*

To understand the trepidation I feel when confronted with mechanical tasks, it's important to confess that it once took me three hours to replace a float valve and flap in a toilet. That's how long some astronauts take on space walks. Headlamp strapped to my forehead, tools strewn, instructions laid across the floor like an ancient codex, I jerked up sweatily when Kathe popped in the doorway and asked, brightly, "How's it going?"

"You do not want to be in here right now!" I barked, and she closed the door.

That night I called my father—chemist by training, data manager by profession—and we had a good laugh.

"I did not get your mechanical genes," I protested. "Or the ones for math."

I thought of this the day I tackled my first telescope disassembly in years, when I removed the main mirror of my ten-inch telescope, cleaned it, put it back in place, and adjusted its collimation. I would think of my father again, under more melancholy circumstances, when years later I tackled another telescope in need of repair, this time an instrument that had somehow returned like a puzzle piece from childhood.

To collimate means "to bring into line; make parallel." For a telescope, it's "to adjust accurately the line of sight." In a Newtonian reflector, like mine, light enters the open aperture, hits the main or primary mirror, which is slightly concave and set at the far end, from which the light bounces back toward the open entry but, ideally, it does so in a focused stream that strikes a small mirror near the top. That secondary mirror then deflects the light up into the hole where the eyepiece sits. With knobs that adjust how high or low the eyepiece is in relation to the secondary, I can adjust the focus to gain a clear, magnified image.

If the mirrors are out of alignment, even slightly, the image degrades. Sharp edges blur like clouds, the view softens as though looking through water. If the mismatch is severe enough, one might only see a white blob where the Moon should be.

Collimation is how we set telescopes aright. It's actually a straightforward process, especially with practice, but when I decided to clean the ten-inch primary mirror of my telescope one afternoon, I knew I had to be careful, given my ineptitude. Remove the screws from the metal ring at the base, lift out the heavy mirror, unscrew the clips that hold it down, then submerge the mirror in water in a very clean sink. There are steps with distilled water, with soapy water, with cotton swabs to remove droplets. There's a lot of rinsing. The primary dried—terrifically clean—while I swiped wet cotton balls across the secondary. Half the battle was deciding to clean the mirrors in the first place, as their dustiness does not necessarily yield poor views. But mine was gray with grunge despite the cap on the telescope's open end. (Later, I learned that one could fit a shower cap on the back of the tube to prevent dust from entering there.) By dinnertime, I had put everything back in place, reluctantly asking Kathe for help just once, when I could not get a ring screw to return to position; she tried her drill but it was fruitless.

"So," I told her, "I do have a screw loose."

To collimate the mirrors—to get the light straight and true—I inserted a laser eyepiece that sends a red beam through the optics, bounding back to a kind of bull's-eye. At the base, I loosened and tightened knobs that slightly flex the mirror till the red beam hit center. It was awkward and time-consuming, but I was done. Or so I thought. I noticed I could not quite get center after all. My secondary was out of whack, and I could not adjust the delicate mirror held by spider vanes. The next day I drove to Starizona, a Tucson astronomy shop so expert they've built equipment that's flown on the International Space Station, where Scott politely pronounced my collimation "close," then deftly adjusted the secondary and finished the job. Adjusting the secondary mirror is the first not last step, he told me, but I'd still need occasional help in the years ahead.

Only later did it occur to me that I was part of a lineage, part of the family of telescopists, that I was descendant and kin, a perpetual, sometimes clumsy apprentice to be sure but one who, with every hour at the reflector, was gaining comfort and some proficiency, whose now-sharper Moon had an ancestor that appeared in a small refractor on cold nights in Italy four centuries ago.

Born in 1564, the same year as William Shakespeare, Galileo Galilei became a talented boy, raised by his musician father and a mother one biographer describes as "lunatic." Passionate about math, music, and building things—the latter unusual for the day—Galileo would also develop a deep love of the visual arts and poetry, especially Ariosto's *Orlando Furioso*, a poem in which a trip to the Moon relates the belief that lost things go to the Moon. Galileo would receive training in the arts, a background that some believe helped him as a scientist. And foreshadowing his son's own contrarian nature, Galileo's father would become embroiled in a public battle over the physics of sound. This same father sent Galileo to a monastery then to university to study medicine. The son would take instead to mathematics, and during his studies earned a telling nickname: "The Wrangler."

What Galileo did not immediately earn was a professorial position, but he made the most of an opportunity when he was asked to lecture on the mathematical geography of Dante's hell in 1588 at the Academy of Florence, where he also taught perspective and other artistic subjects. Such a topic was then of significant literary and theological import, and the talk, along with other mathematics work, helped Galileo finally secure a coveted university job. His academic career began in Pisa and was marked by friendships with intellectual allies and with opponents. Along the way, he made discoveries and was the target of jealousy and enmity. He remained the Wrangler and did not suffer fools. Once, he wrote a poem that mocked the university dictate that he wear his academic robes at all times, including in his private quarters. When he left Pisa, his last lecture was booed by the Aristotelians.

Soon he became the head of the family after his father's death. Galileo made his way to Padua in 1591, where he became a math professor and enjoyed the fruits of a lively community that included a massive library, a liberal atmosphere, and, like all universities then, many chilly rooms. In one such, he lectured from a wooden podium—like a pulpit—about Archimedes, Euclid, and Copernicus's heliocentrism. A professor's days were long, writes Charles Mee, beginning at four or so with prayers, followed by one's lectures, a meal, more lectures and teaching, dinner, then an evening of lessons or one's own research. In Galileo's case, the work would court controversies that would spark the beginning of modern science and result in threats of torture and the final humiliation of his house arrest at the hands of the Inquisition. Though he had children out of wedlock with a mistress (not unusual for the day, especially since he was of higher social standing), Galileo was a devout Catholic, and two of his daughters became nuns. This religious devotion, however, did not stand in the way of Galileo's project of upending received wisdom from Aristotle and other ancients. The boos in Pisa may have made him smile. He took them to heart in Padua.

While there, for example, Galileo found that all falling objects accelerated at the same rate throughout their trajectory and solved a long-standing problem of, in James Burke's words, "Why falling objects do not fall to the ground to the west of their starting-point on a turning earth. Galileo argued that as the earth turned, everything on it turned too, so the falling object moved east with the earth . . . it was like dropping an object from the top of a ship's mast. It hit the deck, because both the ship and the object were travelling together." Da Vinci himself already had shown that cannonballs described arcs; this was contrary to Aristotelian dogma that only in the high heavens could there be motion describing curves and circles. Galileo conducted work on acceleration and motion in 1602, just two years after Cardinal Bellarmine had burned Giordano Bruno at the stake, for, among other heresies, claiming that there was an infinity of worlds beyond Earth and that they could be inhabited by intelligent beings, though his gravest error was suggesting that Christ was not divine.

The Church, with its modified Aristotelian edicts, was increasingly on the defensive, and with the advent of global exploration, begun by Magellan, the idea of other worlds—nonperfect spheres in heavens—was becoming less astonishing. The problems with calendar time—reconciling lunar and solar years—also threw unwelcome light on the Church's schizophrenic embrace of both geocentricism (as the preferred model of reality) and heliocentrism (whose math worked better for organizing calendars).

Even something as seemingly simple as linear perspective was changing the zeitgeist. Galileo was trained in linear perspective, a technique that art historian Samuel Edgerton says helped usher in the Scientific Revolution in part because it requires standpoint, a subjective point of view encountering an objective world and gives the illusion of presence, even mastery.

In the last decade of the sixteenth century, England's royal physician and the man who discovered magnetism, William Gilbert, drew the Moon. Gilbert's Moon was not a symbol in religious art, nor was it a da Vinci–like specimen study. This was a map, the first-ever of the Moon. Gilbert's was by necessity a naked-eye rendering, since there were as yet no telescopes. Covered in grids, the map shows the major maria, and then Gilbert, in full cartographic style, named the things he saw. This was a watershed moment. The physical Moon was now the object of naming, and naming's purposes include notation, description, and, at least in the Western European tradition, colonization. That Gilbert's names were largely banal—such as Regio Magna Occidentalis (the Great Western Continent)—is overshadowed by the name he gave Mare Crisium: Britannia. Gilbert puts England on the Moon, suggesting all those other landforms are there for the taking.

Worlds upon worlds—real places, here and beyond—rose into our consciousness.

❋

When did I first see a world beyond? I have no memory of looking through my childhood telescope, that three-inch-wide reflector made of shiny white fiberboard. It was set upon a metal mount atop a tripod of V-shaped wooden legs. The viewfinder was long and narrow, like a pirate's spyglass, and the eyepiece—tiny, smaller than those used today—slid up and down against a red rubber lining. There was no mechanical focus. Inaccurately, the telescope was called a Space Conqueror, and it came from Edmund Scientific, whose catalogs my father gave me and over which I spent countless hours of pleasure, looking at microscopes, lasers (a person could buy a laser!), gyroscopes (I got one!), that white-black spinning flag inside a lightbulb-shaped glass that was a vacuum (it spun because of heat! I watched it!), chemistry supplies, test tubes, and more. In 1968, the telescope cost $29.95.

The Space Conqueror looked cool and alien, and I was intimidated. Like so much from my skittish childhood, memories have been eclipsed by time and the traumas of our angry house. I don't recall when I got it, or whether it was from both parents or just my father after he left us. I suspect the latter because I never asked for his help—this I know—when I had trouble aiming the cumbersome and unsteady tube. I'd later realize the viewfinder was neither well-aligned with the main telescope nor easy to adjust. And the eyepiece had a narrow field of view. I know I saw the Moon a few times after much reckless scanning, but I don't recall epiphanies or even moments.

What I do remember is feeling frustration and shame when, probably sooner rather than later, I just could not see anything through the Space Conqueror. I told no one. At least once I stuck my angry hand into the tube and yanked the little secondary around. This I vividly recall. It solved nothing. I had a telescope. It did not work.

The Space Conqueror stood in my bedroom on Sawyer Street. My sister, who remembers everything, recalls that I still had it as late as junior high. Eventually, though, it vanished. I wouldn't get another telescope until I was married to my first wife and living in Kansas. A different design, it doubled as a spotting scope for birds, and I looked at the Moon sometimes, but found more challenge and gratification from seeking little green wisps of galaxies over a neighbor's roof or from the prairie, where I tended my loneliness like an ill-gotten gift. Still, the Celestron C90 was the first telescope I fell in love with. And because of its design I never had to take it apart. Especially portable, the C90 was a gem. Why I sold it when Kathe and I moved to Tucson I still don't know.

When Galileo heard of something called "the perspective tube," he was intrigued first by its potential usefulness to the military (and, therefore, to improving his financial position) then, more crucially, by its potential for scientific discovery.

By 1608, a handful of lens makers in the Netherlands had made crude versions of the telescope, more or less at the same time, resulting in the government refusing to issue a patent. Very quickly telescopes appeared all over Europe. They showed up in Germany at fairs, were sold in Paris, and soon enough arrived in Italy. Galileo called the device an *occhiale* or, in Albert van Helden's translation, a "spyglass." Craftsmen had been making magnifiers and spectacles for a long time, so the invention of the telescope appears inevitable in retrospect despite difficulties that included what kinds of lenses to use, at what distances to set them from each other, and the sheer difficulty of grinding and polishing glass. Early telescopes only magnified a few power, were blurry, and were fringed with color and difficult to steady. Yet they worked.

It's commonplace to say Galileo was the first person to look at the sky—especially the Moon—using a telescope. I recently purchased an updated night-sky guide and set of charts, and the section on the Moon repeats that claim. Historians know, however, that Galileo was not the first person to record looking at the Moon through a telescope. That honor, so far as we know, goes to Thomas Harriot.

On an English summer evening, on July 26, 1609, Harriot turned his six-power perspective tube to the Moon, which was less than a week old. Harriot drew a thin arc of the terminator, though he did not shade the nightside portion of the lunar globe, and sketched in a few hasty maria, including Crisium, Tranquillitatis, Fecunditatis, and Nectaris. A scientist and naturalist, Harriot had been part of the Raleigh expedition to Virginia. "The first impulse of discovery," Scott Montgomery writes, "was to record . . . a picture, not to translate this vision into words." Galileo biographer John L. Heilbron notes Thomas Harriot's excellence in math—he was more expert than Galileo—but that Harriot kept his work unpublished due to a stolid reticence. Harriot set the lunar sketch aside. It was, Heilbron says, "a meager thing without a hint of landscape." The drawing looks like a few cat scratches and a part of an arm and elbow. That lack of landscape is key to Harriot's meager rendition, for it would take Galileo's special gifts to transform our understanding of the Moon.

☙

In 1609 Galileo's Padua household had the following grocery list: "Glass, forms for grinding, and abrasive compounds, along with . . . lentils, chickpeas, and rice, and winter clothes for his infant son Vincenzo," writes Albert Van Helden. By the time this list was compiled, Galileo had handled one of the early, janky telescopes being sold as curiosities. He found it wanting. So he made his own.

"These were busy days for Galileo," adds Van Helden. "He had to give his lectures at the university, privately tutor the sons of noblemen in subjects such as fortifications, teach the students to use a proportional compass of his design, run a boarding house of sorts for some of these students, and spend . . . hours grinding and polishing lenses." The lenses he made were superior to any others "for at least a year." Galileo had become "a member of important intellectual circles in Padua and Venice, a man interested not only in philosophy and practical mechanics, but also in language and the visual arts, a convinced Copernican who wanted little to do with Johannes Kepler's public advocacy of that theory." Kepler was nonetheless a towering figure, the most important astronomer in the world. At forty-five, Galileo was on the make, and for him to devote himself to an entirely new task—how to fashion glass to make a superior telescope, a tedious, painstaking, and error-filled enterprise—confirms he saw the instrument's promise. If he got it right, the bearded and bulbous-nosed Galileo would be famous, secure, and on his way to discoveries no one could imagine.

After he'd created a scope of nine times magnification, Galileo presented it to the Venice Senate. Distant ships could be spotted hours before seen with the naked eye. The strategic advantage was obvious, and Galileo got his immediate reward: a lifetime appointment at the university and a raise that doubled his salary. Eventually he'd move on, but this was a plum.

By autumn, Galileo had fashioned a twenty-power refractor far better than anything else on the planet, and starting on the cold night of November 30 and for the next several weeks, he pointed it to the sky.

"The instrument," he later wrote, "must be held firm, and hence it is good, in order to escape the shaking of the hand that arises from the motion of the arteries and from respiration itself, to fix the tube in some stable place. The glasses should be kept very clear and clean by means of a cloth, or else they become fogged by breath, humid or foggy air, or the vapor itself which evaporates from the eye, especially when it is warm."

I once saw a 1931 Italian replica of this telescope at the Franklin Institute in Philadelphia. It was shockingly long—some forty inches—and gorgeous, and made, like the original, of two wooden tubes fastened together by copper wiring. It was covered with paper decorated in gold flourishes: a technological instrument and work of art.

Galileo continued observations off and on till just before the publication of *The Starry Messenger* on March 13, 1610, his revolutionary book about the

Moon, stars, and the discovery of four little orbs around the planet Jupiter. This small volume helped to usher in the modern world. It demonstrated that with the application of human reason to a new tool—a new technology—one could reveal things and processes about the universe that empirically contradicted received wisdom. The telescope was an instrument of discovery, and it showed to anyone who could use it that there was more than we had seen before.

✥

"My god," I muttered, pleasure returning under the headache of disbelief.

Many years ago, my therapist in Logan suggested I go shopping to aid my recovery from a concussion. A few weeks prior I had bent over to push open the blocked door of our shed, only to have cast-iron pipes crash into my skull. Months followed of depression, near-vertigo, insomnia, and anxiety, the latter especially in stores. "It's called agoraphobia for a reason," Dr. Christian had said.

So one gray, cold, wet spring day in Cache Valley—the winters are long there—I drove to DI, or Deseret Industries, the Mormon version of Goodwill. Kathe and I often shopped there, finding used records (Fran Warren's *Come Rain or Come Shine*, Howard Hanson's *The Lament for Beowulf*), cheap clothes for yard work, and old pots and pans for camping. I had walked by stonewashed jeans, elbow-patched blazers, shoes scuffed or polished, into the aisle of DVDs and VHS tapes of Mormon stories, then back to the collectibles area that typically featured nothing more interesting than a bowie knife or a lensless 35mm Nikon.

That day, in the glass case, was a three-inch Edmund Scientific Space Conqueror. *My telescope.* One just like it, but still, my telescope. Laid flat, the optical tube in perfect condition, wooden legs bound together by rubber bands, the top of the tube covered with a Skippy peanut butter lid, the telescope lacked only the original sixty-power eyepiece. I swayed and held onto the glass cabinet, almost crying. I briefly imagined that it was my very own, somehow making its way from Indiana to Utah, just as I had. Circumstance can be a gift especially in a time of need. Yet I hesitated. I didn't buy it then and there. Why I waited I cannot say. A few days later, the Sun breaking through the cold clamp of reluctant cumuli, the telescope was still there—thank goodness!—and I brought it home and set it up in the living room, where it looked quite spiffy.

Happily, I told my father about it, finally admitting that I'd found the scope pretty hard to use as a child. By then, I'd acquired my capacious ten-inch telescope, so the Space Conqueror became, as it had been in my childhood, more icon than instrument. Later in the desert, as my love affair with the Moon became more serious, I acquired some smaller-than-standard eyepieces for the three-inch scope and, despite its dusty mirror and uncollimated views, now and then I'd try to center the Moon, would succeed briefly before inad-vertently tapping the jittery tube and losing sight again, and then having it spasm back into view.

This would not do. I decided that I'd resurrect the telescope of my child-hood. Clean and collimate it, tighten and oil the mount and tripod, adjust the finderscope as best as I could. For a while, it remained an aspiration. Then I read of Galileo's struggle with his telescope. While mine was more powerful and better made than his, I, the unhandy poet, thought, *This will be my Galileo telescope.*

We can still feel the thrill of discovery that Galileo felt when he looked at a waxing Moon on November 30, 1609, steadying his perspective tube and per-haps inhaling with a caught breath as the Moon became larger and strangely rugged. Well-known amateur astronomer Leslie Peltier says that when he looked at the Moon for the first time, through a two-inch scope, he felt the same "incredible delight as Galileo did." Most people *do* gasp when they first see the Moon through a telescope. When we do so, we're part Galileo. Some of us continue to be impressed long after.

Those cold nights in Padua, Galileo stayed up late, observed, drew, and wrote, no small feat for the exhausted head of a busy household, a scholar prone to illness. Unrecorded in Galileo's Moon records are whether he had a carpet by his bed on which he might have rubbed his cold feet, whether a servant had arranged mulberry beneath the mattress to rid it of fleas, or whether, in the wee hours, he ate some bread and cheese before bed. Could he even sleep?

Soon Galileo also had deduced the lights around Jupiter were tiny moons of that planet. It was the first addition to the skies in eons and it showed that there were multiple other motions in the sky, contrary to geocentric doctrine. The discovery of moons around a planet that had been known since antiquity was the Renaissance equivalent to Edwin Hubble's twentieth-century discovery that the universe is expanding. And when Galileo turned his lenses to stars themselves, he observed more stars than we could see with just our eyes. The universe was growing.

His investigations began, though, as did *The Starry Messenger*, with the Moon. His descriptions and deductions are still fresh after more than four hundred years. One is brought to the strange new place, hovering in an uncanny, familiar sky over an uncanny, familiar terrain.

Galileo begins modestly, saying, "It is most beautiful and pleasing to the eye to look upon the lunar body," a pleasure enhanced by discovery that, in his words, "the Moon is by no means endowed with a smooth and polished surface, but is rough and uneven and, just as the face of the Earth itself, crowded everywhere with vast prominences, deep chasms, and convolutions." This was the difference between Harriot and Galileo. The latter was ready to see the variations of light and dark as features of topography. This was profoundly revolutionary. Reality in the sky had revealed its blasphemous and awesome imperfection.

This was also reasoned inference. Drawing on scholar Roger Ariew, Scott Montgomery writes that "Galileo did not *see* mountains and valleys on the Moon. He saw other things—a jagged outer edge, evolving and irregular patterns of light and shadow—and the meaning he gave them called upon a certain eye of faith in a non-Aristotelian lunar truth." He observed "salients and entrants of light, emerging points in seas of shadow that could only indicate an uneven, complex surface." (With his eyes only, Kepler had deduced the same.) Galileo, then, was advancing an empirical version of the "Moon-as-another-Earth idea" by using an instrument that literally magnified the boundary between light and dark.

Part of that empirical version of the Moon as another Earth had to include the maria, what Galileo called "these darkish and rather large spots . . . obvious to everyone." He called them "the large or ancient spots" and contrasted them with other ones "smaller in size and occurring with such frequency that they besprinkle the entire lunar surface, but especially the brighter part." In other words, craters.

Galileo drew one of them—"a certain cavity larger than all others and of a perfectly round figure"—which he compared to Bohemia as surrounded by mountains snagging light before it falls on the middle of the cavity, the flat floor of a valley or crater. We now believe the crater he spoke of was the eighty-five-mile-wide Albategnius.

Galileo's drawings of the crater and the Moon are beautiful, if not entirely accurate, which is in part to be expected given the relatively poor quality of the optics. Albategnius appears in one drawing like a shadowed octopus sucker. Some say Galileo exaggerated the crenulated terminator, made some craters larger than they should have been, and simply left others out entirely. But was the telescope, with its extremely small field of view, the only reason? Perhaps not. It's possible, some say, that Galileo was more interested in *rendering* the Moon to emphasize the essentials of its appearance than transcribing its appearance precisely. That is, he was being an artist. Recall, too, he was working outdoors with candles in the cold over many nights.

Albategnius was, in any case, different from what he saw in the maria, where the light did not suggest profound irregularities. Rather, the Sun little altered their appearance—"now a little darker, now a little lighter" with "their boundaries mingling and running together." Not so with the cavities (he never used the word *crater*) because the cavities in "the brighter part of the Moon"—that is, in the highlands—were "like sheer cliffs sprinkled with rough and jagged rocks . . . divided by a line which sharply separates shadow from light . . . This does happen beautifully in the other, smaller spots . . . day by day these are altered, increased, diminished, and destroyed, since they only derive from the shadows of rising prominences." Thus did Galileo describe the changing light on the Moon's craters, perhaps the most aesthetically pleasing and accessible vision of the sublime that we can find in astronomy.

He also reasoned why, if the Moon was rugged, he saw its limb, the edge, as smooth: range after range would even out their appearance at this distance, as "in a billowy sea the high tips of the waves appear stretched out in the same plane, even though between the waves there are very many troughs and gulfs so deep that not only the keels but also the upper decks, masts, and sails of tall ships are hidden." The water metaphor was telling, for Galileo also suggested the large, dark, ancient spots were actual seas, an interpretation he later disavowed for lack of evidence. It would be another half-century before more capable telescopes would reveal the disk's edge against space as imperfect as the face of the Moon itself. It would take a lot longer to discern the real nature of the maria.

Such imperfection was not a sign of ugliness, not for Galileo. On the contrary, his prose brims with exuberance at the beauty of the lunar surface. He spoke of "innumerable contortions, that is, elevations and depressions" of the Moon "sprinkled all over with prominences and depressions" and, again, "some peaks within the dark part of the Moon appear drenched in light, although very far from the boundary line of the light . . . It is evident, therefore, that the lunar prominences are loftier than the terrestrial ones."

The Moon was not only *not* ugly; it exceeded the Earth in its mountainous glory. Over a few winter nights, using a new invention he had improved and recording what he saw, Galileo changed our world forever.

Ninety seconds after I mailed a card to my father saying I loved him and that I was sorry growing old was so difficult, he called me to tell me he had cancer. I'd been reading of Galileo those days and eyeing the white Space Conqueror gleaming in my Tucson study. After the held-breath shock passed and we talked and I hung up, I knew: I had to fix it. Now. I would clean the mirrors and align them to remedy its imperfections and perhaps a few from the past. My father, nearly ninety, had a tumor in his esophagus. He would not battle it. Of suffering, he'd had enough: he'd been abandoned to his grandparents as a child by a schizophrenic mother; he'd been sent back to her alone on a train from South Dakota with his name pinned on his jacket, his beloved pet chicken left behind; he'd gone AWOL from the Navy to help her during one of her episodes, as when she walked naked down a highway or when she proclaimed herself Christ while anointed with her own feces or when she had shock therapy. My father was not much older than I am now when he told me these stories and we finally learned how to talk. His alternations between withdrawal and rage finally made sense. His now-healed estrangement from my sister made sense. We had collimated our lives.

I moved the Space Conqueror so it stood between the stereo and the sofa in my office at Casa Luna. It was a summer day. In the bathroom, I had supplies of distilled water and dry cloth, a soft lens brush, compressed air, lens cleaner, drying tissues. On the sofa, printed instructions on how to clean a mirror because I hadn't remembered the exact steps. I exhaled, anxious but determined to clean the entire scope—mirror and viewfinder, inside and outside the tube. This was for the child I had been, for my father about to die, and for a scientific hero.

First, I brushed the whole thing off with a feather duster, humming. Then I cleaned the glass of the slender finder scope. The gray cloud of disuse on the front of the lens disappeared. Next I loosened screws holding the black back cap in place, where the three-inch mirror nestled, and I carefully twisted it out, the mirror in its metal mount suddenly in the palm of my hand like an artifact from a prior civilization. At each step, I took photos in case I lost my way and to share with my father, who that day was recovering from a week of tests. With tiny bolts, springs, and clips the mirror was held and I undid them all, putting the pieces in reverse order on the white carpet where I worked. Carefully holding the bespotted mirror with a towel beneath it, I put it under a stream of tap water, which gushed in doubled currents across the surface, beading and running.

Turning off the tap, I uncapped the distilled water and poured it across the mirror like mercury. This telescope, this telescope was *mine*. What happens when we tell that feeling of disregard that we care? Things come clean. Daubs with cloth corners and cotton swabs at reluctant drops of water. The mirror shined, almost glowed, I put it back together. With the gentle turning of the back screws, the brilliant mirror moved in increments, my eye, my vision centering on itself, and, already tweaked, the secondary was fine. I oiled the tripod legs. It was clean. It was collimated.

I ushered Kathe in to admire my work and texted photos of the Edmund repair mission to my father, then set the scope up outside. Burchfield, our biggest and most poignant cat, walked about the tripod legs, meowing and rubbing them. I squatted and pushed my face into his fur and stood up to find the magnified Moon, but not before I wiped away tears.

The viewfinder slid back and forth to focus, and the waxing Moon was there, crisp in the glass and bright in the late-day sky. I dropped in an 8mm eyepiece for a reasonable magnification, moved the eyepiece up and down against the gripping rubber of the draw tube, and there was a whole Moon in sight, so much clearer than I had expected that I exclaimed. Used to my big scope's views, in which I feel like I'm in close orbit, I had to adjust my perception to this View-Master-like Moon, a diorama Moon, this little orb that hung in the sky like a scene from some future history. But I could see what I knew: Mare Crisium, the craters Hercules and Atlas, even a tiny gash that marked the deep Vallis Alpes cutting through an entire mountain range that from here looked like gnarled fragility.

Still I struggled. The mount was rickety, the legs shaky, moving the scope a tricky operation.

"Damn it!" I yelled, as I applied yet again too much pressure to move the optical tube, realizing that I had to wrap both hands around it to move it while bracing my feet against the tripod.

I exhaled. "No wonder I got so frustrated as a kid," I said.

Kathe put her hand on my shoulder. "Not many kids would be that patient." My dad would tell me how much it meant to see me with the scope.

Puzzling out the chain of craters near and along the terminator, I found Albategnius, just south of Hipparchus and west of the easily recognizable trio of Ptolemaeus, Alphonsus, and Arzachel. I marveled that I could see the big secondary impact on the rim of Albategnius, which gives the feature a kind of loose eyeball effect. There are several craters like this on the Moon. So this was the region that figured so importantly in *The Starry Messenger*, a book that put science on a collision course with faith. A crater is a story. A single telescope is a genealogy.

Renaissance scholars believed they were changing they world, and they were. Charles Mee notes that "by the middle of the fifteenth century, 'humanist' had come to mean an individual who believed in the beauty and goodness of a [more secular] human-centered world." Furthermore, the arguments over the workings of the world and our species, their relations to God and to the unfurling discoveries of classical manuscripts in new translations—these were not esoteric matters. Royals, merchants, and religious authorities spelled out their preferences in learning and did so in learned ways. The mind mattered more than ever. Professors and scholars were given more latitude than most in order to convey knowledge and opinion. This often mixed classes, for scholars came from all over, and this elite, liberating, and rowdy movement could run afoul of tradition, custom, and the Church. About twenty years after Italy's first printing press began operation in 1465, the Church alone designated who could print books. By the early sixteenth century, the Church was an active censor. Times were being troubled. Exploration—and exploitation—beyond Europe began. And the static sky wasn't static: Galileo had lectured on a supernova that appeared in 1604.

The Starry Messenger was an instant bestseller, with all five hundred copies purchased quickly. Galileo was to have received thirty from his publisher;

he got ten instead. One of his copies—presumed so because there are marginalia in Galileo's hand—is owned by the University of Arizona Special Collections Library. I've seen it: modestly sized, with crisp white pages showing only a bit of foxing, and, opened to the famous drawings—those inky shadows—utterly captivating. It was called by one Englishman, upon sending it to the king, "the strangest piece of news . . . that he hath ever received from any part of the world." Crowds demanded that the book be read aloud if they learned it was being delivered to someone nearby. Poets sang Galileo's praises, comparing him, as Thomas Seggett did, to Christopher Columbus, who "gave man lands to conquer by bloodshed." Galileo's "new worlds [were] harmful to none. Which is better?" The term *telescope* was invented by mathematician and poet John Demisiani at a dinner in April 1611 to honor Galileo. Kepler wrote to Galileo, "Therefore, Galileo, you will not envy our predecessors their due praise. What you report as having been quite recently observed by your own eyes, they predicted, long before you, as necessarily so." Reportedly, a French queen "fell to her knees in the presence of her subjects, astonishing and frightening them," Richard Panek writes, so impatient was she to look through a telescope.

There were doubters. They accused Galileo of painting the planets on his lenses or simply said that because no one could see the Jovian moons with their own eyes that they were a fiction. It was true that telescopes were shaky and prone to all sorts of problems, from errant color to distorted images. Further, people had not looked through such an instrument before, so in a way they did not know how to see with it. To help combat this, Galileo took to showing his audiences some local earthly scene then waiting until evening to turn the telescope to the skies. Despite this, because the telescope was becoming more available, its powers were put into the hands of a wider populace. The doubters would eventually lose.

Some painters responded to this heresy by painting perfectly smooth white Moons in concert with the Virgin Mary. (Not all: Cigoli, Galileo's friend, painted the first post-telescopic, realistic lunar surface beneath the feet of the Virgin Mary in a Roman basilica.) Galileo lectured, sometimes winning over doubters, and he answered many letters, Van Helden reports. More crucially, attacks began in print, first with Martin Horky's *A Very Short Excursion Against the Starry Messenger* in June 1610. (Van Helden shows how cutthroat Horky was in his letter to Kepler, accusing Galileo of creating a spyglass that somehow fictionalized the sky even as he secretly made wax impressions of the lenses.) Initially, Church officials found no heresy against Galileo; his book had, after all, been approved by the censors. Even the man who had ordered the burning of Giordano Bruno, Cardinal Bellarmine, approved. But one man, Christopher Clavius, was disconcerted by the contentions that lunar dark spots were surface features. He steadfastly held to one of the ancient arguments—that the apparent imperfections were the result of changes in the thickness of the Moon. (Clavius, perhaps ironically, has a crater named for him.) Heilbron writes, "Galileo's mountains were philosophical as well as physical monsters. To Galileo's growing impatience, the Jesuit mathematicians on whom he had counted had declined to admit his or embrace another interpretation of the lunar surface." Even if Clavius and others who had reviewed Galileo's findings agreed the telescope was partaking of reality, the findings were still open to interpretation.

Wry, witty, persistent, arrogant, and determined, Galileo wrote a reply to these and other arguments that wondered, as Heilbron puts it, "What is so splendid about sphericity?"

Meanwhile, after getting a copy of *The Starry Messenger*, Harriot promptly went back to the "perspective tube," took notes, and made a map of the Full Moon as seen through the telescope, something Galileo had not done. Unlike Gilbert, Harriot assigned no names but instead appended numbers and letters to lunar features. Finally seeing topography, Harriot wrote notes like this: "A little ragged with a peninsula" and "it shewed mountenous . . . with an opening in the middest & some black passage from it." Famously, Harriot's colleague William Lower also could not at first conceive of a Moon with topography. He had compared the Full Moon to a tart his cook had prepared, "here a vaine of bright stuffe, and there of darke, and so confusedlie al over." After Galileo, whom Lower compared favorably to Magellan, he finally understood: what had been a mysterious "strange spottednesse al over" were the shadows of ridges and mountains.

In Padua, at the Arena Chapel, Giotto's *The Last Judgment* from 1306 depicts that scene with the Sun and the Moon among the mingled angels. Rather amazingly, the Moon seems to include a rendering of a single maria, possibly Mare Frigoris, though the other, more plainly seen dark spots are not included in realistic fashion. It's a Man-in-the-Moon image. Did Galileo have this in mind when he turned the telescope toward the sky in 1609? In any case, the Moon would loom above another judgment, one more personal, as Galileo continued his blazing trail of discoveries in the years ahead.

In 1612, Giulio Lagalla writes *On the Phenomena in the Orb of the Moon*, contradicting *The Starry Messenger*'s conclusions about the Moon's rough surface. The Wrangler had just been in a testy argument about why ice cubes, among other things, float (trivial today, but not then). In 1613, Galileo's work *Letters on Sunspots* asserts, contrary to doctrine and received wisdom, that there are blemishes on the Sun that can be seen with the telescope. This, and Galileo's harsh style, provoke far more hostility from his opponents and the Church than did his first book based on telescopic observations. In 1614, a Florentine priest attacks Galileo from the pulpit. A year later, while Galileo is ill, the first complaint against him is lodged with the Inquisition. In 1616, the chief theologian, Cardinal Bellarmine, warns Galileo to neither believe nor espouse heliocentrism.

Galileo is essentially cleared and even goes on to meet with the new pope, Urban VIII, with whom he has friendly audiences. He keeps writing, even dedicating a new book to the pope, who rather likes that. The niceties cease in 1632. In the popular *The Dialogue on the Two Chief Systems of the World*, Galileo defends heliocentrism and makes his book's proponent for geocentricism a character named the Simpleton. Thinking Galileo means him, the pope is not amused. The Inquisition convenes against him, Galileo travels while ill and is imprisoned in nice homes while threatened with torture. Finally, in 1633 he is coerced to offer an apology for contradicting the Church. Placed under house arrest near Florence, Galileo is old, sick, going blind, and mourning the death of his beloved daughter Maria Celeste. Still he works. He returns to experiments on motion and publishes a book that helps lead Isaac Newton to the laws of gravity. Just before he goes blind, Galileo turns his small refractor toward his old friend the Moon and discovers lunar librations, the daily and monthly variations in how it tilts toward the Earth. On January 8, 1642, Galileo dies.

Not until 1835 does the Church remove pro-heliocentric books from its ban. Not until 1992 does the Catholic Church officially admit error for what it did to Galileo.

I read again what Galileo wrote in *Letters on Sunspots*: "Why should our special hatred of death render fragility so odious?"

Decades ago, my father survived a small airplane crash and never flew again. Except once, after steeling himself with sedatives and beer to board a commercial flight to Utah where I taught at Utah State and we lived in the house along the Blacksmith Fork River. He saw that water, he saw cottonwoods, and close enough, though his hearing was bad even then, he could hear meadowlarks, his favorite bird of a motherless childhood in South Dakota. We drove up canyons, showed him the mountains we love, watched night fall. I sensed it then, how when presence ceases, like a bolt, memories are the only kin. Galileo was right. Fragility isn't odious. It's tender, and both fathers and worlds are born, live, and die.

CHAPTER FIVE

By Any Other Name

Naming lunar features was our first Space Race, as scholars have pointed out. While it didn't involve rockets, this far-earlier contest also embraced politics, prizes, and a quest for a kind of immortality. Naming the mountains, valleys, and concavities of the Moon would gain momentum with the spread of the telescope, with scientific curiosity and pride, and with the spread of power, commerce, and exploitation from Europe around the seas of the Earth. A Moon with names, it turned out, might help sailors navigate on this, our home world.

Early lunar astronomers hungered for the fame of being the first—and only—person to precisely map and permanently name the Moon. To bestow lasting words upon a known world and help us conquer this one—which of these men could resist the lure? And to have your name be one of those marking features of the Moon—well, who could resist that either?

Galileo neither mapped nor named the Moon. But he understood the power of legacy. In *The Starry Messenger,* his high-flown prose sometimes scrapes the floor with flattery. He called the four moons of Jupiter the Medicien stars, and the Medici family may have been pleased. The names did not last.

Galileo argues for the time-defying power of human commemoration:

A most excellent and kind service has been performed by those who defend from envy the great deeds of excellent men and have taken it upon themselves to preserve from oblivion and ruin names deserving of immortality [. . .] images sculpted in marble or cast in bronze are passed down for the memory of posterity; because of this, statues, pedestrian as well as equestrian, are erected; because of this, too, the cost of columns and pyramids, as the poet says, rises to the stars; and because of this, finally, cities are built distinguished by the names of those who grateful posterity thought should be commended to eternity. [. . .] Others, however, looking to more permanent and long-lasting things, have entrusted the eternal celebration of the greatest men not to marbles and metals but rather to the care of the Muses and to incorruptible monuments of letters [. . .] knowing full well that all human monuments perish in the end through violence, weather, or old age, this human ingenuity contrived more incorruptible symbols against which voracious time and envious old age can lay no claim.

In other words, words outlast stone. Perhaps. Perhaps not.

Rather than going to heaven after death, a name abides in the sky. One elaborate version of this was the belief of the philosopher René Descartes, who thought that souls migrated to craters named for them on the Moon. But named by whom? Could I name a crater on the Moon for my dead mother, my dying father? Does the person who names the crater have to believe in souls? Or the person whose name is on the crater? And what if you don't want your soul lurking about a crater that's been named for you anyway? Can you sneak off to an asteroid instead?

In such a view, names are as powerful as things themselves. To name is to create or destroy. Even in postmodernity, names, though suspect as literal referents, have the ability to confuse us in their endless play of meanings. They still have power. They can shock us in certain contexts; some words, words we know, are not to be spoken by most people in most circumstances: taboo.

Vows are words. Contracts are words. To speak is to weave a multivalent web of relationships, self to other, this to that, now to then or next, powerlessness to power or vice versa. To speak is to see.

❦

The names printed on my several Moon maps and guidebooks—I keep buying more—have been used for centuries now. I enjoy rolling their regal Latin over my tongue. Vallis Rheita and Mare Nubium. Rima Cauchy and Rupes Cauchy. Montes Agricola and Lacus Mortis. Oceanus Procellarum. Promontorium Agarum. Dorsa Smirnov. Catena Abulfeda. Sinus Amoris and Palus Putredinis. There is rhythm and elegance in these names, and, ever-embarrassed of nearly failing modern Greek in college, I use this Latin lunar overlay as a modest second language, one I'd begin to teach myself over summers at the cabin.

At the beginning of my lunar obsession, I didn't puzzle over the origins of the Moon's proper nouns. Apart from his vodka, who was Smirnov? Abulfeda? Arzachel? For several months, the generic Latin terms were enough. *Vallis* for valley. *Mare* for sea, the solid lava plains. *Rima* for rille—a stress crack or a channel for formerly molten rock or a graben. *Rupes* for cliff or a scarp. *Montes* for mountains. Other lava plains are *lacus* for lake, *oceanus* for ocean, *sinus* for bay, *palus* for marsh. A *promontorium* is a cape or headland. A sinuous wrinkle ridge of hard lava is called a *dorsum*, while a chain of craters is a *catena*, which, of course, is Latin for a Moon bar, where the Selenites gather to drink.

The craters and other generic topography have their proper names, and of course I knew some of the people they honored. Galileo, certainly, though the crater is modest, and there's a story there. Copernicus, for another, though his crater is grand yet set apart in isolation, and there's a story there too. Mythical craters: Hercules and Atlas, the imposing duo in the northeast quadrant of the Moon. Several famous Greeks: Eudoxus, Aristarchus, others I could not keep track of. Astronomers like Tycho and Kepler. So I knew that the other craters—Werner, Davy, Messier, others—had their biographies. Sometimes I look them up

when I'm observing or afterward, when I sit in a chair by lamplight. Some I
recall, many I don't. They form a palimpsest of actual lives across the strange
world of the Moon. Sometimes I see through that palimpsest. Sometimes I
see it as I see the surface.

Take Hercules and Atlas, two of the easiest craters to identify and remember.
They are set close together, like the bells of trumpets in an orchestra. Nearly
seven thousand feet deep and nearly forty-three miles across, Hercules is
the older of the two. You can tell because it has no particularly interesting
features on its floor. Apart from a big-O interior crater—which smacked into
the lava-flooded floor—one mostly sees just that lava floor. Lava-flooded
crater interiors are common on the Moon and the lava obliterates the kinds
of things that we can instead see in Atlas. Fifty-four miles wide and about as
deep as its twin, Atlas, as a younger crater, compels with its terraced walls and
a complex of mountain peaks, fissures rambling like gossip across the floor
and its spooky dark spots, the origins of all this as-yet largely unknown to
me. But I look. And I think it unfair that Hercules was less interesting since
he was the keener of the two, having tricked Atlas into putting the world
back on his shoulders when Hercules said he'd continue to hold it up forever.
But Hercules needed to pass it over to Atlas—*just for a sec, pal*—so he could
cushion his shoulders. Hercules promptly took off. Surrounding each crater
is the glacis or nimbus, an ejecta blanket like a halo, a circular field of strewn
debris, which is the outcast rock and regolith of the impact. (Regolith is the
Moon's "soil," powdery and rocky by degrees, but as lifeless as terrestrial soil
is teeming.) Perhaps it's okay that Hercules is the less interesting of the two,
because it's older and its namesake wiser.

Though these two craters weren't on my list of observing challenges from
the American Lunar Society, the program laid out by this now-defunct organization was my first systematic introduction to the Moon's surface. Without
it, I'd still be rolling about the Moon with little discipline. For more than
a year, I threw myself into the ALS program as though I were a Boy Scout
earning a badge or a college student working hard to get that A, staying up
late, getting up early, checking off features found, and reading about those I

had not yet spied. I learned the places, at least the most prominent ones, and their names, and how beautiful, how sublime, they appeared in ever-changing lunar sunlight. Because of the Moon's absurdly complicated orbital motions, how any given peak or rille or crater or low volcanic dome might appear on any given night would be subtly different the next lunar month and still again. You could observe a single feature over decades and see it differently each time. Truly obsessive lunar observers and illustrators, such as the British amateur Harold Hill in his lifetime, make such views their grails.

The project largely moved me across the surface from east to west, that is from right to left across the naked-eye Moon, though my telescope flips the image due to its design. I had no trouble making the adjustment, though at first I wondered if I would. Different Moon maps provide different layouts for how telescopes alter images in the eyepiece. I had in part avoided the Moon for fear of such directional confusion. It had been something of a neurosis. My mother inculcated in me an anxiety about knowing which streets to take in our hometown of Indianapolis when I borrowed her brown Chevette, making high school one long worrisome prelude to the Global Positioning System. Finding my way around the Moon proved easier than sophomore-year expeditions downtown to the *Indianapolis Star*, where I was a copy boy. Soon I divvied up my observing goals into four sections.

There, in the northeast quadrant of the Moon, on the wide lava plains of Mare Crisium, was one of the "wrinkle ridges," Dorsum Oppel—an arcuate sweep like the white spray of a stalk and grass seed blown in the wind. Though wrinkle ridges look like waves solidified into place, they are shallow upthrusts formed from surface contractions. As the Sun climbs in the lunar morning, Dorsum Oppel, like all such low ridges, eventually disappears. Years later, I'd get around to looking him up: Albert Oppel, a nineteenth-century German paleontologist. If big, complicated craters grab the easy visual glory, wrinkle ridges, found only on the maria, suggest modesty and quiet.

I had by now memorized the maria and, working through my visual traverses, I began to place craters and valleys and mountain ranges in reference to them, knowing that to the north of Mare Serenitatis I'd find two grand impact scars, Aristoteles and Eudoxus. I had trouble remembering their names until I called another complex of craters to the south and west "the Triple A"—Aristillus, Autolycus, and Archimedes. For the ALS I made detailed notes, including on the occasional use of a magnifying enhancing tube called a "Barlow" and a light-yellow filter screwed onto the eyepiece, which cuts glare and enhances detail. I had yet to internalize the Moon's deep-time epochs—the timeline of surface feature formations—so for now it was about where and what I was seeing. The when and how could come later. I worked.

19 June 2018 clear, still air, very steady, #8 filter, 10 mm with Shorty Barlow = 250 mag [Object] 24. Crater Aristoteles: Excellent example of a complex crater from the Eratosthenian Period. *The crater is 90% in shadow. Western side is v-shaped and sun lit with a black notch at the apex. The northwest wall is very white and seems steep, followed by a black band then to the west is a narrow less-white band. The southwest wall shows a notch near the south side. I can make out one dark pinnacle shadow. There's a white spot on the eastern wall hard beside the visible crater Mitchell. From north of Mitchell to the south side of Aristoteles is quite hummocky above the rim. The nimbus of the crater's northeast rim has another rectangular black notch. All in all, a lovely complicated feature right now.*

25. Crater Eudoxus: Good example of a crater from the Copernican Period. *Eudoxus is 95% in shadow. The western rim shows white/dark broken terracing, including a rectangular black notch on the southwest rim. I see the small crater on the northern rim. Badlands to east extend all the way to Lacus Mortis, the edge of which is straight, like a scarp. The somewhat v-shaped southern rim nimbus debouches toward an unnamed small mare field. V shape looks like a ghost crater rim Eudoxus intrudes upon.* [A ghost crater is a crater so flooded by later lava its circular rim is barely visible.] *The eastern nimbus is well-developed and seemingly orderly.*

I moved the telescope to the southeast quadrant, made up of white highlands, sometimes called "terra." This is heavily cratered and bright country,
hard to traverse by eye (and, I imagine, by boot). Craters upon craters upon
craters, like receding layers of solid clouds, the light changing them not just
over nights but over minutes, sunlight creeping, shadows moving, a slow
procession of profusion. The Great Eastern Chain of Langrenus, Vendelinus,
clocklike Petavius, and Furnerius appears in the southeast when both Mare
Crisium and Mare Fecunditatis are lit. Large and foreshortened, each of the
members of the Great Eastern Chain shows prominently. You can see how
shallow most lunar craters are, like the plate that holds the teacup and not the
cup itself, as one well-worn metaphor puts it. You can see how some craters
are more degraded by time than others. That degradation and overlapping of
craters was the foundation for piecing together the formation history of surface
features. Intrusion of one crater upon another is chronology.

Nearby, distinctly curving across the eastern Moon toward its limb is Vallis
Rheita, a giant canyon formed in seconds by flying boulders sent hurtling from
the massive impact that formed Mare Nectaris. Looking at it gives one a true
sense of the Moon as a globe. I gasped the first time I recognized it. You look
down the canyon as though you might land there. The gouge is remarkable.
About as long as the Grand Canyon, some 270 miles, Rheita Valley formed
nearly four billion years ago when an impactor blasted out the Nectaris
impact basin, digging out a cavity some three hundred miles wide and fifty
miles deep and sent about a dozen rocks—well, fast-flying mountains—in
a straight line that slammed, slammed, slammed, and slammed some five
hundred miles away, gouging Rheita Valley in seconds. The Nectaris impact
event has not been modeled, but researcher Brandon Johnson has studied
a similar-size basin on the lunar far side, Orientale, and found that it was
caused by the vertical in-fall of a forty-mile diameter rock hitting at nine miles
per second or some thirty-three thousand miles per hour. NASA's Barbara
Cohen, a lunar scientist, tells me, "Orientale formed later than Nectaris and
the thermal conditions of the crust might not be similar, but that's about the
best we can go on right now."

I think of the Rheita Valley whenever I'm in or by the great canyons in the American West, those that took millions of water-years to form, the Grand, say, or the Goosenecks of the San Juan. The violence of flying impact-created mountains versus the patience of water. Varying in width, Vallis Rheita looks in its western reaches like it's been scraped by a giant's trowel. As it bends across the lunar surface, the eastern portion thins. Here and there, craters punctuate its surface.

Vallis Rheita is easy to find, south of the Great Eastern Chain and before the vertiginous confusion of the highlands' brawling craters, a confusion that to this day can still devil me. To learn the major craters of the southern lunar highlands meant taking time and patience, traits I'd largely lost in Tucson. My lunar exploration brought them back.

Words place us, and we name places. George Stewart, in his classic book *Names on the Land*, writes of the North American continent before any human settlement: "Once from eastern ocean to western ocean, the land stretched away without names. Nameless headlands split the surf, nameless lakes reflected nameless mountains; and nameless rivers flowed through nameless valleys into nameless bays." Stewart's friend, the fellow Western writer Wallace Stegner, says this "commonplace" idea when "looked at steadily, begins to glow with strangeness and wonder." If Galileo fabled the power of a name to nearly outlast time itself, Stewart and Stegner more accurately saw names as originators of human time. Stegner says of Stewart that he "understood that nothing is comprehended, much less possessed, until it has been given a name." Eventually, as Stewart writes, "the names lay thickly over the land," for settlement had come, Indigenous then invader, and today we live the mixed legacy of those names: "Alder Creek and Cedar Mountain. Long Lake and Stony Brook, Blue Ridge and Grass Valley . . . the thing and the name were almost one." Evocatively, Stewart imagines that deep in the sounds of certain names might be prehistoric resonance: "Some name which first meant 'Saber-tooth Cave' . . . There is no sure beginning," he writes.

With the Moon it was different.

There were beginnings, several, and each attempt to lay designation across the face of this nearest globe embodied the scientific if not political vision of the namer and the cultural currents of the day. Names were given and they vanished, and names were given and some remained. The Moon is an archive animated by the lives of names and namer and witness. In the phrase of environmental historian Michael Rawson, this terrestrial naming of lunar features was a form of "anticipatory geography."

William Gilbert's naked-eye map of the Moon, drawn in the late sixteenth century but not published until 1651, had little effect on how other observers saw and labeled the Moon. The map was of no use in the new age of the telescope, and its names were mostly mundane, though I like his term for the present-day Sinus Medii, which is a small mare in the middle of the Moon's face: Insula Medilunaria or Middlemoon Island. Thomas Harriot's more detailed telescopic map with its letters and numbers also gained no adherents.

Not long after Harriot came a First Quarter map by Christoph Scheiner in 1614, who used letters on a chart. Four other observers made illustrations, though only the work of one of these, Francesco Fontana, saw print; Fontana's Moon is highly stylized, with the maria relatively well portrayed, but the distribution of craters seemingly uniform over much of the surface, almost like polka dots. Fontana's rays of Tycho crater—his work's most impressive feature—radiate thin and pointed like a starfish. Although Fontana's work also had no particular effect on selenography—the now-antiquated term for lunar study—his forgotten generic landscape terms are sweet: he speaks of black hairs, gems, little pearls, and small founts.

The first person to thoroughly map the Moon was Michael Florent van Langren, a Belgian savant who was a child when Galileo first looked at the orb. Known as Langrenus, he worked on this project for years, busy with other tasks, such as mapmaking here on Earth, but in 1645, he published *Philippian Full Moon, adorned with proper names,* the first map of the Moon that is recognizable as such to modern eyes. The features are carefully drawn, with maria stippled in, the craters round and shaded as though the Sun was shining late in the lunar morning, a practice other selenographers would follow. What would not remain would be many of his names, which largely honored Catholic royalty and popes. According to lunar astronomer Ewen Whitaker in his book *Mapping and Naming the Moon*, Langrenus deployed 325 names; 164 of them were eventually moved to different features while four of his remain where he put them. (Also lost to time are most copies of the map; only four exist.) What is today called Lacus Somniorum, the Lake of Dreams northeast of the Sea of Serenity, Langrenus designated as Lacus Scientiae, the Lake of Science, which I like rather more. Today's Mare Crisium he called the Caspian, a name that survived from Plutarch's day and was used by others, including Harriot. Whitaker suggests the term persisted because "it occupies roughly the same position to the Moon's face that the Caspian Sea does with respect to a map of Europe, N. Africa, and the Middle East." Langrenus called the entire southern highlands Terra Dignitatis, the Land of Dignity. The thin horizontal mare near the top of the Moon, Mare Frigoris, the Sea of Cold, is for Langrenus the Mare Astronomicum, the Sea of Astronomy. And the far northern cratered highlands are named for honor. Science, dignity, astronomy, honor. Good names for good things.

Flung upon the Moon in large measure to satisfy Catholic pride during Europe's bloody Thirty Years' War, Langrenus's crater names are a casualty of their timeliness and his hubris. He even threatened a fine—three florins—if anyone copied his map or used different names. I doubt he ever collected. Skilled with drawing and egotism, he named a mare and a crater after himself. The crater remains, Langrenus, near the eastern limb, the first truly monumental crater south of Crisium, one so large it once qualified for a term now out of favor: a ringed or walled plain. Langrenus is the first of those four craters in the Great Eastern Chain.

Apollo 8 astronaut James Lovell, one of the first three men to orbit the Moon, called it "quite a huge crater," and that it is. Langrenus appears oblong near the Moon's limb, but it is essentially circular, about eighty miles wide, ringed with terraces and with central mountains rising some fifty-four hundred feet high from its nearly fifteen-thousand-feet-deep floor. These features—steplike terraces and mountains in the middle—would become a refrain, though hardly boring, as I found other such complex craters on the Moon and as I learned why they looked so different from other holes punched in the surface. It's not far, as the eye looks, from this crater to the bright, white, battered southern highlands of the Moon. Lovell would not have known that Langrenus called the southern highlands the Land of Dignity, but he would have agreed, I have to believe, that it is sweet that something so battered and so ancient received such a befitting name.

✿

Three nights after I usually observe the Great Eastern Chain there appears a huge cliff—Rupes Altai, which stretches across part of the dignified highlands and that reminds me of the Book Cliffs of southern Utah or Cedar Ridge and the Echo Cliffs near the Arizona Strip. Rupes Altai curves in the southeastern portion of the Moon for about three hundred miles and attains in places a height of nearly thirteen thousand feet. It's unmistakable and was, for a time, my stopping place because beyond the scarp there occurs the disarray of dozens of named craters, overlapping like wild dogs at a kill.

Easier was scanning to the southwest, where north-south lines of larger craters had fooled some into believing there were weak faults that had erupted into volcanic depressions. Well past the ends of these illusory chains and just into the Moon's southwest quadrant is the most readily identifiable crater in the southern hemisphere: that of Galileo's friendly critic Clavius. It is so large and so shallow that standing in the middle you could not see the crater's encircling wall. This, one of the biggest and most memorable craters visible on the nearside, spans more than 140 miles. Its interior craters stare like goo-goo eyes, and the crater is a giant in my eyepiece. Once I had an effective magnification of six hundred times—about two times more than my usual highest—but steady air arrived now and then, and Clavius filled the view like a braggart. I could almost imagine that I spied its Moonbase from *2001: A Space Odyssey*.

If I had any misconceptions about all craters being round, it ended when I checked off Schiller, to the west and north of Clavius. An elongated crater, Schiller is a paramecium of stone, and a crater-rimmed broad valley curves about Schiller's southern edge so flat and wide it looks like a vast boulevard stretching toward the Moon's western limb. This is the Schiller-Zucchius Basin, whose first look for my ALS "Moon Badge" did not impress. Only when I saw the area lit by the low sun of a lunar morning did I appreciate that, like Vallis Rheita in the east, this perspective emphasizes in mesmerizing fashion that the Moon is a globe in space. The basin invites walking down its broad gray concourse over the limb, to the mysterious far side whose features I would, of course, only know from maps and photos and of whose names and topography I would have to make a later study.

Langrenus was bested soon after his map by a beer maker. (Apparently brewers could be quite wealthy and interested in science.) Johannes Hevelius of Danzig established the best observatory in the world in the 1600s, according to lunar historians William Sheehan and Thomas Dobbins, and this self-described "muse of the Moon" self-published in 1647 *Selenographia* four years after beginning his careful observations. *Selenographia* is a masterpiece. It is the first lunar atlas. I got to peruse it once at the University of Arizona Special Collections Library.

Crater edges are rendered, in Ewen Whitaker's words, like "termite hills," a convention of the time. They look like rocky spikes and are beautiful and wholly unrealistic. Hevelius shows the rays that extend from the larger craters, rays whose nature would also be debated; he thought of them as smaller mountain systems and rendered them like rows of rocky points. Crater proportions are not all exact to the modern eye, but that's not surprising given his equipment, which included a strange, mastlike telescope with lenses separated by 160 feet of wooden beams and cords, along with pulleys to move the assembly up and down. It shivered in the breeze. "It must have been exasperating to use," write Sheehan and Dobbins. It was all, they say, an effort to reduce the problems of distortion and false color caused by the lenses of the day.

Paging through Hevelius's *Selenographia* is like trying to tune in an old broadcast signal on a television. Sometimes the picture clears—yes, that is the Altai Scarp, rendered as a thick white line bordered in thin black. Sometimes the picture blurs—can that grayish ellipse be Aristarchus? The library's copy is nearly pristine, the white pages thick and easy to turn between the heavy leather covers. The title page combined black type with bright red, including the title itself. Beyond the elaborate frontispieces of clouds, robed men (Al-Hazen and Galileo), eyeballs inset in marble, and more, including a poem by Cyprianus Kinnerus and illustrations of the Hevelius workshop, the book settles into descriptions of all things lunar as well as other solar system objects. It was an of-the-moment summation of astronomy. And not just his—for his wife, Elisabeth, collaborated with him. Hevelius produced his book with gorgeous engravings of the Moon, from nightly phase-change maps to the whole Moon. The maria are crosshatched black, ejecta are white rays. I close Hevelius's book with a reverence for his expertise and recognize a kinship beneath it. He loved the Moon too. And his book was a stunning achievement in the history of astronomy.

Yet for all the ambition and artistic flair, Hevelius produced a kind of backward Moon, one with features named for the geography of ancient Greece and Rome. Hevelius said he experienced "perfect delight that a certain part of the terrestrial globe and the places indicated therein are very comparable with the visible face of the Moon . . . [I cannot] get over wondering how all this turned out so well." That was a stretch. The terrains are a poor match—and he was disingenuous. Hevelius knew of Langrenus's selenography, but said he was the first to apply names to the Moon. Often clunky—Lacus Hyperboreus Inferior—most of Hevelius's names did not carry forward, but the two most sublime mountain ranges on the Moon—Montes Alpes and Montes Apenninus—do.

Other mapmakers took on the Moon—or busied themselves forging Hevelius and Langrenus—but four years after *Selenographia* appeared, the third of three early titans of lunar astronomy appeared on the scene: Giovanni Riccioli, who published an excellent, highly accurate lunar map drawn by a fellow monk named Francesco Grimaldi. In his *Almagestum Novum*, Riccioli affixed an entirely new nomenclature to the Grimaldi map, this time honoring mostly ancient and deceased astronomers and philosophers with crater designations and, believing with many that the Moon influenced the Earth's moods of weather, he gave the maria the names we know today, Latin for Dreams, Rains, Storms, Mists and Death and Time and more. Riccioli also clustered together the names of related figures—Plato is relatively close to Aristotle, for example—and the older a figure is the farther north he is on the Moon. The southern highlands were reserved for intellectual titans closer to Riccioli's day. (These were all men, of course. As of now, there are about three dozen craters named for women on the Moon out of nearly sixteen hundred that are labeled. To address this, the International Astronomical Union is prioritizing naming lunar features for women.) Riccioli was not above the politics of the day. As a putative geocentrist he exiled the heliocentrists in Oceanus Procellarum, the Ocean of Storms. That's why we find Aristarchus and Kepler and Copernicus there, and it was he who gave Galileo a strangely tiny crater on that lonely expanse. (Whitaker believes that Riccioli was being coy, that he was a heliocentrist and that, apart from the crater Galileo, these other features named for heliocentrists are so striking that their placement enhanced their namesakes' reputations.)

A later commentator called Riccioli's Moon "the cemetery of astronomers," but nearly all of his Moon names eventually bested the system proposed by Hevelius. Both men's maps were used for more than a century, but it is Riccioli's names, by and large, that we find on today's Moon. Once again left behind was Langrenus, who complained, "Again, here is Father Riccioli, professor of Bologna, who has changed everything even though he had nothing but praise when I sent him my selenography."

☾

Of all lunar "water" features, Sinus Iridum—the Bay of Rainbows—is the one that most looks like water. At 150 miles wide, it is unmistakable in the Moon's northwest, below and beyond the western terminus of Mare Frigoris. The Bay of Rainbows was one of the first—and most gorgeous—lunar features I learned. Wrinkle ridges or dorsa seem to lap like waves, time-stopped, into this huge crescent bay, which opens to the massive lava plain of Mare Imbrium, the Sea of Rains. The whole bay is an old crater that tilted down so lava filled much of its interior, leaving at its back an arc of surrounding mountains, the Jura, the old crater rim that rises like a jagged scythe. When the Sun first rises on this scene, it lights the low wrinkle ridges as though white with foam. The tops of the Jura, at more than thirteen thousand feet, shine like lighthouses. At one end of the Jura, Promontorium Laplace is a striking and hulking headland poised above the flat gray of Imbrium, a gray dusted with white.

Nineteenth-century British lunar astronomer Edmund Neison waxes rhapsodic about this part of the Moon, calling the bay "splendid . . . well called . . . the most gorgeous and magnificent of lunar formations" and the Jura Mountains "stupendous cliffs of one of the loftiest great mountain highlands . . . [with] noble peaks towering fifteen to twenty thousand feet above the still dark plain at their base." He overestimated the elevation but their nobility remains. After all, as astronomer Peter Wlasuk points out, because the Moon is twenty-five percent the diameter of the Earth, a ten-thousand-foot lunar mountain is equivalent to forty thousand feet on the Earth.

The damp illusions of this part of the Moon seem all the more appealing when observing from hot and dry Tucson. I struggled. I struggled to keep my feet on this desert ground and to keep the Moon from being another form of escape. *The Moon,* I'd tell myself, *is shining on me here, and it's getting closer to full.* In the old adobe yard, I would admire the little bosque of mesquite trees by the whitewashed house, lime plaster made silver in Moonlight. It was a good yard. At our current house, I admire how Moonlight blanches the white blossoms of Texas Mountain Olive trees along a metal fence. I've gotten better at acceptance, but the truth was the Jura were closer to me than Tucson's ring of ranges, the Catalinas, the Rincons, the Santa Ritas, the Tucson Mountains. (Yet the streets of the barrio in which we live are right here.) The Moon is less inconstant than the poets suggest. Its steadiness, its emerging knowability, has become a compass. It doesn't point to escape. It points to home.

From the Jura, I sail south and west on Oceanus Procellarum, whose relative dullness accentuates a few good landmarks for navigation. Not long before the Full Moon, there are the Marius Hills, a creepy range of low volcanic domes that in low light rises like a blister that shows up overnight. Whenever I walk in the Marius Hills, looking down on their gentle slopes or looking up their gentle slopes, I breathe more deeply.

Across the Moon, the eye is drawn. Back in the northeast, there's the gash of Vallis Alpes running through its namesake mountains near the wide, dark crater Plato. I didn't know, the first time I began to observe in that region, that there was another, exceedingly subtle channel in the middle of the Alpine Valley, a rille that would come to obsess me, a thin passage for ancient lava that would elude me every time I'd look for it. Somehow knowing the Moon meant seeing that rille—somehow, someday, a new place in a familiar landscape. Then, far to the west, is far Gassendi, in Mare Humorum, with its crater floor grooved with lava rilles, a crazy scratching all about its central mountains. And all by its lonesome in the Ocean of Storms is the double-cratered, valley-cut, and subtle yellow-green of the eerie Aristarchus Plateau, a place I'd come to again and again, both with the telescope and with documents concerning strange lights sometimes seen there.

What to the untrained eye becomes a monotonous bounty of rocky bits, quilted lines, and many, many, many holes is, in fact, mapped and named, storied, and, as such, navigable. (There's even an observing project to spy a thousand named features on the Moon!) Early on, I could make myself lightheaded by turning from eyepiece to map, map to eyepiece, my finger in the air like a conductor tracing a shape while I spoke aloud its name and distinguishing features: "Clavius, '2001,' that curve of craters." I was scouting for a later version of myself.

One night after rain clouds cleared in Tucson, after talking with my father about how his day had gone—not well—I set the scope up quickly, not bothering with maps or a list of features to observe. I just wanted to look. I needed to be drawn out. I didn't realize until I was done that I had been moving across the surface of the Moon from a new source: from memory.

❧

In the seventeenth century, Jeremiah Horrocks described the lunar motion about the Earth more accurately than others had, though it was still exceedingly difficult to create so-called ephemerides showing where the Moon would be against the stars on any given night.

This mattered, for knowing where the Moon would be against the stars meant it could serve as a kind of clock. Figuring latitude—one's location to the north or south of the equator—was straightforward enough. On a clear day or night, one could determine how high the Sun or Polaris or the Southern Cross hung above the horizon. But longitude—distance east or west from an artificial meridian—required knowing the time at that meridian and the time in one's location away from it. Time, in effect, is distance. Without a mechanical clock that could allow such timekeeping—and it would not come until the pioneering work of James Harrison in the nineteenth century—sailors relied on their wits and their prayers. Also, they crashed a lot. The motivation to solve the longitude problem was not theoretical. The King of Spain, Philip III, even dangled a purse of six thousand ducats to anyone who could solve it.

The Moon moves precisely, every day and night, covering thirteen degrees in twenty-four hours. If "one has an accurate star chart and ephemerides of the Moon's motion worked out in advance for the standard meridian," then navigators could use this so-called lunar-distance method to calculate longitude, according William Sheehan and Thomas Dobbins in their sweeping (often gripping) book *Epic Moon: A History of Lunar Exploration in the Age of the Telescope.*

Those ifs, however, defeated "the lunars" method. The authors quote Sir John Herschel: "A Lunar at sea seems a rather bungling business." Despite Amerigo Vespucci's use of the lunars in his exploration, the method was doomed because the Moon's position could not be accurately predicted nor were star charts of the day sufficiently complete.

But what if one could eliminate the star charts altogether and dispense with worrying about predicting the Moon's movement in the sky? The first person to observe the transit of Mercury across the face of the Sun, a heliocentrist and theologian who corresponded with both Galileo and Kepler, the Frenchman Pierre Gassendi wondered if the mountaintops of the Moon could help us find our way at sea. He put the idea to the test in 1628, comparing in two locations the timing of different parts of the Moon's surface moving into shadow during a lunar eclipse (when the Moon passes through the shadow of Earth and darkens slowly). It was clear from this try that a highly detailed map would be needed to make this method practical, and Claud Mellan, observing with a telescope obtained from Galileo himself, set to eventually creating beautiful and precise images that even today are aesthetically pleasing in their luminous softness. They were light-years more detailed than previous drawings but, even so, Mellan, along with Gassendi and others, never finished a proper map of the entire lunar face. The ongoing lack of such maps was one reason this method also failed to catch on. Another was libration.

The Moon's motions are insanely intricate, and some of them are called libration. Libration amounts to different kinds of wobbling in which the Moon reveals portions of the far side. In fact, it's possible to see fifty-nine percent of the Moon's surface because of libration. One form of libration is a nodding north and south or libration in latitude caused by the Moon's being tilted a few degrees in its orbital plane relative to the Earth's. The east-west wobbling or libration in longitude occurs because of differences between the Moon's varying orbital speed and its steadier rate of spin. Even where one stands on the Earth can affect how much wobble the Moon seems to have. Libration complicated how to time the transit of light or shadow across surface features, so the Moon's brief navigational career never really took hold.

At least Gassendi's generic surface terminology of mons, mare, rupes, and vallis remains. Gassendi's proper nouns were, however, as ephemeral as the unfinished map, but they are poetic. The southern crater Tycho is called

Umbilicus Lunaris, the navel of the Moon. What we call the Sea of Serenity was, for Gassendi, the place called Thersites, a hideous and furious general of the Trojan War. We don't know to what Gassendi was referring with his names Shady Valley, the Salt Pits, Snowy Cliff, or Bitter Mountain—but they sound like places to explore.

The names first applied to the Moon in the seventeenth century precede all other extensive naming of objects in the solar system. As Scott Montgomery and Ewen Whitaker, among others, explain, lunar mapping and naming was a complex welter of the cultural and the personal. It was not a tidy movement from Galileo's descriptions of the Moon to the names on maps today.

Throughout the bestowal, the selenographers may have wondered: What did the inhabitants of the Moon call *their* mountains, *their* lakes, *their* craters? Lunarians, Selenites surely walked the Moon, and we could talk to them—they could tell us their nomenclature—for the Moon was a real world. It was a globe in the cosmos. Many were those who considered it a place of seas and forests and beings, for God could not waste a world. He would not leave it barren. Our names are a kind of life set upon the land. But the Moon was land that surely itself was alive.

Consider now how barren the Moon must seem, how devoid it is. If Earth's the right place for love, what about the Moon? We're wiser—well, we know more. There's no life on that rock. It took awhile to become amenable to that idea. The centuries preceding, in which astronomers populated the Moon with trees and insects and rational hominids, remind us of our essential loneliness, how it filled the first graves with flowers and cups for the next world and how we have looked up from the lone house of the Earth to make the sky—the Moon, first—a place where our needs must have something like angels.

CHAPTER SIX

The Gardens of Eratosthenes

At the southern end of the Montes Apenninus, the Moon's most majestic mountain range—a long, rugged, sweeping crescent on the shores of Mare Imbrium—the crater Eratosthenes stops the eye with its fortuitous placement, like the period in a calligraphist's exclamation point. Eratosthenes is a complex crater, the kind with a rampart rim, amphitheater terraces, and a mountain in the middle, high peaks that, with the rim, catch light before the Sun floods the crater floor with morning glare. As with all such craters, sunlight and shadow crawl and spike on the surface as hours go by. Eratosthenes is beautiful. Some thirty-five miles wide, it's a smaller version of Copernicus. And once it was a garden. Vines climbed its walls. Insects scuttled across its interior plain. For one man—the eccentric and derided early twentieth-century astronomer W. H. Pickering—these changes of light I see in my telescope were the changes of life itself. Pickering was one of the last to envision a living Moon.

He was far from the first. Some ancients had imagined life in the sky, and, though cautious on the question, Galileo's findings had given new impetus to dreams of a living Moon. Visions of lunar gardens, lunar livestock, and lunar bat-people would fill the pages of books fictional, philosophical, and scientific in the centuries between *The Starry Messenger* and the first barren photographs taken by probes in the 1960s. Perhaps William Pickering would feel partially vindicated to learn that microscopic tardigrades, or "water bears," may in fact survive on the Moon—the ones we launched there, that is.

Galileo's contemporary, Johannes Kepler, was a Selenite enthusiast. Selenites: those who live on the Moon. Long before Pickering there were astronomers who were certain they had seen forests on the Moon and the structures of a lunar city among so many various wonders. In one infamous episode, a nineteenth-century American journalist faked sensational reports of lunar life, fooling even credulous scientists. Pickering himself may have inspired H. G. Wells's vivid, fantastic scenes of Moon plants and more in his 1901 novel *The First Men in the Moon*.

This desire for Selenites of some kind—any kind—was less a matter of attempting repeated, scientifically valid observations of living things among the craters. This was hope. This was longing. Once Christianity had accepted that there are other physical globes in space—the Moon, Jupiter, Mars, and more—the question then was: What had God intended to do with them? Would God make empty worlds? Or would they be populated and, if so, by whom and in what state of grace? Were the worlds above as fallen as this one? The nineteenth century in particular was full of enthusiasm for and arguments about extraterrestrial intelligence or what was then called the plurality of worlds.

"Buddhism . . . accepts the idea of a plurality of worlds," writes Karl S. Guthke in *The Last Frontier: Imagining Other Worlds from the Copernican Revolution to Modern Science Fiction*:

> The Pythagoreans believed that the moon and stars were peopled by humans or similar beings; in the Middle Ages thinkers as different as Nicholas of Cusa and Teng Mu reflected on the individual characteristics of the inhabited planet-worlds. But in all these cases the idea was a matter of faith . . . For centuries, Platonists, Aristotelians, and Christians . . . refused to accept [this idea] . . . But only when the existence of other planets was *scientifically* established did

the dream of other inhabited worlds cease to be merely a product of the imagination.

That dream—with its ultimately untenable alloy of reasoning and theology—informed the eyes brought to eyepieces searching for signs of extraterrestrial intelligence.

Signs they did find, first on the world closest to us. And the possibility of lunar life—and life beyond—received its first popular treatment in the late seventeenth century in an unlikely, wildly popular book. Which was why, one cool, windy summer day, I did something I'd never done before. I climbed a fence in France to trespass onto the grounds of a charming chateau.

☽

Château de la Mésangère is where Bernard Le Bovier, sieur de Fontenelle, wrote the 1686 *Conversations on the Plurality of Worlds*. A fast-failing lawyer, Fontenelle turned to writing and was only twenty-nine when he published *Conversations*. It became an instant bestseller with multiple editions and translations over the years. A fetching fictionalized dialogue between him and a marquise, his lovely hostess at the lovely chateau, the book plumbed the depths of cosmic solitude and the possible plenitude of planets and life upon them. The marquise, based on the real-life widow of the chateau's owner, was only a little younger than the author. It's not a stretch to think they were lovers. After all, this is a man who, in his ninth decade, met an attractive woman and remarked that he wished he'd met her when he was younger—in his eighties. (Despite his roving eye, Fontenelle was centuries ahead of his time in the portrayal of and belief in the equality of the sexes.) The book was a major success for the aspiring man of letters, who, it must be said—in the common aside whenever his name his mentioned—died just weeks shy of his one hundredth birthday and who attributed his long life to eating strawberries every day.

I had come to his birthplace city, Rouen in Normandy, to find some traces of this remarkable figure. Rouen is a gritty metropolis, with little of the appeal—authentic and commodified—one finds in Paris. For a few days in June, Kathe and I stayed in an industrial loft whose entry stairwell was littered with pigeon feathers and broken glass. The museum that would have given me information about Fontenelle's life was closed despite it being high season. But in the old town, as an inset to a squat archway tower, the famous Great Clock hulked in elaborate lines and curves and shapes of blue-and-gold-and-black-and-even-more-gold, showing phases of the Moon, the day of the week, and the time of day. Born in 1657, Fontenelle, in his childhood, surely would have been transfixed by the fourteenth-century clock, one of the many such public timepieces coming into existence across Europe as increased trade and specialized urban labor required a shared and announced temporal regime. And I did see his birthplace from the outside only. It is a modest three-story Tudor painted beige and green and marked with a large plaque in his honor. A gateway led to an arch-covered passage and a courtyard chock-full of bins and bikes. The courtyard was worn around the edges and smelled of rubbish, a bit like the Middle Ages, I guessed.

One day, Kathe and I tried to arrange visits to the Château de la Mésangère. After laborious web searches and several unanswered calls, including to a local historian who previously had encouraged me to get in touch, Kathe, who is fluent in French, found a website that listed a government study of the chateau's grounds. The study included the name of the author, who, lifting my irritation, returned our call. Although he e-mailed the study—a history of the chateau's grounds and recommendations for its preservation—neither he nor anyone else could or would put us in touch with the owners. They were, he quietly said, divorcing.

"I didn't come all this way not to try to find it," I said, so we set out the next day from the streets of Rouen to farm fields, wrong turns, iffy cell service, and narrow byways, eventually pulling up on a sinuous lane by a little ditch running with water, a fence, and a wooden gate that said PROPRIÉTÉ PRIVEÉ, behind which was a grassy opening in the trees and a statue. *A statue.*

"This has to be it," I declared, for the study had proclaimed the estate's beauty in part because of several mythologically themed statues. When

Fontenelle visited and wrote his book in the 1680s, there stood, one source said, these stone figures of the ancient world, markers of the time Fontenelle was, in fact, writing against. He would side with the moderns in the then-heated debate on whether the ancients or the moderns were superior, according to Nina Rattner Gelbart. Science, though he did not call it that, was testament to the modern outlook of observing and testing reality, Fontanelle believed, rather than relying on mistaken didactic assumptions from the superseded past.

In *Conversations on the Plurality of Worlds*, Fontanelle put forth a modern outlook on the universe, one in which human significance is radically decentered and in which we begin to grapple with the sheer scale of it all—themes less explicit but still present in *The Starry Messenger*. The primary point he makes is that there are other worlds: We had discovered them in our solar system, so why not around other stars? Those worlds, if we thought about it logically, ought to be crawling with life. After all, the recently invented microscope had shown us what seemed empty—a drop of water—swarmed with tiny creatures then called "animalcules." What seems empty must be full. His modernity outstripped the mythic personae of the past, but, surely, the statues at the chateau imbued him with a sense of mystery and awe. His outlook owed more than a little to some ancients, such as the pre-Socratic atomist Democritus and the Epicurean Roman poet Lucretius.

Fontenelle writes in his preface that "it seems to me that nothing could be of greater interest to us than to know how this world we inhabit is made, if there are other worlds which are similar to it, and like it are inhabited too." (He admits that only "those who have thoughts to waste can waste them on such things; not everyone can afford such unprofitable expense.") He emphasizes the use of "verifiable physical tenets" and avoiding the "totally fantastic." Crucially, he claims that any sentient beings beyond the Earth are not "descendants of Adam," sidestepping the thorny theological question that would haunt the plurality of worlds debate for at least two centuries: If God has created an infinity of worlds—which his omnipotence suggests he did—would he allow his son, Jesus Christ, to be crucified ad infinitum? That would seem a cruel stroke.

Gathering my notebook and camera, I told Kathe I wouldn't be long and that I'd play the innocent tourist if someone were home. One website had said the chateau was open to public visits, but we had come to the back of the estate. And there was a wall. So, convincing myself that jails here couldn't be too bad—nice cheese at lunch?—and grateful for Kathe's French, I clambered over a thicket and the crumbling stone wall and a halfhearted section of chain-link, thus officially trespassing in another country for the first and last time in my life.

The swaying trees—linden, beech, and hornbeam—transfixed me. Under this tree-dark tunnel, on a grassy path overgrown with goldenrod and dandelions, I walked, stopping at a circular clearing where I touched a dingy white statue of two figures set atop a high pedestal shaped somewhat like a balustrade. The god of the south wind, Zephyrus, whose wings were broken, still guided his wife, Flora, the goddess of flowers. His left leg was missing, a hand as well, and the path beckoned in shade, sun, and dapple. Robins and blackbirds moved among the trees that lined the broad path toward, I glimpsed it, the wide lawn behind the chateau, where, centuries before, Fontenelle and the marquise had walked and pondered the night sky. Clouds scudded over kiting hawks, and there were seconds, minutes, times of washed sun and dark sprinkle, wind-tossed drizzle, windswept hawk, reeling around me as the understory ferns quivered.

When the path debouched into the semicircular lawn behind the chateau, I stopped. The house seemed empty so I untensed. What a night sky they had, circled by trees! Statues in shadow dotted the perimeter, and I orbited the garden, dazzled. This was where they spoke of worlds.

The conversations began very much on Earth and very much in the body. "One evening after supper," Fontenelle writes, "we went to walk in the garden. There was a delicious breeze, which made up for the extremely hot day we had had to bear. The Moon had risen about an hour before, and shining through the trees it made a pleasant mixture of bright white against the dark greenery that appeared black." The night, he says, provokes freer thinking than day. After reviewing the errors of "Antiquity," with its myriad pronouncements on nature that empirical observation had shown to be incorrect, he proclaims that "whoever sees nature as it truly is simply sees the backstage area of the theater." This is preferable to "a false notion of mystery wrapped in obscurity." The primary falsity had been believing in the centrality of humans, so much so that we have thought the universe "is destined for our use." As he finishes a discourse on the new heliocentrism, Fontenelle and the marquise soon turn their attention to the nearest part of the universe in the sky—the Moon.

"The Moon turns around the Earth and never leaves her in the circle the Earth makes around the Sun. If she moves around the Sun, it's only because she won't leave the Earth," he says.

"I understand and I love the Moon," the marquise responds, "for staying with us when all the other planets abandoned us."

After Fontenelle correctly explains why we see only one side of the Moon, he says he

> sometimes [will] imagine that I'm suspended in the air, motion-less, while the Earth turns under me for 24 hours, and that I see passing under my gaze all the different faces: white, black, tawny, and olive . . . At first there are hats, then turbans; woolly heads, then shaved heads; here cities with belltowers, there cities with tall spires and crescents; here cities with towers of porcelain, there great countries with nothing but huts; here vast seas, there frightful des-erts; in all, the infinite variety that exists on the surface of the Earth.

By establishing the Earth as a world, living and diverse, from this god's-eye perspective, Fontenelle suggests the power inherent in the human ability to reason and the likelihood that the Moon and worlds beyond, are more, not less, like the Earth.

Although the selenographer Giovanni Riccioli had said explicitly on his map that the Moon was not inhabited, Fontenelle, as with many others, didn't get the message. It's on the second of the five evening walks that the marquise and Fontenelle plumb the possibilities of lunar life. First, he says he believes the Moon is inhabited but could be convinced he's wrong. (In its way, this is as revolutionary as René Descartes's famous skepticism.) Fontenelle compares our knowledge of the Moon to the knowledge of a Parisian looking at a distant town; she would see buildings but not people.

"We want to judge everything," he says, "and we're always at a bad vantage point. We want to judge ourselves, we're too close; we want to judge others, we're too far away."

Some things are clear. He explains that the Moon reflects sunlight bouncing off the Earth to the Moon and back to our eyes. He explains librations and eclipses.

But now "our learned men . . . travel [to the Moon] every day with their telescopes," he replies, revealing "lands, seas, lakes, soaring mountains, and deep abysses." After the marquise inquires as to how water has been discovered, Fontenelle repeats the convention that the dark patches are lakes and seas, the brighter areas are land.

It doesn't take long for Fontenelle to flirt again—there's been a lot of flirting—as they joke about the claim that all things lost are gathered on the Moon, including, he offers, "the sighs of lovers." Still, if there are creatures up there, not just lovers' sighs, we cannot call them human, Fontenelle confesses, but beyond that, we cannot say much. After all, could such creatures imagine us, with "our mad passions and such wise reflections; a life so short and views so long; so much knowledge devoted to insignificant things and so much ignorance of things more important; so much love of liberty and such an inclination to slavery; such a strong desire for happiness and such a great inability to achieve it?" Who could imagine?

Easier to imagine the landscape of the Moon, and, to the marquise's dismay, Fontenelle suddenly backtracks on his contention that the Moon might have seas. He puts modern skepticism into practice. This rocky Moon might be desiccated or have "dews" in air "a little different from our air" and vapors "a little different from our vapors." What is needed is an "interior motion of the Moon's parts, or any produced by external causes," which would make the Moon a dynamic world, like ours, that could produce "fruit, grain, water" and lunar beings "about whom I don't know either."

Yet the possible lack of water may not be a hindrance. He points out that "even in very hard kinds of rock we've found innumerable small worms, living in imperceptible gaps and feeding themselves on the substance of the stone." This may well be the first reference to the possibility of endoliths—rock-loving creatures—on other worlds.

Thinking of Fontanelle and the marquise, I looked at the walls of the chateau as though they themselves might be full of life, endolithic or otherwise, and admired this rather modest brick affair of sagging and worn elegance—altogether alluring—with high windows and corners punctuated by white quoins, tall shuttered doors beside a small balcony, a mansard roof, and wide stairs leading up to a fence of pointed iron. There, a footbridge guarded by a useless cannon, crossed a grassy moat to a terrace of broken-up white stone, a chip of which, covered in black lichen, I tucked into a pocket. There were vines. There were clouds of roses, red and yellow, more lindens and splashes of asters and daisies and flowers and trees I didn't know, a veritable tumuli of chlorophyl, heaps of life by old stone. I could almost hear Fontenelle and the marquise talking quietly as nightingales called over Moonlit statues.

I wanted to spend the night here. I could look at the Moon from here. I wanted to live here, with windy days and still nights. I wanted to talk with the never-gods. Here, with messenger Mercury, in winged headdress and winged shoes, towering above me in the leafy cosmos. With delicious Bacchus, leaning insolently, holding grapes by his head and a cup before his chest. With an armless, toeless, helmeted Mars. With Apollo, who kept a hand on a lyre, Sun-god brother of absent Diana. The sky darkened. By the stairs, atop their pedestals, Hippomenes and Atalanta sprinted toward the grove to escape the rain. Allied with Fontanelle, I too side with the moderns. But who could fail to be entranced by this little green universe of myth in stone at the edge of Fontenelle's mind, he who imagined talking with Selenites, who believed that flight "will be perfected, and some day we'll go to the Moon"?

There we'd experience the Moon as the Moon. Fontenelle accurately describes sunrise at the terminator: "You're in deep darkness, and suddenly it seems as if someone draws a curtain; your eyes are struck by the full brilliance of the Sun. Then you're in vivid, dazzling light . . . Day and night aren't linked by a transition which partakes of them both. The rainbow is another thing that's lost to the Moon."

The lack of air and the heat of the Sun would compel lunar beings to live underground. Again, he relies on analogy: "Subterranean Rome is nearly as big as Rome on the surface . . . A whole nation exists in a crater, and from one crater to another there are underground roads for communication between peoples." It is the connection, that communication, which Fontenelle emphasizes, giving his speculations a human heart.

"A little weakness for that which is beautiful, that's my sickness," he says. "The other worlds may make this one little to you, but they don't spoil lovely eyes, or a beautiful mouth; those have their full value despite all the possible worlds." All these worlds—Fontenelle has described those of the solar system and is about to raise the likelihood of those around other stars, even though all stars die, extinguished like candles in a bedroom.

"Ah, Madame," Fontenelle says, "rest assured, it takes time to ruin a world."

"Nevertheless, isn't time all it takes?"

It is, he says, "a still longer test." For we are mistaken if "we make our lifetime, which is a mere instant, the measure of some other . . ."

The measure of this remarkable book and figure is considerable. "The ideas he was bandying about were bold, controversial, even forbidden," explains scholar Nina Rattner Gelbart. Yet they "suddenly became the rage." Bruno had been burned at the stake for heresy. Galileo had been placed under house arrest. And the Church banned *Conversations on a Plurality of Worlds* in 1687, a year after its first publication and the same year as Newton's *Principia* appeared. Fontenelle kept publishing new versions anyway.

I stepped across the stone terrace that flanked the chateau's back, walking above the garden lawn, and peered into windows of wavy glass. Evidence of life: books on a table and reading glasses. Rain sprinkled and stopped and I recoiled from my voyeurism—this was a *home*—and I nearly tripped, hustling off the terrace and back the way I'd come, across the lawn and down the wide green path and past gray Cupid, back to Kathe and the tiny car. More breathless than remorseful, I hauled over the fence, spearing my thigh and tearing my shirt. Then Kathe drove us around to the front, where we stopped, got out, and walked up to another fence and its ornate gate of gold paint.

Which Kathe pushed open. "Or," she said dramatically, "you could go through here."

I groaned. So we ambled up the long drive, saw a goose—a living one—and a drying pond and cords of wood and outbuildings—the ruin of a pigeon pen from 1604—and crates of slates for roofs in need of restoration. There was a small herd of horned cattle and a distant church. The back garden with its trees and roses and grass and statues had been a temporal ecotone. Now this was just a house with people trying to live their lives and maybe fix what was broken. They were absent, I thanked them, I wished them well, and we walked to the car. As we left, I thought of Fontenelle, who, by starlight, by planet light, by Moonlight, by candlelight, created, by the logic of the day and the talent of his pen, a universe starting with the Moon, as full of life as the chateau trees had been of wind.

☙

Fontenelle was not sui generis. With logic and panache, he brought ideas about extraterrestrial life and intelligence to a much wider audience, but others had wondered along those lines for a long time.

"As far back as 1588, candidates for master's degrees at Oxford had . . . to debate the question of 'whether there are many worlds,'" writes Karl Guthke. Orpheus spoke of dwellings on the lunar globe. Democritus and the atomists—whose cosmology called for infinite worlds—had been pushed aside by Plato. (Historian Steven Dick writes in *Plurality of Worlds: The Origins of the Extraterrestrial Life Debate from Democritus to Kant* that in ancient Greece "worlds" were not planets in the sense we use today but rather "other universes entirely beyond the range of human senses.") Lucretius was widely read, but his advocacy of atomist ideas of multiple worlds had not sparked a revolution. In Plutarch's "Concerning the Face which Appears in the Orb of the Moon," he has characters speak for the atomist view, as well as the Aristotelian objections, on everything from habitability to illumination.

A couple of centuries after Lucretius, a boy named Lucian was born in the town of Samosata in present-day Syria. Raised into a hardscrabble life, Lucian was set to be a stonemason until he escaped to Ionia, where he learned Greek and the verbal arts, eventually becoming a public speaker, educator, and writer. It is with Lucian that the Western tradition of the lunar voyage commences, tales often relying more on terrestrial satire than cosmological reasoning. He wrote two, the second of which, "A True Story," was so influential it "inspired a long chain of literary descendants" according to science fiction author, editor, and critic James Gunn.

In Lucian's tale, a storm sends a ship hurtling to the Moon where the crew finds a surface "inhabited and under cultivation," while in the sky "was another land mass with cities, rivers, seas, forests, and mountains: we guessed it was our own Earth." Soon the narrator is caught up in a war between the peoples of the Moon and those of the Sun. There are troops made of buzzards, the "Saladbird Cavalry" (the bird is covered in lettuce, not feathers), Peashooters, and Fleaborne Bowmen, among other oddities. Armor is made of beans and lupine husks. The solar troops are equally ludicrous, with an Ant Cavalry, for example, and they are ultimately victorious. Out of spite, they set up a wall to cut the Moon off from all sunlight. Eventually, a more equitable peace is sworn, and our narrator is free to learn more. There are no women on the Moon. Young men carry babies in their calves. The men can cut off a testicle, plant it, and when this penis tree bears fruit, "men are hatched from them." These men use ivory or wooden penises for penetration, "never die of old age but dissolve and turn into air, like smoke," eat only frog, do not pee or poop and, lacking an anus, have intercourse behind the knee. One-toed, long-bearded, cabbage-rumped, milk-sweating, with removable eyes and a belly pocket, these lunar men "wear clothes of flexible glass" or, if "poor . . . of woven copper," for the Moon is well-endowed with copper. More poignantly, Lucian describes a mirror suspended over a well; with it one can hear conversations and see the sights of distant Earth.

Many centuries stretch between Lucian's story and the next important imagined lunar voyage, written by Johannes Kepler, a reflection of the fact that until the telescope began to reveal the Moon's potential nature, stories of the Moon largely continued to be long-held myths and fables. It took the telescope to unlock not only the scientific Moon but also a more boldly imaginative one, and, as we'll see, the line between fact and fancy did not sharpen until well into the nineteenth and twentieth centuries.

Careful to keep his lunar-life enthusiasms mostly to himself, Kepler was afraid of crossing the authorities since his mother had been accused of sorcery and he had worked hard to free her from prison. His *Somnium, or Lunar Astronomy* was passed around privately after Kepler wrote it in 1610. (James Gunn says this manuscript might have been one of the reasons for his mother's arrest in 1620.) She died soon after her son freed her, and Kepler himself died ten years later, and the book appeared to the public in 1634.

Stephen Dick points out that Kepler had observational foundations for his lunar text. He used a camera obscura "to form an image of the moon on paper; the image contained dark spots no matter how he moved the paper . . . the spots were real and not a product of eyesight." He found "that the apparent mountains seen on the moon during a partial lunar eclipse were not shadows projected from the Earth's mountains" but native to the Moon itself.

To Galileo, Kepler was bold. He spoke of how a lunar atmosphere would be conducive to life and wondered if the circular features—the craters—were made by sentient beings. He recalled how a former teacher of his had observed rainfall on the Moon. Kepler claims that on Palm Sunday 1605 his teacher Maestlin watched a lunar eclipse during which dark rain clouds covered a portion of the Moon. (It was nothing of the kind, of course, being just an effect of the Earth's shadow through which the Moon passes in a lunar eclipse.)

Though the book is called "a dream"—Kepler's narrator arrives at the Moon in a magical sleep—it is a comprehensive scientific account of the lunar orb.

He calls the Moon "Levania," and, in getting there, the narrator experiences "extreme cold and impeded breathing," but "up there," he says, "we are granted leisure to exercise our minds in accordance to our inclinations." Kepler names the Earth-facing hemisphere Subvolva and the far side Privolva, with the Earth itself called Volva. After dispensing with the journey, Kepler focuses on "the nature of the region itself, starting like the geographers with its view of the heavens." Kepler's "dream" is a compendium of the physics and timing of the Moon's complicated motions, motions that suggest different seasons and environmental zones on a Moon with air and water. His narrator observes mountains, valleys, short-lived but giant creatures, towns, deserts, forests. Creatures are crepuscular and cool their water in caves. In a gesture surely made with his teacher in mind, Kepler conjures rain clouds on the nearside of the Moon.

I've seen no weather on the Moon, but one night I watched for the first time something ephemeral and gentle, what counts, I guess, for the weather on the Moon: a Sun-rising light slanted across something nearly submerged, as though rising to the light, something wended and convoluted. I watched dawn break on the gray braided ridges of Lamont. Splashy craters get all the attention. I had identified and admired several by this point. Now, looking for other features in my quest for an American Lunar Society certificate, features often quiet and easily missed, I gazed transfixed upon odd Lamont. Lamont may be a caldera swamped by its own lava; the wrinkle ridges might be the surface sign of magma dikes. But the common view is that sometime long ago a rock hit the Moon there, punching a basin down and impact rims up. Later lava flowed across the region like a rising sea, filling the basin nearly to its top but leaving the kelp-strands of Lamont in place, an interior oval partly surrounded by an incomplete wider encirclement and touched by wide radiating ridges, like a round, flagellate creature. Lamont's ridges break the surface in low light. I stared for a long time as that light moved across the ridges. This stirring of the underneath surfaced like uncertainty. Lamont is briefly seen. High Sun obliterates the textured relief of a few hundred yards' height. It is a giant ghost crater, with an outer ring diameter of some seventy-five miles.

I found Lamont early in my studies for the observing certificate. This was a couple of years before my father's diagnosis. Even then, I knew and felt the finality of our time, mine and his, all our loved ones, and knew and felt the wisdom in a quote from Albert Camus I'd find the day we put down one of our cats: "The misery and greatness of the world: There is no truth in it, but only objects for our love." My father, my wife, our cats, living and dead, the fossil traces of Lamont. Does it seem strange that the Moon can become so companionable? As eerie and pleasant as it was for Fontanelle and the marquise?

There are other low ridges on the Moon, the wrinkle ridges that are little conflicts in the crust, where the contractions have pushed up long lines of lunar rock. Many have names, and I identify some for my program, though one stands out in particular. North of Lamont, on the eastern reach of Mare Serenitatis, running north to south for 140 miles, is Dorsa Smirnov or Serpentine Ridge, its better-known and more poetic name. Undulant and seemingly braided, the Serpentine Ridge terminates its twistiness in a y-shaped split at its northern end. If you want to get a feel for the Moon's low ridges and rims, whether of ghost craters and ghost basins or of dorsum, walk a mountain trail and trace the journeys of the tops of exposed pine roots in the dust. That's what they look like. Or, literally at hand, look at the blue veins between your wrist and knuckles. Your body as Moon.

Four years after Kepler's book was released, the English bishop Francis Godwin published a fictional and popular book called *The Man in the Moone*. In it, we see the developing perspective that extraterrestrials will be better than us, smarter, and happier. As well, says scholar Michael Rawson, the book "popularized the idea that the earth and moon might share species and constitute what we would refer to today as a single ecosystem." That same year, in 1638, another cleric, the English bishop John Wilkins, the first secretary of the Royal Society, published a speculation about the Moon called *Discovery of a New World*. In 1657, Cyrano de Bergerac's *A Voyage to the Moon* came out two years after the death of the rakish figure.

Wilkins's book, which was nonfiction, was central to this new literature. He considers questions of gravity (then called "sphere of attraction"), the chilling attenuation of the atmosphere, and the physical needs of space travelers. On the latter question, he reasoned that since bodies in free fall to the Moon will not be exerting themselves, neither food nor sleep will be necessary. (The astronauts would find otherwise.) He argues that the Moon is solid, separated into sea and rugged land, and shines by reflected sunlight. This habitable Moon, he says, has an atmosphere. This question of lunar air would remain vexed.

The lively Moon became also the province of poets, who previously had used the orb mostly to light bittersweet romances in gardens or to symbolize faithlessness, love, or both. In 1737, Thomas Gray, best known for "Elegy Written in a Country Churchyard," composed a poem in Latin hexameter called "Luna Habitabilis," in which the speaker declares the old modes of transport to the Moon—all those fictions!—were obsolete. *All one needs is a telescope.* Gray is writing in the spirit of Jeremiah Horrox, who wrote of the telescope in 1639, "Blest with this, / Thou shalt draw down the moon from Heaven." In a prose translation, Gray writes, "Just apply yourself to the little tube (you have reached a good position and are looking aloft from a hillock); as soon as you enter the bottom of the tube with gaze thus sharpened, the lofty mansions of the sky will be revealed. Instantly, when you have ventured to gaze upon the realms of the moon, you will walk upon the earth but place your head among the clouds."

He is sure he sees oceans and "shadows cast by groves of trees." It is a land of dew, clouds, and rain. Gray asks, "Can you believe that a world so vast lacks some kind of inhabitants? These beings till their fields and found cities of their own. No doubt, too, they wage war, and when they are victorious celebrate triumphs: here too glory has its fit reward. Fear and love and mortal chances affect the minds of these creatures." Here the predominant conceit of superior intelligence on the Moon and planets is set aside.

In her 1948 book *Voyages to the Moon*, literary scholar Marjorie Hope Nicolson writes that the Moon and other possible worlds became a passion because "perhaps the spirit of melancholy and despondency that marks so much late Elizabethan literature had some basis in [the] fact . . . that the unknown was known, the remotest regions of the world written down on maps." This is a bit of an exaggeration—consider the lateness of true polar exploration—but it goes a long way in explaining the excitement that met the revelations of the telescope. Nicolson wrote in an article preceding her book that this interest in lunar mapping was not strictly a specialist's pursuit. "Such maps were pored over by laymen, interested in the terrestrial analogues they disclosed," she says, adding that globes were also objects of fascination. There were even "playing cards" with astronomical information for sale in seventeenth-century London.

These interests in extraterrestrial life merged with ancient dreams of human flight—and reasonable speculations of the same—by such figures as Roger Bacon and Leonardo da Vinci. Still, imagined lunar voyages usually required some fantastic means of transport—dreams, birds, angels, storms. More practically, wings could be attached to the body or, as Wilkins put it, you could go "by a flying chariot." These last two approaches supplanted the supernatural or absurd. (Nicolson points out that "the sophisticated modern reader" may scoff at tales of bird-drawn humans flying to the Moon, but, in the seventeenth century, "ornithology was still in its infancy. Birds hibernate somewhere. Why not in the moon?") Between Cyrano's suggestion, in Nicolson's words, of using "fireworks or some other forms of gunpowder" and, late in the nineteenth century, Jules Verne's use of a cannon, Nicolson notes, only a couple of other authors made use of this motive force.

In many stories, once visitors reach the Moon, they witness more of the usual giant creatures, luxuriant plants, and even Utopian cultures. All of these beings would, of course, become British subjects. In *Annus Mirabilis* the poet John Dryden wrote that the British would voyage first to the Moon and claim it for the monarchy. There they would see the Earth and, in his memorable line, "view the ocean leaning on the sky."

I didn't know of Cyrano's suggestion to use fireworks to reach the Moon until I began researching this book, but it reminded me immediately of that summer evening when my sister and I—mother working the night shift, father having left us—fired bottle rocket after bottle rocket from behind our tipped-sideways picnic table. We set the rockets in an empty Coke bottle and lit the fuses. We didn't get to another world, but it was thrilling.

We launched our craft with a childish combination of nervous giggles and awe. I remember the color of the arc and explosions as a kind of electric green. They flew so high! The hiss and sizzle and crack! This was the backyard of the house I remember best from childhood, on Sawyer Street. The treeless yard had a concrete patio by the back door and a wide expanse of not especially welcoming grass. I have no memory of setting up my Space Conqueror there, but from the second-story windows the Indianapolis sky seemed huge over the roofs behind us.

Roofs under which angry neighbors called the police. Terrified when the officer arrived at the front door, I stood as high as his polished leather belt. Having advised me to say we were sorry and that we didn't have any more bottle rockets, my sister surely chided me when I handed over every remaining one. The next day I might have "hid" in the space between the front of the house and a giant juniper bush to ease the tongue-lashing that surely occurred. I loved that front yard with what seemed a massive tree—maple? oak?—I once climbed to shake a branch under which a hornet's nest was stirred. It went badly.

Yards are our little living territories, whether beside a chateau or behind a cheap American tract house or on a fictional Moon, and those of my childhood were rife with struggle. Fearing the angers of the house and suffering from a "lazy eye," more than once I burst through glass doors to escape into the back. My parents would pluck shards of glass from my skin and hair. It wasn't until my ex-wife and I had a house in Kansas, whose backyard was lined with tall cedars and full of hackberries, one with a cavity in which northern flickers nested, that I felt in a yard some measure of calm. Ever since, the flicker's been a favorite bird, and I see life everywhere, though not on the Moon. Kathe and I had a small backyard in our first rental in Logan, Utah, and you could see the mountains from there—you can see the mountains from a lot of yards in Utah. There and in Tucson, the yards became sacred ground for the ritual of hauling out the scope, setting up a table, a chair, a place for charts and maps. Both refuge and point of departure, the telescope-gifted yard is monastic and communal. I am usually alone but I am brought to feeling the kinship that Sigmund Freud himself said he never experienced but that he named, beautifully, *the oceanic*. I imagine cutting strawberries for Fontanelle while he looks, dumbfounded, through my ten-inch reflector.

Our cabin in Logan Canyon is edged in the back by a rocky slope of maples and junipers. To the east, there is a small, rising grassy meadow with straggly aspens. To the west, our little patio and more maples and lilacs beside an old stone fireplace. The cabin's front opens to half a lawn—the other half is dirt in which we cannot get a thing to grow. There's the small dirt drive and beyond that, another lawn, the firepit meadow where I boiled blood and wine and an eye and a frog. It's weedy but open. I can see western tanagers sitting on the tops of willows leaning on the sky.

It is from there that I saw my first catena, a chain of craters pinging one after the other as evidence of multiple impacts (or volcanic outbursts). Catena Abulfeda is named for the crater from which it spins off, like sparks from a flywheel, like rapid-fire bottle rocket explosions. I used the unmistakable Theophilus as a marker, moved west and a little south. As a crater, Abulfeda isn't much, one of the myriad degraded craters lacking the grand features of younger complex impacts like Theophilus. But it's easy to see with its catena jetting off from its rim past Almanon and Tacitus. Chance dropped the multiple impacts in a 130-mile-long chain as a comet or asteroid dropped its debris directly or as the secondary ejecta from elsewhere. In places, it's whitish, like firework smoke.

How many craters are in this chain? Indeed, how many at all? As with so much on the Moon, a seemingly simple question yields more than one might think. "The number of lunar craters depends on the minimum size chosen," lunar scientist Charles Wood tells me:

> A recent Chinese paper tabulates 1.3 million all around the Moon larger than about ten meters. A Full Moon count of craters [bigger than] twenty kilometers [about twelve miles] yielded about five thousand, and a catalog I worked on . . . in the '60s counted about eleven thousand [bigger than] 3.5 kilometers [about two miles] on the nearside. The International Astronomical Union is only concerned with named features and has a list of about seventeen thousand craters . . . that have a name or letter designation.

As to how many make up Catena Abulfeda, Wood told me, more or less, take your pick.

When I saw Catena Abulfeda for the first time, I exclaimed—I did that a lot as I saw lunar nuances for the first time. Now I smile in recognition and feel the way I feel when I see infinite fields of golden Mule's Ear in bloom or when I hear mountain chickadees with their burry notes dropped into the air, one after another, a line of sound from one creature to another, countable but, really, countless.

Critics of lunar life—indeed, of science in general—did not stay quiet in the face of what they considered to be dangerous speculation. There were readers who would find John Donne's 1611 verses far from reassuring: "And new Philosophy calls all in doubt, . . . / The Sun is lost, and th' earth, and no mans wit / Can well direct him where to looke for it. / And freely men confesse that this world's spent / When in the planets, and the Firmament / They seek so many new . . . / 'Tis all in pieces, all cohaerence gone; / All just supply, and all Relation." In 1623, the poet William Drummond articulated fears of a seemingly broken compass: "Thus Sciences . . . have become Opiniones, nay Errores, and leade the Imagination in a thousand Labyrinthes."

As the work of the first scientific organizations got underway—such as that of the Royal Society, founded in 1660—interest in the Moon was high. Our increasing knowledge of it, our improving telescopes, and all the rest—Newton's laws, the world opened up beneath our eyes through microscopes (red blood cells! bacteria!), the invention of the barometer, the world of static electricity, the seed drill, calculating machines, calculus, artificial water filtration—opened up to the Enlightenment—which, if its faith in itself was at times misplaced and if its biases still need correctives—gave us both the humility of our tiny place in the universe and the zest to improve that lot by embracing our smallness, by empowering human reason to know more and to act from wisdom, even compassion, based on facts.

In his 1686 essay, "A Free Inquiry into the Vulgar Notion of Nature," Robert Boyle spoke for the modern case against the ancients as he discussed lunar mountains. He wrote,

> And the moon which was anciently a principal deity, is so rude and mountainous a body, that 'tis a wonder speculative men, who consider'd how many, how various, and how nobles functions belong to a sensitive soul, could think a mass of matter, so very remote from being fitly organiz'd, should be animated and govern'd by a true, living and sensitive soul. Indeed, these deifiers of the celestial globes, and the heathen disciples of Aristotle, besides several of the same mind, among the Christians, say great and lofty things of the quintessential nature of heavenly bodies, and their consequent incorruptibility; of the regularity of their motions, and of their divine quality of light, which makes them refulgent. But

the persuasion they had . . . seems not grounded upon any solid physical reason, but entertain'd by them for being agreeable to the opinion they had of the divinity of the celestial bodies; of which, Aristotle himself speaks in a way that hath greatly contributed to such an excessive veneration for those bodies, as is neither agreeable to true philosophy, nor true religion.

This is a direct attack on superstition, the inertia of Aristotelianism and Christian dogma about the material world. And Robert Hooke, in the 1674 "An Attempt to Prove the Motion of the Earth by Observations," got his blows in as well, rather more comically by complaining of those who "have confined their imagination and fancies only within the compass and pale of their own walk and prospect . . . that suppose the Sun as big as a Sieve, and the Moon as a Cheddar Cheese, and hardly a mile off."

As the natural philosophers continued to observe and reason from the data as best they could, many poets made use of the ever-renewed finding that the Moon reflected sunlight instead of generating its own. It seemed a handy metaphor, as Marjorie Hope Nicolson points out, for a lack of originality. If someone was "Moonlight," they were borrowing ideas from elsewhere.

In a sense, then, the astronomers who, like the poet Thomas Gray, believed they observed evidence of life on the Moon—and not just made deductions about its likelihood—were themselves "Moonlight." The idea of lunar life was already well-established by the eighteenth and nineteenth centuries. What changed was the application of the telescope to this idea.

A government official who wanted to draw and map the entire lunar surface, Johann Schröter was an early claimant for observational evidence of lunar life. For his telescope, he developed a kind of binocular viewer with one eyepiece for his eye and one for projecting an image on a drawing surface. It was clumsy but he put it to work, and in 1790, he published *Selenotopographische Fragmente,*

then a follow-up book in 1802. Ewen Whitaker calls his illustrations "rather naive" and his "artistic talents . . . somewhat limited." Like others, he drew crater rims that looked like they were circled with trees, but he was the most precise of the first wave of nineteenth-century selenographers, and his work helped to spark the German and English passion for lunar astronomy. In fact, Charles Wood notes that Schröter's "drawings had more details than previous maps, and he discovered mare ridges, numerous rilles and domes, tall mountains at the South Pole, and he provide the first details of various limb regions." He discerned that the larger a lunar crater the shallower it is.

Schröter was convinced that the Moon had an atmosphere and rational life. The conviction was again based on the popular idea that God would not waste worlds by leaving them bereft of flora, fauna, and sentience. The 1771 *Encyclopedia Britannica* said the resemblances of the Earth's surface to that of the Moon's "leave us no room to doubt, but that all the planets and moons in the system are designed as commodious habitations for creatures endued with capacities of knowing and adoring their beneficent Creator."

Despite his shortcomings as an illustrator, Schröter was an assiduous observer who backed up his philosophical and theological beliefs with what he considered to be accurate visual evidence. He saw changes in lunar surface colors at a very fine scale, sure that alterations in brightness and the appearance and disappearance of "many a small spot" was attributable to these places being "constructed habitation[s] of the rational inhabitants of the moon; and perhaps in that and in . . . industries . . . lies the explanation." Smoke and fog were, after all, a part of urban existence on Earth. So too was agriculture. "I at least imagine," he wrote of the craters Plato and Newton, along with all of Mare Imbrium, that they were "just as fruitful as the Campanian plain [in Italy]. Here nature has ceased to rage, there is a mild and beneficial tract given over to the calm culture of rational creatures, who . . . give thanks for the fruit of the field and perhaps only fear Mont Blanc and . . . cratermountains [that] may cause new disorders through new eruptions and overflow many moon cottages." Alas, all this turned out to be illusory overinterpretation of slender visual data. There are no fertile plains or cottages on the Moon, at least not yet.

Schröter had an ally in Sir William Herschel, a German musician, teacher, and composer who moved to England in the mid-eighteenth century. Best known for his telescopic discovery of Uranus in 1781, the first new planet since the ancients, Herschel burst onto the scientific scene with work that included ascertaining lunar mountain elevations. Privately, he enthused about the Moon as a place of habitation, even though many selenographers considered it airless, dead, and dry.

In a letter to Astronomer Royal Nevil Maskelyne, Herschel conveyed an enthusiasm for Selenites that startled and worried the senior scientist, so much so that he responded with measured tones. Herschel's initial correspondence betrayed his own doubts about how his lunar views would be received: "If you will promise not to call me a Lunatic I will transcribe a passage (from . . . observations on the Moon . . .) which will shew my real sentiments." Acknowledging that he has to argue from "the analogy of things . . . beyond the reach of Observation," he forges ahead with bold claims. The Moon has light, warmth, and soil, so it must have inhabitants. It is also sublime. Even if it lacks the foundations that make life as we know it possible, the Moon, for Herschel, is to be preferred to the Earth.

"What a glorious View of the heavens from the moon!" he wrote to Maskelyne:

> How beautifully diversified with hills and valleys! No large oceans to take up immense plains, fit for pasture &c: Uninterrupted day on one half, and on the other a day and night of a noble length, equal to many of ours! Do not all the elements seem at war here when we compare the earth with the moon? Air, Water, Fire, Clouds, Tempests, Vulcanos &c: all these are either not on the moon, or at least kept in much greater subjection than here . . . For my part, were I to chuse between the Earth and Moon I should not hesitate a moment to fix upon the moon for my habitation.

Herschel would have built himself a Moon-cottage.

In his notebooks, sections of which historian Michael J. Crowe published for the first time in 2008, Herschel began to be convinced that the Moon was a real place with real life. Using "a new ten foot Reflector [aimed at] the Moon with a power of 240," Herschel "was struck with the appearance of something I had never observed before," though he admitted that it might be "an optical fallacy." What he thought he saw was *a forest* in Mare Humorum, one of the last major maria that becomes visible as the Moon nears full. He even sketched little hairy dashes in the area, including the ancient crater Gassendi, arguing that a vast forest rather than a different-colored surface accounted for the darkness of the maria. Mare Humorum is more than 260 miles wide, with fissures and cliffs and craters and mountains and solid lava of varied shade.

The trees must be, he thought, "at least 4, 5, or 6 times the height of ours." In intervening years, he observed what he believed were pyramids, canals, and byways, arguing as well that the craters—whose nature and origins would be the subject of fierce debates in the coming two hundred years—might well be "works of Art"; that is, "Metropolis, Cities, Villages."

Crowe notes Herschel's remark to a fellow astronomer that he had magnified the Moon by 932 times in his telescope. This is a magnification so high that it distorted and falsified Herschel's views, and he would later admit this. Distortions inherent in certain kinds of optics and in high magnifications is a theme that recurs in modern lunar controversies.

Herschel wasn't alone in his enthusiasms. Cotton Mather was a believer in extraterrestrial intelligence. So was a well-known French poet, Paul Gudin de la Brenellerie, who in 1801 wrote a long pro-pluralist poem that deftly claimed that on a waterless and airless Moon its inhabitants did not have to drink or breathe. Crowe tells us that at least "a dozen" books on extraterrestrial life were published between 1710 and the middle of the eighteenth century. These authors would be joined by many others in the nineteenth century, including Thomas Dick, who was a popular author of astronomy books and seemed to be at the vanguard of a veritable army of pluralists arrayed against their lonely opposition, the anti-pluralists. One of the most prominent of the latter was

William Whewell, a titan of the times, who published a book called *Of the Plurality of Worlds* that devastated the pluralist argument along several lines, but the idea of a lively cosmos, from the Moon outward, was too entrenched to succumb to his rigor.

Thomas Dick even called for a kind of citizen science project to obtain proof positive of intelligent Selenites. The program would involve "vast numbers of persons, in different parts of the world" to observe the Moon—all its features "accurately inspected"—in order to "lead to some certain conclusions." Citizen science has no more romantic an origin than in Dick's belief that vast numbers of persons in different parts of the world all equipped with telescopes could find the evidence he presumed they would: Moon-cottages and so much more.

☙

Perhaps the most notorious of those who would reach "some certain conclusions" was the astronomer Franz von Paula Gruithuisen, son of a falconer employed by Bavarian royalty, and who, like a bird of prey, was sharp-eyed. After serving as a combat doctor's assistant at age fourteen, he too was in royal employ and able to buy a telescope to look at the Moon while using the competing lunar maps of Riccioli and Hevelius, as William Sheehan and Thomas Dobbins relate. The falconer's son became a scientific polymath teaching in Munich. In 1811, a bright comet appeared and Gruithuisen wrote "On the Nature of Comets," wondering, like others before him, if they might contain living beings. It would be his lodestar. He would go on to say, "We are still in love with the beautiful Moon, and dry reports of observations are better able to hold our attention if we can somehow keep alive the possibility of Selenites."

Gruithuisen was also a doctor who "invented a surgical device for crushing bladder stones," Ewen Whitaker writes, and became a keen lunar observer who drew many features and published "an aesthetically pleasing lithograph map of the whole Moon" in 1825. A later version, a copy of which I've seen, uses two columns of text on either side of the orb with lines running from names of features to their location. It looks like the Moon has been captured by a very precise spiderweb. His separate drawings of features are gorgeous. Gruithuisen used a handful of small refractors to observe the Moon, from, irony of ironies—Munich's Sonnenstrasse or Sun Street.

It was not long into his lunar observations that he began to speculate about natural and artificial changes on the Moon. With his good eyes, he found the five craterlets on the dark flat floor of Plato—spotting those craterlets is still a test for good optics, good observing skills, and steady air—but he wondered if they'd appeared to him because a fog had cleared over these possibly water-filled depressions. Significantly, Schröter had not reported them.

Gruithuisen believed maria to be seabeds long since drained of water that now were forests and farms. After all, plants on Earth appear blacker the farther one is from them, so "how much more must this be the case on the far-off Moon!" He could even spy yellows and browns—the autumn world of the far-off Moon—and clearings he thought were roads. He was mistaking changes in contrast with color, it turns out, and being tricked by the mind's desire to connect points into a line. The neurobiologist Mark Changizi and others say that our brain scoots ahead about a tenth of second to conjure an expected visual image. Even our brainy vision doesn't want to be in the present; it gets light from the past and tries to foresee what's next.

Gruithuisen believed the channels or rilles (he discovered many of them) and valleys of the Moon were "produced by a higher order of animal life, traveling frequently between these regions." They were flanked with trees and "could be completed only with shrewd planning and concerted effort"—foreshadowing Percival Lowell's forthcoming arguments about Martians planning their canals to save a drouthy world.

Obsessed with the towering figure of Schröter, Gruithuisen named an area of the Moon the "Principality of Schroeter," i.e., Sinus Aestuum (or Bay of Billows), a plain of exquisite basaltic deposits that look like ink brushed upon the Moon or a mandala taking form as a shaman slings dark sand.

As Sheehan and Dobbins write, on July 12, 1822, he scanned the land-scape named for his exemplar and burst with disbelief, then, of course, with belief. "O Schroeter, here is that for which you always searched in vain!" He called it the *Wallwerk*, a system of three long ridges bisected by three major slanting ridges, rather like, he said, "the veins of an alder- or a rose-leaf. At first sight of this object, I fancied I was looking down from the height of a steep mountain, through all the seething ocean of the air, and had the bird's eye perspective of a city before me." No mere cottage. A city. Finally. A city on the Moon.

In his 1824 article, "The Discovery of Many Distinct Evidences of Lunar Inhabitants, in Particular a Colossal Artificial Structure by the Same," he wrote:

> No one, no matter how fanciful, would consider it possible that Nature alone could bring forth such a structure. Could crystal druses 5 geographical miles in breadth arise on the Moon, when on Earth they excite the greatest wonder when they reach 5 feet? . . . Could the structure have been built by termites? Certainly, for all we know to the contrary, giant wingless insects, with their instincts for such construction, might dwell in the Moon. But would they have the understanding of a man? The answer is surely no.

Sheehan and Dobbins tell the story as though it were a scientific adventure, which, in its day, it was. For Gruithuisen believed the city was covered by plants, was likely still being built, and went on to find an adjacent temple laid out like a star. The Selenites of course would make a temple in the shape of a star because the clarity of the air over the Moon would reveal stars in such wonder. He pondered the life of the Selenite—probably, mostly underground—and he awaited his fame.

Astronomers mostly derided the announcement, even when they were more polite to Gruithuisen's face. Some people claimed they saw the city and temple too. Others did not. Gruithuisen visited Schröter's former assistant and showed him the city through the master's telescope and even dropped by to see the great Goethe, according to *Epic Moon*. All this earned Gruithuisen an astronomy professorship at what is now Ludwig Maximilian University in Munich, but the scientific community soon turned against his vibrant lunar pluralism. Eventually, he resorted to publishing his own work in his own journals, including the "discovery" of another lunar city. He died in 1852, his reputation in tatters. This, say William Sheehan and Thomas Dobbins, obscures what he did get right, including hypothesizing that comets delivered water to the Moon and that the craters were formed by impact, an idea whose time had not yet come.

T. C. Elger, writing in 1887, in particular praised Gruithuisen's observations of lunar rilles. "His records are by no means to be depreciated," Elger said. The astronomer and writer Rev. T. W. Webb, also writing in the nineteenth century, said of Gruithuisen that he "assuredly thought, and published, an uncommon amount of nonsense." (He includes impact craters as such nonsense.) "Yet," Webb continues, "this man made good use of a keen eye and sharp instrument, and saw much, and if he had spared us his inferences, would have been accepted as an observer of no little weight." When Webb first examined Gruithuisen's city, he recorded his thoughts: "The whole object looked coarse, and though curiously arranged, would never have given me the idea of an artificial production."

Today, some amateurs still seek it out. I have too, using my three-inch reflector and my ten-inch. It lurks north of the crater Schröter, but I've not yet seen it. The complex is just a series of low ridges. Someday to witness it would be like living for a moment in Gruithuisen's skin.

The Wallwerk was on the mind of an enterprising journalist in New York City, who, just over a decade after Gruithuisen's monumental announcement, had similar, even more fantastic news to share. Unlike Gruithuisen's, this news would seduce the public en masse, including many scientists.

In late August and early September 1835, the *New York Sun* boosted its circulation so quickly that on one day it sold some nineteen thousand issues, topping its local competitors and, indeed, all other newspapers around the globe. The reason? A series of articles by Richard Adams Locke with the news that Sir William Herschel's son, Sir John, had discovered life on the Moon. Translations appeared within months, all recounting how Sir John, from his observatory in South Africa, had seen not just changes in colors or spots of light or mere cities but witnessed conversing bat-men and blue unicorns. By developing a special and humongous new telescope—a "hydro-oxygen reflector" capable of forty-two thousand times magnification!—Sir John was able to find volcanos, including those his father allegedly had seen, along with a nice negative touch: the cities that Gruithuisen and Schröter had found were just natural rock formations.

Presenting his text as the journal of one of Herschel's assistants, Locke recounts the discoveries in an unfolding panorama of tension and wonder. Times of the evening are noted, initial scenes described: rocks first, a "greenish brown" basalt, then a covering of "a dark red flower," like the Earthly "rose-poppy," a lunar flower that was "the first organic production of nature, in a foreign world, ever revealed to the eyes of men," as another collaborator put it.

It would not be the only miracle. Here is a partial list of the discoveries: "a lunar forest," "a beach of brilliant white sand, girt with wild castellated rocks, apparently of green marble, varied at chasms," "grotesque blocks of chalk or gypsum and feathered and festooned at the summit with the clustering foliage of unknown trees," "water . . . nearly as blue as that of the deep ocean . . . in large billows upon the strand," "giant amethysts," lunar bison of small size with, as "nearly every lunar quadruped . . . a remarkable fleshy appendage over the eyes, crossing the whole breadth of forehead and united to the ears . . . [which is] a providential contrivance to protect the eyes of the animal from the great extremes of light and darkness," and that blue unicorn ("an agile sprightly creature . . . with all the unaccountable antics of a young lamb or kitten)." The astronomers named its haunts between Mare Nectaris and Mare Fecunditatis the Valley of the Unicorn. There was a hut-building bipedal beaver, eight other mammal species, and thirty-eight of tree. Most compellingly, several kinds of bat-men or "Vespertilio-homo." They walked

"erect and dignified," as "lens H z" was able to show, along with their cop-
pery hair, beards, four-foot height, and animated conversations. In an isolated western crater, Bullialdus, there was "an equitriangular temple, built
of polished sapphire, or of some resplendent blue stone." Near Langrenus,
an even more dignified species of Vespertilio-homo was found, "eating a
large yellow fruit like a gourd" while holding "rural banquets" and making
ritualistic gestures. They found not a single carnivore. Unable to resist the
lure of lunar naming, Locke's astronomers bestowed their Moon with
the Land of Drought, the Land of Hoar Frost, Vale of the Triads, and the
Ruby Colosseum.

It was, of course, a fake, including the three-hundred-mile-long quartz
crystal.

Upon hearing word of the articles—which had fooled many readers,
including some scientists—Herschel was bemused and only later annoyed.

The success of the Great Moon Hoax speaks to the abiding need to con-
firm our status in the cosmos. Having been demoted in space—we orbit the
Sun, not vice versa—and having been demoted in time—geologists were
finding evidence of a very old Earth, one we were apparently not always
part of—we asked and still ask, "Are we alone?" (Charles Darwin was on
the *Beagle* the same year Locke's articles appeared.) Idle curiosity and novelty played a part in the Great Moon Hoax—I mean, blue unicorns on the
Moon, of course, who could resist? But this charade played out in a time
obsessed with understanding the physical nature of the universe around us, a
universe whose forces and indifference were becoming all too clear. Just the
year before the Great Moon Hoax, in 1834, Benoit-Pierre Émile Clapeyron
articulated the first version of the Second Law of Thermodynamics, that of
entropy, how disorder and decay increase in closed systems. The implication was that all clocks ultimately wind down. If we were alone in such a
cosmos, we were—are—appallingly unique. Our need for blue unicorns
on the Moon, for bat-men on the Moon, our need, lately, for real scientific
signs of intelligence beyond the Earth, or even just traces of ancient creatures elsewhere in the solar system—a thorax jutting from Martian sand—it

means, I suddenly realize, that we are afraid: when we ask if we are alone, we are asking why we have to die.

I have this realization while looking at the only feature on the Moon I might have ever mistaken as artificial: Rupes Recta. Like a scalpel's mark, Rupes Recta looks like a slice in the southeastern region of Mare Nubium (the Sea of Clouds). It is the first maria visible on the Moon's southwestern quadrant. Rupes Recta—the Straight Cliff or Straight Wall—is another of those standout features for which confused lunar beginners are grateful. It's not a cut but a scarp of varying altitude, from about 650 to nearly 1,000 feet high. The slope is a steep-enough twenty degrees, and the scarp runs seventy miles long and is so straight it used to be called the Railway. Rupes Recta is a black blade terminating at its southern end in the so-called Staghorn Mountains, which also look like a sword's handle. A dramatic illustration from a 1960s juvenile science book shows the Straight Wall as nearly vertical, with thin layers like sandstone depositions from long-lost oceans. Presumably such sediments would have marine fossils. On the preceding page, the scarp is scored with vertical gullies, as though it had been a victim of erosion. Both are wonderfully inaccurate and redolent of a living Moon. Researchers speculate that Rupes Recta was in fact created by a thrust fault or the settling of the mare lavas to its west.

Schröter, Herschel, and Gruithuisen could not imagine a dead Moon, so the only alternative was a living one. They could not abide a lonely cosmos, so they populated the Moon with temples and trees. They lived in two worlds at once. This I understand. As I spent so many evenings, in Tucson and in Logan Canyon, looking up my ninety features for the American Lunar Society, I also lived in two places at once: wherever I was on Earth and whatever I saw on the Moon.

If I had not come of age with a lifeless Moon, would I, too, have seen, if not a Valley of the Unicorns, then a copse? A freshet? A squall? I believe I would have. Sometimes I stare at Rupes Recta and see a dark highway built across the Moon.

Born into a Brahmin family in 1858, W. H. Pickering grew up to become, like his brother Edward, an astronomer. He discovered a satellite of Saturn called Phoebe in 1899, speculated about the possibility of plate tectonics on Earth, and helped found the Appalachian Mountain Club, but for most of his career he was an observer of the Moon possessed with what can only be called a wildly imaginative and cantankerous personality. Working in various capacities for Edward, both in Massachusetts and in Peru—Edward was director of Harvard College Observatory—W. H. Pickering was so vociferous in his claims about life on the Moon that eventually Edward would have nothing to do with him. William then managed to eke out a lonely professional life in Jamaica, where, till his death in 1938, he maintained that he alone among twentieth-century astronomers understood the nature of the Moon. By the time W. H. Pickering was publishing article after article in *Popular Astronomy*, the *Century Illustrated Monthly Magazine*, and other venues, no professional journal would touch his work because it was understood that the Moon was dry and airless and more or less static. In one article titled "Is the Moon a Dead Planet?," Pickering answers firmly in the negative: "In a study of the daily alterations that take place in small selected regions . . . we find real, living changes that cannot be explained by shifting shadows or varying librations of the lunar surface."

William was a devout visual observer who recorded his sessions in notes and drawings. He had good eyesight, and it wasn't for years that lunar photography could catch up with what the human eye and brain and hand could process and record. While in Peru, pursuing his own lunar obsession, William ignored his brother Edward's telegrams to "photograph with the thirteen inch" that Harvard had placed there for stellar research and to "restrict . . . [yourself] more distinctly to the facts in this as in other cases." This was a reference to William's contention of having found lakes on Mars, according to Sheehan and Dobbins, who tell Pickering's stormy story in *Epic Moon*.

In time, William left Edward's pleas—and Peru—behind. He moved to an old plantation in Jamaica, built a telescope, and entertained local and visiting dignitaries, naval officers, and scientists, though if any were astronomers they likely were not impressed with Pickering's fanciful lunar notions.

Those began, according to an article in *Sky & Telescope*, with what he thought was some kind of eruption in Schröter's Valley, the positively gigantic former lava channel that cuts through the strange landform in the Moon's northwest reaches called the Aristarchus Plateau, a place utterly different from its impact-basin surroundings. A diamond-shaped elevation built up of mile-high layers of volcanic ash and lava, the plateau also features that bright crater, its namesake Aristarchus. Because the plateau is so big and so unusual, my eye goes to it every time the Sun is rising or setting on it, as it emerges like a weird island from a pulp sci-fi movie.

Pickering's 1905 piece in *Nature*, "Changes Upon the Moon's Surface," laid down the gauntlet, saying that anyone who argued for a static Moon—"a burned out cinder, upon which nothing ever happened"—had never *looked* at the Moon. Every chance he could, he argued his points, backed up with historical records and his own observations. Pickering called his approach to lunar studies—his advocacy of lunar life—the "new selenography." He was convinced, wrongly, that when the Moon passed in front of Jupiter, the distortions of the latter were caused not by his optics but a lunar atmosphere. There were, he said, eruptive forces on the Moon, albeit on a small scale; new craterlets were appearing, such as those on the floor of Plato, something observers had tussled over for years. (Pickering was incorrect about the air and the new craterlets appearing in Plato.) Further, he reasoned, if there is volcanic action, then that would include a lot of water—it does so here on Earth—so there had to be water on the Moon. (He was right, sort of; there is water ice trapped in shadowed regions of the poles, but this wouldn't be known of for decades.) For Pickering, the valleys and channels even then being attributed to lava flows in ages past were obvious evidence of copious surface water. There was, he claimed, enough moisture to support the formation and falling-out of snow in craters like Messier, Linné, and on the mountaintops in central peaks in Eratosthenes and elsewhere. (Alas, no snow falls on the Moon.) With snow and hoarfrost in the sunlight, you had melted water. With melted water, you had wet soil. With wet soil, you had plants. That would explain what he called "variable spots." (Like those before him, he was tricked by optics, unsteady air, contrast, and his mind's eye.)

"The lunar vegetation is not green. It is gray like our sage brush and some of our cacti, and black like our lichens," he writes in one of his articles on Eratosthenes, the crater whose interior was, for Pickering, a kind of garden:

> The lunar vegetation is scattered, generally in rather small patches . . . None is found near the poles. The only greenish spot . . . is the floor of the great crater Grimaldi. By far the greater part of the lunar surface appears to us to be simply a desert waste. The vegetation, where found, is often associated with minute craterlets, as in Alphonsus . . . Sometimes it is associated with rills as in Atlas. In any case its growth and decline must necessarily be very rapid.

To my observing programs, I might have added, "Detect lunar sagebrush." I squint and there I spy a hint of grayish green growing larger on the crater wall. When I'm on the trail in Utah or just by the cabin, I never fail to run my hands through sage and bring the branches to my face, inhaling a local scent. Would that I could do so on the Moon.

H. G. Wells captured the poetry implicit in Pickering's vision in the chapter "A Lunar Morning," from his 1901 *The First Men in the Moon*, a novel whose descriptions of sunrise on that surface impressed no less than T. S. Eliot. I knew this book before I came to the Moon, then read it again before I came to the travails of W. H. Pickering. Wells scholars note that the book was reviewed prior to publication for scientific accuracy, and Wells, with his science education and interest in popular education, could hardly have missed the long, dreamy history of lunar life, especially as it was culminating with Wells's contemporary, Pickering. Yet even by then most astronomers scoffed at the assertions of air and water on the Moon.

In "A Lunar Morning," humans from their landing vessel see "scattered here and there upon the slope, and emphasized by little white threads of unthawed snow upon their shady sides . . . shapes like sticks—dry twisted sticks of the same rusty hue as the rock upon which they lay. That caught one's thoughts sharply. Sticks! On a lifeless world?" Then: "Among these needles a number of little round objects. It seemed to me that one of these had moved . . .

"One after another all down the sunlit slope these miraculous little brown bodies burst and gaped apart, like seed-pods, like the husks of fruits; opened eager mouths that drank in the heat and light pouring in a cascade from the newly risen sun." There were cactuslike plants, orange puff-balls, and human astonishment.

> Imagine it! Imagine that dawn! The resurrection of the frozen air, the stirring and quickening of the soil, and then this silent uprising of vegetation, this unearthly ascent of fleshiness and spikes. Conceive it all lit by a blaze that would make the intensest sunlight of earth seem watery and weak. And still amidst this stirring jungle, wherever there was a shadow, lingered banks of bluish snow. And to have the picture of our impression complete you must bear in mind that we saw it all through a thick bent glass, distorting it as things are distorted by a lens, acute only in the centre of the picture, and very bright there, and towards the edges magnified and unreal.

Magnified and unreal, indeed. Pickering must have loved Wells's novel, or, at least, this passage, the last, great vivid evocation of imagined lunar life that had begun so long ago. For the astronomer, Wells's poetry was grounded in the truth he saw, as in one of my favorite Pickering articles, the romantic "The Snow Peaks of Theophilus," in which "these peaks seem to be but the vestiges of a once great smooth central cone, whose sides were grooved by glacial action, which scored them with a series of U-shaped valleys into their present deeply eroded form" and whose "névé or upper snowy area whence they sprang is still readily visible with even a small telescope." On Theophilus, as elsewhere, "from moisture given out by a small invisible crater" there drops "freshly fallen snow." (In fact, an entirely snow-covered Moon was one popular theory to account for the lunar surface, as Peter Schultz writes in his helpful consideration of lunar-impact theories.)

I adore this vision, so blazingly inaccurate and so frostingly inviting. Ever since moving to Tucson, I have missed the seasons of the Central Rockies, especially winter when I strap on snowshoes and huff up mountains empty of humans and silent as the pauses in a poem. Pickering lets me imagine opening my spacesuit helmet on the Moon to stick out my tongue and let selenite snowflakes prick the flesh with dots of cold. After traversing snowy Mons Pico, I could hike far south to intersect one of the white streaks emanating from the crater Tycho, whose bright ray system, like others on the Moon, Pickering attributed water vapor rising from cracks and forming "ice crystals . . . like those terrestrial cirrus clouds to which we give the name mare's tails." For a time, astronomers thought the Moon was terribly cold, even in sunlight. Now we know it is terribly cold in the dark and terribly hot in the Sun. At mid-latitudes the Moon's night is minus 200 degrees Fahrenheit while some 250 degrees Fahrenheit in the day. I close my imagined helmet. I save snowshoeing for the Wasatch.

Later, Pickering would attribute some changes on the Moon—such as those in Eratosthenes—not to plants but to insects, suggesting there were hordes of them moving slowly in search of moisture, the garden of Eratosthenes plagued by lunar locusts. It's a fetching science, Pickering's mistaken Moon, a failed science made more dramatic by the astronomer's tragic life. He, like Gruithuisen and others, earned scorn. Scientists can be deeply stubborn, unwilling to let go of cherished theories despite strong, even overwhelming, contrary evidence. Science grinds such ego down. It takes time. For all its flaws, science tests, verifies, and rejects. It slowly builds an understanding of reality in which the Sun does not revolve around the Earth and in which neither locusts nor crickets migrate across the floor of Eratosthenes. If that crater remains a garden, what grows in it is neither vegetable nor animal; it is emotional. It's humility, something that Pickering never spotted.

For his troubles, Pickering neither changed the course of lunar science nor endeared himself to his brother, who cut off support both personal and professional. Edward's successor at the Harvard Observatory, the famed astronomer Harlow Shapley, also spurned William. To top it off, his sure-fire formula for winning stock market bets turned out to be a dud. The aftertaste was bitter. By the time of his death in 1938 at age seventy-nine, the Moon was of no interest to professional astronomers. Pickering's claims were considered laughable, and, as some have pointed out, in a final irony, his brother Edward has a lunar crater named for him. Historian Howard Plotkin recorded the complex tensions between these two men, with the older Edward feeling responsible for William, even as he was embarrassed and rebuffed by him. For his part, William resented Edward's condescension, power, and prestige.

Pickering even had the gall to compare himself to Copernicus, saying "the observer will triumph over the man who depends exclusively on his reasoning." (Ironic since Copernicus was not an observer but a theorist.) And Pickering suggested to amateur astronomers they, too, could be on the vanguard, as they could see for themselves "proof of the existence of vegetation, as indicated by the variable dark spots, [which] is open to every possessor of a three-inch telescope." It was a faint echo, whether William knew it or not, of Thomas Dick's call for a global brigade of amateur astronomers to find signs of lunar intelligence.

Pickering's critics were merciless. One review of *The Moon: A Summary of the Existing Knowledge of Our Satellite, With a Complete Photographic Atlas* noted Pickering's assertions as "startling" and "sensational." The reviewer complained that the drawings and photographs were "so excessively enlarged that the details are mere blotches." The images meant to prove vegetal change were useless. The reviewer poked fun at Pickering's snowy spots, calling them, in a brilliant turn of phrase, "survival of the brightest."

Yet the dead Moon has its lively nature, if only on Earth.

Consider: the complications of the Moon's orbital motions mean that in 2025, its tidal pull will begin to strengthen for about two decades—just as climate change is speeding sea-level rise—and the effect will be to further threaten coastlines and their communities.

In such seas, there are lives whose rhythms are lit by the Moon. A study in the *Journal of Experimental Botany* has shown that crustaceans, when deprived of other cues, time their biological cycles to tides from where they had been living. California grunion heave their bright, whiplike bodies on shore to mate, timed to these lunar watery rhythms, then the hatchlings are born, taken back to the sea by lunar high tide.

Moonlight—which is one thousand times brighter than the stars—influences coral growth and bird migration. An Indian Ocean petrel species only nests on a Full Moon. In Africa, wildebeests huddle in safety from lions on the darkest nights and grow bolder as the Moon waxes. African buffalo gather on dark nights, seeking protection in numbers. Certain fish do the same. A fish of New Zealand's reefs, the common triplefin, grows more readily when the Moon is bright, presumably because the light helps them find plankton. And Arctic zooplankton will make vertical migrations according to Moonlight during the sunless winter. When the Full Moon is up before dawn, birds sing earlier than on other mornings.

And perhaps the old folk beliefs about avoiding sleeping in Moonlight do have a basis in science after all. A 2023 study found that "deaths by suicide are significantly increased during the week of the full moon," possibly due to how the light affects "circadian clock genes." (The study also found increased suicides in the afternoon and during autumn.)

Despite that grisly twist, life has benefitted from the Moon, not only in giving the Earth the axial tilt to produce seasons and thus provide the environmental complications that evolution favors, but also from its early proximity to Earth when tidal mixing was more forceful: a stronger stirring of the prebiotic soup. The Moon's former magnetic field also shielded the early Earth from radiation that otherwise would have stripped the nascent planet of its atmosphere. The Moon-forming impact, about which I'd learn more, may also have helped start plate tectonics, another factor implicated in life's evolution on Earth. Pickering would've liked that, I'm sure.

When Apollo 12 landed on the Moon in late 1969, the mission demonstrated pinpoint navigational skills, putting down within spitting distance of a robotic Surveyor lander. Retrieving parts of the machine to bring back to Earth for analysis, the astronauts seemed to have inadvertently found the first known signs of life on the Moon: bacteria from this planet that apparently had hitched a ride on Surveyor, surviving, it seemed, in dormancy after more than two years of exposure to the harsh lunar environment. Later research showed, however, that the bacteria probably had been introduced to the Surveyor equipment after it was returned to Earth.

Pickering would have cried.

We do know that in the twenty-first century the Chinese sprouted cotton leaves inside an automated capsule—the cotton died in the lunar night—and they sent silkworm eggs to the Moon. An Israeli lander, which crashed, probably scattered microscopic, hardy tardigrades, along with human DNA samples, across Mare Serenitatis (in violation of international norms, as it happens).

Someday someone may try to find and revive the little water bears. Tardigrades have survived in desiccated dormancy for a decade. Lately, a couple of researchers have even suggested that the very early Moon might have had warm oceans of water, the perfect place for life to evolve, though such would have long vanished. Of their work I would also learn more.

Whether that possibility, like cotton leaves and silkworms and water bears, would have provided any consolation to William Pickering, who can say? But his new selenography was the old selenography—having more in common with Lucian than Copernicus—and it died because the Moon itself is dead, a fact whose austerity masks so much that is rich and moving, something that the dreamers of a living Moon had missed all along.

CHAPTER SEVEN

"A corpse in Night's highway . . ."

I n 1866 the British poet Coventry Patmore described the Moon seen through a telescope. He was not impressed. Where reverent selenographers saw grandeur despite the Moon's austerities—or because of them—Patmore saw ugliness, the grotesque. The Moon was, he wrote, "a corpse in Night's highway, naked, fire-scarr'd, accurst." In his poem "The Two Deserts," Patmore wrote that "View'd close, the moon's fair ball / Is of ill objects worst." The Moon, he recognized, was dead. In this narrow fact, Patmore—a minor Victorian sentimentalist—was correct: Nearly three decades had passed since the publication of a magisterial German book on the Moon arguing that our satellite was, contrary to the enthusiasms of Franz von Paula Gruithuisen and others, devoid of air and water and, therefore, of life. The Moon was barren.

As to its accursedness, I disagree. For during the long autumn of my father's dying, I kept vigil with the Moon, and its sublimity was like bursts of steady air: the calming sharpness of craters wreathed with ridges fining in slopes from rims down to lava plains and the subtle grays of interplanetary antiquity. Mountains made of a white mineral I would learn to name rose to heights I could never scale, and their peaks would flare in the dark as the Sun struck them before passing to the lowlands. I traced the paths of cracks and valleys and serpentine waves of buckled rock. Light and nuance and knowledge are the hallmarks of careful looking at the Moon, the only way to see beyond stasis to history and wisdom and to change, even if that change is only crawling Sun and receding shadow.

By then I'd completed my Lunar Study and Observing Certificate from the American Lunar Society, framing the diploma that a man named Eric had sent me. It lauded my "excellence in the study of lunar geology and excellence in the art of lunar observing" and made me unreasonably happy. Without that program, I might still be whipsawing my telescope from one catchy crater to another, missing subtleties and what they can teach me. I sent a photograph of the certificate to my father, whose cancer he, at eighty-six, was not battling. On the phone, his voice was weary though I could still make him laugh, sharing antics of our cats or a terrible joke or two. Kathe and I would fly to Indiana to help him and my stepmother move into an adult-care apartment, and later I'd return for his early December birthday, where on a cold night I stopped in the parking lot, tears in my eyes, caught short by a cloud-shrouded Moon and by the image of my gaunt father looking at me from his chair as the door closed. It would be the last time I saw him.

The Moon's wilderness became my terrain. Far from excellence, I had much to see and learn—so that fall I began to traverse the surface for another observing project, a much more difficult one sponsored by the Royal Astronomical Society of Canada. I started to revisit old friends—massive Clavius, brooding Pythagoras—and found ones I would come to know: a saucerlike shallow visible only when the Sun is fresh in the crater Albategnius, the very place that Galileo drew like an eyeball in *The Starry Messenger*. Little craters, weird ones, fallen blocks in big ones, sloping and relatively rare volcanic domes seen only close to shadow, some with secretive pips at their tops where the lava hadn't erupted but oozed. That calcined stillness was my terrain.

Over the most recent backyard in Tucson, over the scrawny Texas Mountain Olive trees, looking at the Moon, I thought of my father, of love and mistakes of injury and reconciliation, of philosophers and mountains. I felt the difficulties and gifts when entropy becomes personal. I imagined the chaos of collisions, whole valleys formed in seconds by flying hypersonic rocks the size of peaks I hike here on Earth, the eruption of fire fountains, beads of glass that fell like rain, and the hardening of molten magma. Fire-scarred, indeed, and more awesome for that.

Patmore's attitudes toward the lunar surface were more akin to those of the seventeenth century, when the prescientific Moon was that smooth sphere entrained in sacred motion, and to the early eighteenth century, when terrestrial mountains were God's reminder of Satanic excrescences on Earth. In those days, when English travelers crossed the Alps, they shuttered their carriage windows to hide the view. The sublime as a figuration was not yet fully articulated, though our need for it had been satisfied in other ways, as in bowing beneath the streaming light of stained-glass windows. The sublime: that need to be astonished, even overwhelmed, and, though tinged with the possibility of mortal threat, to be, even for a few seconds, possessed by an awed speechlessness that mutates back to fumbling expression, to admiration, to delight and even kinship. From the desert, the Moon is my closest sublime terrain, as it would be for the great lunar observers of Patmore's century.

For that poet, the microscope was marginally preferable to the telescope because it showed living things, "a torment of innumerable tails." Best of all, however, was "our royal-fair estate / Betwixt these deserts blank of small and great," what Patmore calls the "obvious ways," from which "ne'er wandering far." The poem's comfort is the present that we sense, which I laud, and ignorance about it, which I do not. Science makes the present glow. There's more than a whiff of fear in "The Two Deserts," a fear of death, and a loathing of the scientific view, which is shot through with death past and death to come. Perhaps for Patmore the craters I admire looked like human wounds (they do), the varied channels carved by lava or cracked by stress like the incisions of some deranged surgeon. The poet was widowed. He took comfort where he could.

☾

As it happens, the Moon's mountains helped whet the terrestrial sublime. All the way back in 1638, in his *The Discovery of a World in the Moone*, John

Wilkins set down one of the foundations of the Western sublime (another being a middle class and elite hungry for travel and recreation). Wilkins was cutting against the grain, for in the seventeenth century there was no ardor for mountainscapes, as Marjorie Hope Nicolson writes. To use the title of her influential study, we went from "mountain gloom" to "mountain glory." Yet even before that shift, which came full-on with the Romantics, Michael Rawson, citing Nicolson, notes that "Kepler's description of the lunar mountains as simultaneously forbidding and majestic presaged the transition to seeing terrestrial mountains as sublime." Wilkins carried that forward.

"If well considered," he wrote, mountains "will be found as much to conduce to the beauty and conveniency of the universe, as any of the other parts." His use of the word *beauty* also foreshadows the looming philosophical debates on the nature of it and of the sublime, a term yet to be in vogue though that is clearly what Wilkins is describing. The Latin roots of the word *sublime* mean a high lintel, a door on high. What could be higher in the sky than the high mountains of the Moon?

"You must know that there is not meerely one ranke of mountaines," Wilkins writes, "about the edge of the moone, but divers orders, one mountaine behind another." Range after range upon the lunar edge, like a crown.

The fascination that was developing for the Moon, aided by the telescope, also "had something to do with the growing admiration for terrestrial mountains [in the eighteenth century and later], as consideration of the physiography of the moon and planets certainly led scientists to a new interest in the physiography of their own globe," Nicolson explains. The difficult and obscure task of mapping lunar mountains brought us closer to our own. The deadness of our companion globe enlivened our appreciation of this one.

So in the minority is Coventry Patmore's Moon, which fit for him the definition not of the sublime but of the grotesque, which "arises," says critic Geoffery Galt Harpham, "with the perception that something is illegitimately in something else." Everything hinges on that adverb. Who gets to decide the legitimacy of a phenomenon? Patmore was a late advocate of a kind of hallowed Moon, appalled that it was a place with contentious topography. He is a caution to me now. Never lift a wish for reality to the level of reality itself.

The nineteenth century was a time when science and philosophy began to take the encounter with reality very seriously. It had to. The facts revealed disturbed conventional faith in purpose, meaning, and human destiny as sanctioned by God. Geology took center stage. By the time of Patmore's poem, the Scottish natural historian James Hutton had been dead for decades. Hutton had founded geology as we more or less know it today, proposing that we use rocks and landforms of the present as key to interpreting the past. And it was a deep past, millions of years past—billions we know now—all built up in rocks, some composed of the shells and bits of deceased things, a past far older than Bishop Usher's Earth, which he calculated from Biblical references to have been created in 4004 B.C.E., and so more amazing because of the denizens of its diverse ages. The next great figure in earth sciences, Charles Lyell, would die nine years after Patmore's poem—I wonder what he thought of it—and his foundational work *Principles of Geology* posited fossils as markers to the slowly changing chronology within the Earth's strata.

A then-unknown naturalist named Charles Darwin took Lyell's book with him when he boarded the *Beagle* and left Plymouth Sound on December 27, 1831. Twenty-eight years later he published *On the Origin of Species*, which, along with the work of rival Alfred Russel Wallace, established the fact of extinction as well as the mutation and adaptation of living species. Innumerable tails stretched into swampy recesses, a fact that neither poetry nor faith had established. Indifference to death became a feature of the cosmos.

As to the future, what Darwin, Lyell, and others were suggesting was that it would be different, very different. We might not be here at all. Why should the human race last when others did not? No less than H. G. Wells projected the implications in his late Victorian-era novel *The Time Machine*, seeing a future in which humanity diverges into two unflattering species, the Morlocks and the Eloi. Time, he saw, is titanic. Time eats its children.

So too space. Astronomers began to see how disordered the solar system had once been: asteroids were discovered where astronomers thought a planet might have existed, and this discovery would become part of the long-standing debate over how all those craters formed on the Moon. What was the fate of all those flinging rocks? Do they ever land? Do they still? And careful measurements of stellar parallax revealed unthinkable distances. Our Sun remained a mystery. Lacking a full understanding of atomic theory, nineteenth-century physicists were troubled by the stunning amount of energy the Sun generated. But secrets were revealed: radio waves, X-rays, the electron. Physicists did understand—and the public mind seized on—the fact of entropy: the Second Law of Thermodynamics meant the entire universe eventually would suffer what was popularly called "heat death." Today the most likely model of the future universe suggests a version of the heat death that haunted the Victorian intelligentsia, everything "condemned," as physicist Hermann von Helmholtz put it, "to a state of eternal rest." The Moon was a premonition, a future Earth.

At least to the Western mind. As lunar studies progressed with the advent of the telescope, it is a striking fact that this work stayed largely within an Anglo-European tradition. In his 2018 *Geosciences* review of "Scientific Knowledge of the Moon, 1609 to 1969," Charles Wood is "unaware of any non-Western discoveries or hypotheses about the lunar surface, and none that impacted Western science." That may be only slightly an exaggeration, as Japanese and Russian scientists were writing about transient lunar phenomenon during the twentieth century. But, as he says, "The lack of lunar observations and discoveries even after Westerners brought the first telescopes to Muslim lands, India and China in the early seventeenth century differs greatly from Europe, where the Moon was one of the first objects to be telescopically investigated." Lunar studies outside the Western context have only really taken off in the twentieth and twenty-first centuries.

It was in 1620 that Jesuits brought to Goa, India, the new telescope. To what extent they were used by Indians for lunar observations is largely unknown, but appears to have been of little interest there and elsewhere. The eighteenth-century Indian observatory of Jantar Mantar was entirely nontelescopic, preserving traditional—and quite sophisticated observations—from what may have been seen as an intrusive technology. The reasons for non-Western disinterest must have been many, ranging from a focus on the Moon that was calendrical or astrological (not astronomical) to a reluctance to use a tool associated with colonialism and set against Indigenous cosmologies.

According to Carmen Pérez González,

> Lunar cartography and photography was basically a European concern. Why was this so? A possible answer would be that the required technology (the telescope) did not arrive early enough in non-western countries. But this is not correct: the telescope spread around the world within a few decades of its invention in 1608 . . . The reason is subtler than mere technological limitation, for non-western (or, more precisely, Asian) astronomers lacked the obsession of their western counterparts with producing realistic drawings of the moon.

There is an early nineteenth-century Japanese lunar map, and in India the Maharaja Jai Singh did use a telescope for Moon studies, a telescope given him by the Jesuits, among other non-Western historical tidbits unearthed by the researchers in the book *Selene's Two Faces*. "Japanese astronomers had long shown no interest in directing telescopes to celestial bodies to know their true character," writes Tsuko Nakamura, citing poor optics, expense, and a focus on astrology in that country. Telescopic lunar observations were for a long time few and far between, though one such resulted in the arresting description of the Moon's surface as like "human skin suffered from small pox." Pérez González writes that that the first photograph of the Moon in Iran was taken around 1864. So the wider non-Western use (or avoidance) of the telescope is a global history to be written.

It was Victorian science that relentlessly instrumentalized the astronomical and lunar revolutions begun by Copernicus and Galileo. Our centrality was neither a given nor even a frailty. It was a lie.

☾

The Moon is the fact of rock. It is dead. Its clean, magnified surface shows this, and lunar topography revealed itself in detail never before recorded in *Der Mond*, published in 1837 by the banker Wilhelm Beer and the astronomer Johann Heinrich Mädler. This volume was the culmination of intensive observations and was so full of descriptions and mapping and was so good and so thorough that it had the profound effect of stultifying lunar studies for decades to come. The Moon seemed, finally, ultimately, known.

In his researches, Mädler had bested another ambitious selenographer named Wilhelm Gotthelf Lohrmann. Lohrmann's mapping skill was, in the 1820s, unprecedented. What sections of his lunar maps he published were praised as the best yet done. I marvel at their beauty. But Lohrmann had saved the most difficult sections of his incomplete and large map for later drawing sessions when he'd become exhausted from his day job of surveying. Both surveying and lunar mapping afflicted him with vision problems. Those last, hard sections of the map were full of mountains, the most challenging terrain to draw especially by a faint lamp in the open night. In 1839, he managed to complete and print a much smaller whole Moon chart, using the so-called hachure approach to showing topographic elevations, which involves drawing tiny parallel lines with spaces indicating slope. The closer they are, the steeper the gradient. Before contour intervals, this was the best surface relief method for mapmakers. If you blur your vision a bit, the hachure slopes look relatively realistic; otherwise, they look like microbes. By 1840, Lohrmann was dead, a victim of typhus.

The parents and kindly uncle of Johann Heinrich von Mädler had themselves died of typhus in 1813, forcing the bright nineteen-year-old to become a schoolteacher instead of a university student. But this would not hold him back. Mädler would do for the Moon what other explorers had done for the Earth. He would render it.

William Sheehan and Thomas Dobbins rightly make much of this Earth-Moon connection. In fact, the most famous explorer of the day, Alexander von Humboldt, was responsible for introducing the wealthy Beer to the scholarly and energetic Mädler. Humboldt was a giant, one of the last great explorers and certainly one of the last to imagine he could compile, as he did later in his life, a kind of encyclopedic cosmology, his *Kosmos*. Beer and Mädler would name a barely visible lava plain—Mare Humboldtianum—after him. The two men began their work in 1830. Lohrmann was still alive but his work was languishing. He was an inspiration to his rivals, however, down to the kind of telescope they used and their map's scale. Deciding to map the Moon in quadrants, they set to.

"Mädler's first observation for the lunar map was made on April 29, 1830," write Sheehan and Dobbins. "Over the next six years, he spent six hundred nights observing and sketching the Moon . . . employing a magnification of 140X, though with frequent recourse to 300X." He also "measured a network of reference points" to help keep distances and scales correct when the actual mapping began in 1832. "In preparation for work at the telescope, he took eight-by-eleven-inch sheets of paper—there would be 104 of these sheets in all—and on them marked the positions of the points of the first order." He created a grid of nearly two hundred triangles to help place important landmarks before filling in the rest. "He worked at great speed and completed his last drawing on March 19, 1836."

Not only was the work incredibly accurate—a huge stride forward from earlier maps—it added more rilles and showed the elevations of not quite a thousand lunar mountains and the sizes of 150 craters. Edmund Neison, the British astronomer, noted that the Germans made nearly a thousand measurements and named almost 150 new features. They used a refractor of not-quite-four-inches aperture, a size considered very small these days, at least for visual use.

So I have it easy in my Royal Astronomical Society of Canada lunar program. I am using a ten-inch reflector to locate, identify, and observe some 150 features, as well as observe and log changes in phase, motion, and location of the Moon. There are big, easy craters—like Gassendi and Tycho—and more difficult landmarks, like a lava channel near the crater Marius. Some of those difficult landmarks are part of the "challenge" portion of the effort, and ultimately I want to get quite a few of those as well. It's a fairly daunting set of objectives for a beginner, but absorbing. One evening could bring, say, a crater not noticed before: Damoiseau, with its graceful curves imposed on another larger crater, curve within curve. Or another cold night, a shock of black basalt, unnamed on my map, south of the crater Vieta, and above that a massif that often looks white-hot. It makes me appreciate the precise obsessiveness of nineteenth-century lunar observers, as well as the pleasures of central heating.

Der Mond's German keeps me from reading it, but it doesn't take long in skimming to see how mathematically thorough it is. Beer and Mädler, I learn, forewent fragmentary descriptions, flights of fancy, and frequent invocations of God (as past lunar mappers had indulged) in order to just comprehensively describe the Moon. From Sheehan and Dobbins, we learn that *Der Mond* tells us that the maria cannot be seas for they have marks upon them, that there are slight color differences among various maria (which I am beginning to discern), that craters range in size with features common to them according to width, from the smallest punctures to the truly enormous craters that Mädler and others called "walled-plains." The Moon was dry and airless. The rilles and rays were mysteries; the former were not rivers for they did not always run down mountain slopes into the maria, and the latter had no visible surface relief. Lunar mountains appeared to be razor-edged (they were not, to the later disappointment of many). As to the craters, they did not look like holes caused by volcanos on Earth, a fact that would lead to much controversy. And as to life, if there, it was necessarily quite different from Earth's—and the sober implication was, therefore, it was nonexistent. The Moon was "no copy of the Earth."

Unable to access a full-size original copy of their illustrations, I resort to digital versions, still awed by the level of detail, so much of it and so thickly rendered with unfamiliar stipple and hachure, that I am dazzled and a bit lost. The technique lacks a crispness I am used to in modern maps—not just photographic, but contoured—and so the craters and ranges and, well, everything, seem a strange mélange of cells, another stream of innumerable tails. I recognize outlines but they look curiously soft. Here and there, a flash of recognition: the white crater Aristarchus, the black crater Plato. The outlines are there but their rendering, clearly majestic, is not one I can easily read. I encounter their work feeling both impressed and illiterate.

�❧

I print and bind my Canadian checklist and mark off objects as I go, highlighting those on pages that remain and poring over the modern maps in Antonín Rükl's *Atlas of the Moon*, to which the checklist is keyed, and, as a supplement, the sharp photos and useful descriptions in Charles Wood and Maurice Collin's *21st Century Atlas of the Moon*. The day after an observing session, at the kitchen counter, I give Kathe updates on my progress, and she begins to tease me that she is reporting my work back to the Canadians.

Soon enough I'd see the rest of the craterlets in Plato: like distant whirlpools in that great black lake.

Certain places on the Moon are grails. The first was the Goclenius Rilles, which had eluded me for several weeks while I worked on the American Lunar Society certificate. When I had a clear, steady view for the first time up at the cabin, I hooted with excitement and waved my arms about. No one saw. They stream about an area of the Moon that includes the lobster-clawed crater Gutenberg, a favorite. The current grail—and it will be a hard one to see—is that very narrow Rima Alpes, the slender lava channel which runs along the bottom of Vallis Alpes, the Great Alpine Valley. The valley is as prominent, dramatic, and as easy to see as the rille is demure, subtle, and hard to spot. I will need a night of perfect seeing or perhaps a much larger telescope. Or both.

Less difficult but still challenging is Rima Hadley or Hadley Rille. This nearly hundred-mile-long, now-empty lava channel threads hard by the steep scarp of the western face of the Montes Apenninus, the Apennine Mountains. Not only is the entire length of this sinuous rille not easy to see, but also it is one of the most inspiring places on the Moon. It's where the Apollo 15 mission landed, requiring its crew to fly over mountains nearly twenty thousand feet high then descend close to the rille, which is some thousand feet deep and a mile wide. That was dangerous. Apollo 15 was perhaps the most scientifically ambitious and fruitful of all the missions, and later I would meet two of its crew and, on my own, with my conversations with them in mind, I'd visit an Earthly gorge where they trained to simulate their expedition to Hadley. I love Hadley Rille. I try to see it every month.

It begins as a wide angular gouge by the southern Apennines, fishhooks briefly, then takes a hard northerly course, narrowing as it curves along a flat land—all relatively visible—until it sneaks between the Apennine scarp and a fresh bowl crater. Then it becomes a shape like a winged raptor seen straight on, seeming to disappear. Finally it makes a long course mostly northwest before slitting next to another scarp. This I had seen in photographs. This I had not seen at the telescope.

I knew enough by now to study, to train the eye's ability to register the mind's knowledge. On cloudy days, the maps were out. I read. I consulted online atlases and photographs. I'd even obtained a USAF lunar chart map of the Apennines with the same contours, colors, and paper stock of the old "quad maps" I once used to plan backpacks. It was not a far step to imagine myself there in 1971 with astronauts Jim Irwin and Dave Scott. There, I told myself, putting my finger on the map, there, that's where I'd set up my tent. So I knew exactly how the rille coursed along, but the air of our planet had conspired, it seemed, for several lunations to be cloudy or bubbling with unsteadiness. After the fishhook turn, after a bit of the rille's flatland run, it always vanished.

Until it didn't. One night I blinked. There it was, all of it, the entire length of the most beautiful thing on the Moon, all of Hadley Rille, seen even where it curved past the crater, where Apollo 15's *Falcon* Lunar Module had landed, where Scott and Irwin were standing by the face of the eleven-thousand-foot Mons Hadley Delta, hopping along with bags of rocks, discovering things, describing it all with a geologist's precise diction and a child's giddiness. I sat back from the eyepiece. It was ten thirty P.M. in Tucson and the olive trees rustled and I felt as though I was locating that channel like some latter-day John Wesley Powell, driven by vistas to succeed where other desires had failed. My skin tingled. There was no danger in my looking, yet it was the sublime I felt, and it passed into a satisfaction a thousand feet deep. Beside the telescope was the old metal typing stand I used as my field desk for lunar nights, and I opened my checklist, marked Hadley Rille, then studied Rükl and proceeded to fly over the Apennines, identifying the major peaks. I was here and I was there, facts like wings, soaring.

By then I knew the differences among the three types of rilles you can see on the Moon. Both the Alpine and Hadley rilles are "sinuous," mostly living up to that name, and having been lava flows. Some were formerly subsurface channels that have wholly or partially collapsed. Those that remain closed—called lava tubes—have been proposed as sites for human settlement since their roofs would provide protection from solar radiation, cosmic rays, and micrometeorites. Rilles wind around the surface like exposed nerves. Some are wide, some narrow, some easy to see, others hair-lashed into visibility only when the air calms. Some sinuous rilles have calderas, where lava pooled then drained off, leaving a bowl.

My favorite one is so big it's a valley: my oft-thought-of Vallis Schröteri. It bisects the Aristarchus Plateau, cutting down to three thousand feet deep, with its source at the caldera called the Cobra Head, just north of the bland crater Herodotus and the bright crater Aristarchus. The six-mile-wide Cobra Head then tapers to about half that width over the course of its some hundred-mile very snakish length. According to research cited by Charles Wood, lava poured down this barely sloping channel at unbelievable rates: from twenty-two million pounds to ten times that amount "erupting every second!" This sent fire fountains into the sky to fall and land on Oceanus Procellarum, building up the plateau itself.

There are constellations of sinuous rilles around Triesnecker, around Goclenius, around Gutenberg, there are the Hypatia Rilles a stone's throw from Apollo 11's landing site, there is the Hyginus Rille with its gently curving form interrupted in the middle by what's been called the Moon's largest volcanic crater.

Another kind of rille on the Moon is one well-known on Earth: grabens, where the land expands and cracks open, leading a block of the crust to drop down. On the Moon these might be called a linear rille or rima, a valley or vallis or a scarp or rupes. One of the most famous lunar grabens is the afore-mentioned, dramatic Vallis Alpes, which cuts its 120-mile-long, six-mile-wide swath through the Alps bordering Mare Imbrium, the surface of which is cut by the demure lava rille I want so desperately to spot. (A linear rille that's a large graben-valley is distinguished from those valleys that were gouged out by flying rocks, like Vallis Rheita or Vallis Snellius.) In low light and magni-fication, the Vallis Alpes graben appears chisel-edge straight. It looks like the first cut in a tree trunk left by an ax. Other linear rilles include the graben of Petavius, which forms the clock-hand of that crater.

The last of the lunar rilles are arcuate, the gracile or spiky lines left after lava flowed into a crater, the floor cooled, and the weight of the material at the center settled to form cracks at the edges or sometimes across much of the floor. They curve along the crater floor of Hippalus and they crackle along most of the floor of Gassendi, adding interest to an old crater that lacks the fresh terraces and central mountains of younger, more arrogant craters.

Whatever the type, rilles are a dramatic counterpoint to the eye-catching lunar stars—those craters like Copernicus—and it takes some imagination to move from seeing them to visualizing them. Nineteenth-century selenog-raphers James Nasmyth and James Carpenter say this about one of the rilles that make up Rimae Archimedes, terrain not far from Hadley:

> If the reader will only endeavor to realise in his mind's eye the ter-rific grandeur of a chasm a mile wide and of such dark profundity as to be, to all appearance, fathomless—portions of its rugged sides fallen in wild confusion into the jaws of the tortuous abyss, and catching here and there a ray of the sun sufficient only to render the darkness of the chasm more impressive . . . he will . . . learn to appreciate the romantic grandeur of this, one of the many features which the study of the lunar surface presents to the careful observer, and which exceed in sublimity the wildest efforts of poetic and romantic imagination. The contemplation of these views . . . are . . .

vastly enhanced by . . . the unchanging pitchy-black aspect of the
heavens and the death-like silence which reigns unbroken there.

I contemplate these views. I think: *I am exploring another world.* At the
telescope, I have whispered "You should see this" more than once to my
distant father.

&

Edmund Neison's tome, *The Moon and the Condition and Configurations of its
Surface*, the first great book in English on the Moon, was published in 1876,
a decade after Patmore's poem. A book whose weight and reputation suggest a
majesty both antiquated and delightful, it is hefty in its re-bound hardcover
from the University of Arizona Library. The frontispiece is an ink and sepia
illustration of Copernicus, its rim and surrounds—again, the "nimbus" as
the British call it—splayed open like a cell and its interior looking like some
fossil footprint. I enter the pages of Edmund Neison with something of
Keats's wonderment at Chapman's Homer: a version of the Moon obsolete
and familiar swims into view.

First, I am surprised to find an argument for a lunar atmosphere and,
peppered throughout the book, suggestions that the Moon might be
active—somehow—and that visible changes have occurred on its surface.
The dead Moon of Beer and Mädler is not Neison's. Neison notes records
of "misty appearance[s]" and places "after sunrise [that] appeared hazy and
indistinct." He admits, "The observations mentioned above are in each case
of considerable delicacy and of by no means a decisive nature." These and
other such views would vex lunar astronomy even to the present day, in the
controversies of so-called transient lunar phenomena.

Against the arguments made for an airless Moon—by, correctly, for
example, Roger Boscovich in 1753—Neison is unequivocal: "The existence of
an atmosphere to the moon must . . . be regarded as certain, resting as it does
on the evidence presented in so many forms by the present physical condition

of the surface, and all that remains uncertain is what density it possesses." This certainty, as well as the certainty that volcanism was a driving force on the surface of the Moon, would be proven wrong in the decades ahead. It seems Neison was the fountainhead of a tradition of mostly British amateur astronomy that was poetically, if didactically, attached to these views.

Neison and others detected what looked, incorrectly, to them—and do still to me—like the coastlines of ancient oceans. For Neison, the relic seas retreated below the surface, and, from time to time, with the glaring heat of the sun—he spends much time ascertaining temperatures—might release "aqueous vapour" that could account for "some of the otherwise inexplicable appearances that have been observed." The maria were once seas, Neison claims, citing "the formation of diluvial deposits . . . detected by many selenographers." The most extreme argument of this nature was made by D. P. Beard some decades later, who argued that lunar surface features were coral atolls!

Volcanos exist on the Moon as on Earth, Neison writes, though "the greater craters . . . [are merely] low-lying spaces surrounded by mountain regions or disturbed highlands." Small craters—and *craters* was then a term associated with volcanos—were merely "shallow hollows such as are not uncommon on earth, rather than the craters of lunar volcanoes." The latter could be found "rising steeply from the surface, with a precipitously falling conical aperture of small dimensions, whilst all around lies apparently ejected matter, visible often in long streaks radiating in various directions to the lower lying districts."

From these broad contours of interpretation, Neison then resorts to an impulse of so much nineteenth-century lunar study, cataloging detail after detail with highly specific categories, a kind of epic taxonomy. In descending order of width and complexities there are "walled-plains, mountain-rings, ring-plains, crater-plains, craters, craterlets, crater-pits, crater-cones and depressions." There are, in descending order of elevations, "the great ranges, highlands, mountains, and peaks . . . hill lands, plateaus, hills, and mountain ridges . . . hillocks, mounds, ridges and landswells." Neison's confidence turns to curious uncertainty in describing the mysterious rays and delicate rilles; he has no easy explanation for either.

There are "bright plains," the cratered highlands of the southern Moon. And there are the maria, which in places, "resembl[e] . . . the rolling lands of America." He asserts a veritable landscape painter's palette to the colors of the maria, shades I mostly don't see and some I can barely detect: "In all these greys a double degree appears to exist, namely a pure grey sinking to a very dark cold steel gray, and a brownish-grey sinking to a blackish-brown color, intermediate tints between these two often appearing." He declares that "the surface of the moon exhibits every kind of variation of pale yellow, grey, and white, and in many places the yellow merges almost into a pale brown." In places there occurs a "thin bluish white of little intensity, and almost as if it were semi-opalescent."

Neison describes positions, works out elevations and slopes, details the difficulties of estimating these during ever-changing sun and shadow, and determines surface brightness. His descriptions are so precise that I quickly just ride the language—the ring-plains are fine, magnificent, the rilles delicate, the masses labyrinthine, the mountains lofty. I appreciate the diligence of a passage that in part reads, "The 4° bright crater I and the 5° bright crater e, with the curved mountain-ridge near the last, are likewise distinct, and the peak ε is very steep. F, in +4°7′ lat. and + 40°3′ long., is 5° bright and easily seen, and so are the two small 5° bright peaks west of F, though the rest of the mountains of this region are only 4° bright."

Then I find in Neison a passage free of numbers. Here is the edge of Mare Crisium: "Broad deep bays, partly winding, partly delta-like, separate these projections, and penetrate deep into the southern highlands, the whole forming an indented coastline of most interesting character." A flash of recognition that ties me to Neison and to the evenings I've spent cruising the coastline of Crisium.

Once I read aloud Neison's long description of sunrise over Clavius, "a walled-plain," to feel his voice within me as we both watch what we have seen:

The extreme summit of the lofty wall of Clavius . . . becomes distinct, fine streaks of light begin to extend across the dark mass of shadow on the interior . . . from the light breaking through some passes . . . and these streaks widen near the centre and form illuminated spots on the floor . . . strongly contrasting with the now brightly-illuminated crest of the lofty . . . wall and the great circular broad rings of light formed by the small ring-plains within . . . The illumination of the interior . . . now proceeds rapidly and forms a magnificent spectacle: the great brightly-illuminated rings on the interior, with their floors still totally immersed in shadow; the immense steep line of cliffs . . . are now brilliantly illuminated, though the entire surface at their base is still immersed in the shades of night; and the great peaks . . . towering above . . . are thrown strongly into relief against the dark shadow beyond them.

This is the language of the sublime. Not only am I exploring the Moon from tens upon tens of thousands of miles away, as well as the closest contours of life, I am, by yoking place to experience, worlds to meanings, exploring the limits and promises of language.

Perhaps it's strange to speak of the sublime when nestled in city backyards beneath familiar trees and steps from a door and cozy house. At the cabin, however, where I've killed rattlesnakes and where mountain lions walk in winter, there can be a chilling edge to night, Moonlit or not.

The philosopher Immanuel Kant spoke of two versions of the sublime, the mathematical and the dynamic, the former devoted to sheer physical scale, the latter devoted to forces in nature. Clavius would encompass the main length of Utah's Wasatch Front, from the small town of Nephi up to Brigham City, where we turn into the mountains to get to Cache Valley and Logan, then on to the cabin in the canyon. The Wasatch peaks would not reach the top of the Clavius rim. They are not much higher than eleven thousand or so feet while the crater floor drops down sixteen thousand. This is the sublime of dimensions, of the very wide, the very deep, the very steep, and the very tall. From afar, like from my backyard, sunrise over Clavius, with its light and dark, accentuates the fact of its "placeness." I am not on the Moon but the facts I bring to vision emplace me here and there.

There is also the sheer force of sunlight landing on the Moon, that light peeling frigid dark apart, light that is heat, the thermal gradient spiking like a statistical mountain from two weeks of night that's three times colder than the Earth's South Pole to two weeks of light that's like standing in a slow cooker. On Earth, such force is conveyed by storms of air and agitated earth—hurricanes, tornadoes, wildfires, eruptions, earthquakes—the dynamic sublime. On the Moon, it's sunrise over Clavius. Edmund Neison got that right. But can words convey that? It would be one thing to stand there suddenly sun-whacked in a spacesuit, coolers spinning up, replacing the whir of heat to keep a human body alive in the Moon's dark. It's another to read of it, mediated through language that at once expresses and attenuates experience.

We have sublime encounters with things (canyons, storms, even human constructions like a rocket), we represent those encounters and call the representations sublime (paintings of canyons, tornado videos, a grainy TV broadcast of the launch of a Saturn V), and we mediate between the two, not only offering up representations after the fact but also absorbing representations *before* we encounter the thing itself. The historian David Nye has written about this final factor, describing how first-time visitors to the Niagara Falls often felt underwhelmed because their anticipation had been shot through with descriptions, verbal and visual, of the place they had not yet experienced. The falls were smaller than they expected, there were tourist traps, there was sunshine instead of storm. Familiarity from a distance can breed disappointment. It can take time to let those mediating texts go, and, when they do, visitors to sublime landscapes often feel the rush of the originating emotion. One is not always initially stunned. The sublime can take patience.

However, perhaps ironically, it can be cultivated and sharpened by foreknowledge. Someone passing through Cache Valley might notice its small towns, its many Mormon churches, the large houses sprouting in developments that used to be cow pastures and hayfields. They might notice the peculiar foothills, called "benches," especially on the east side of the valley, in the Bear River Range. They might be driving north to Yellowstone, a sublime landscape with its relatively predictable geysers—Old Faithful has a schedule and is fronted by benches for tourists—and, in doing so, speed by Red Rock Pass. At Yellowstone, they'll encounter traffic jams, possibly wildlife, and likely the former caused by the latter. They may try to take a selfie with a moose, which is not sublime, just a bad idea.

Had the tourists known, they might have matched the landscape with the history: that some fifteen thousand years ago Cache Valley was underwater as part of Ancient Lake Bonneville, as big as Lake Michigan and three times as deep, and that the benches mark the varied shorelines of that ancient ecosystem. They might have considered the force of water as it once burst through Red Rock Pass in one of the world's most catastrophic floods. Carrying the past into the present, we prepare ourselves for future reverie. Knowing this makes Red Rock Pass a kind of sublime—a physical remainder and reminder of the temporal sublime, the past inscribed in the land.

The flood is past, there is no danger. Right beside the sublime is mortal threat, the difference between witnessing a flood—I have several times—or being in it. Mortal threat is terror. The sublime requires a sense of safety, though often quite nervy, a jittery distance from which terror hails rather than obliterates. Edmund Burke writes, "When danger or pain press too nearly, they are incapable of giving any delight, and are simply terrible."

It's possible that my first taste of the sublime was when my father lurched into our camper-trailer on a summer trip to the Great Smoky Mountains. My mother, sister, and I were terrified by crashes and shouts outside. I must have sat up wide-eyed as my father came back to the camper's flimsy door. "I jumped out of my skin," he said of seeing a bear. I took this quite literally and puzzled for a long time how the process of fear-induced organ-skin reversal and recovery took place. I went from fear to the sublime to bafflement.

Burke famously made a distinction between the sublime and the beautiful, where beauty is calm charm, pleasure, quiet loveliness. His discussion is deeply and problematically gendered, and his bifurcation between the two overly stark. Anyone can experience either or both as in a steep slope of mountain bluebells quavering in the wind as a thunderhead builds overhead.

As in such a moment, the Moon is both sublime—in a way—and beautiful. Across time and cultures, we have focused more on the latter. Nocturnes of Moonlit paths, Full Moons behind plum blossoms, love poems. The Moon's sublimity, for the Earthbound, arises at a fair distance, to be sure, but magnified in the eyepiece, the lunar wilds rise up like a white wave. They don't threaten me, but knowing what they are and how they came to be reframes the moment: there's the back door, a few steps away from an ancient impactor cruising soundlessly toward a crash that will launch rocks as big as city blocks. My lunar sublime crosses space and time. Facts bring the past into view, and my body feels the prick and tang of cosmic danger.

Do I become inured sometimes? Of course. But glimmers rise, the deep breaths, the held breath, a crystalline view, a new moment: with Kathe and our friends, we have bundled against a late-spring snowfall while, unbelievably, meadowlarks sang. Around us, a mountain valley stretched forever with golden flowers. Virga became snowflakes on our shoulders. Or the bumps and ridges of the lunar limb, just glimpsed, the far side world sloping over the curved Moon, a real place, made of the real: rock and fact and light and shadow, the steep sides and the shallow dips and the cracks that dance across a crust so dark alone and so bright in the dark sky that my contact with the deep past, with my own, dissolves redemption into kinship with its lovely certainty of death. *I* experience *that*. Just to be awake is sublime.

The most relentlessly sublime region of the Moon is the southern highlands, that helter-skelter explosion of overlapping rim and bowl and rim and bowl, nearly all worn down and battered. If brutalist architecture had curves, it would look like the southern highlands. It's a confusing lot of craters that apprentice observers such as myself find daunting. My observing programs have allowed me to navigate with more precision, eye looking in the eyepiece then turning on the headlamp to match the map's depiction, my finger tapping the map as though transmitting a wireless message, "This is the crater Vlacq. It is surrounded by six other craters of similar size. It is near the southern limb. It is south of Janssen. Call it the Vlacq Complex from now on."

The southern highlands look like a rough sketch of destruction. But facts also render the highlands a rough sketch of creation. This is the oldest region of the Moon we can see from Earth—about four billion years old—as it is unaffected by the later lava floods in the northern maria. The white mineral that makes up most of the highlands is, I now know, called anorthosite, whose role in the Moon's history I would come to appreciate. The highlands rise nearly two miles above the northern maria's average elevation. So while battered, the highlands are the nearside's most primordial country.

Among the crater cornucopia of the highlands—Andrew Planck compares the region both to a "blizzard" and to Utah's Maze district in Canyonlands—are three now-familiar landmarks: Janssen, Maurolycus, and Stöfler. I kept confusing their names until I realized that they appear in alphabetical order as the Sun rises over each in turn. Janssen is big, about 125 miles across, but with all the overlapping craters—including the younger Fabricius—its domain appears even more extensive. One can discern a roughly hexagonal shape to Janssen, a function of underlying basalt structures affecting post-impact appearance. Appearing a couple of nights later is hefty Maurolycus, with a vast elevated ledge and a few terraces from its more youthful days, followed by nearby Stöfler with some ten intruding craters, including Faraday on the southeastern rim, which has left a jumble of deposition in the main crater's floor. Stöfler is about the same size as Maurolycus, but has a darker floor from basalt deposits. They're eloquent in their old age. The same, but different. There is more to a crater than just being a crater. Its size and features depend on impact energy and when the impact occurred.

They likely were once complex, like the ones I've been drawn to again and again: Copernicus, Theophilus, and Tycho, the latter a complex crater of the western southern highlands young enough to have kept its splash of bright ejecta rays from darkening in the solar wind. These are the showstoppers with stair-step walls and interior central mountains. (Randomly, guidebooks I have in front of me call Copernicus "incomparable," "prominent," and "magnificent.") The impact digs out subsurface material that rebounds and piles up the central peak. The overburdened rim collapses, leaving coliseum terraces in its wake. These craters are typically wider than about ten to twelve miles, and the glacis and rays of complex craters are splayed like a mansion's foyer chandelier. The rim of Copernicus, of any complex crater, would be dizzying.

Over time, once-complex craters degrade from further impacts, landslides, and floor-fracturing. Scientists and observers often consider floor-fractured craters their own type, pointing to a place like Vitello, a sploshy, cracked crater opposite its bigger brother, the fractured Gassendi, in Mare Humorum. Floor-fractured craters are those that, after their formation on the edges of the lunar seas, had their floors lifted by underlying magma like a blister. Fissures linear and curved—arcuate rilles—cracked open, lava oozing forth. The rilles of Gassendi bear a superficial resemblance to dried mudflats or, I think one night, the hieroglyphic records of a shipwrecked giant.

Rarest of all are concentric craters, in which a crater is formed smack inside another, a double bull's-eye. They're small, and the only one I see with regularity is Hesiodus A, which looks like a bagel sitting in a bowl. The smoothness and symmetry of concentric craters are at odds with the usual irregularity of the rest of the terrain. They are as improbably neat as a lapel pin on a threadbare jacket.

Less immediately compelling are intermediate craters, those that are older and lack central peaks and can have a ring of peaks or other formations in the middle. As their walls subside to the Moon's weak gravity, they can form scallops, and such craters can display ponds of impact melt, like darkened spilled mercury solidified in place. Eighteen-mile-wide Kepler is a good example, with its frothy flanks rising to the rim, its heaps of fallen wall and its mounded central hills. Rays still project from Kepler, however, providing a bit of youthful panache.

The last category in lunar crater type, broadly speaking, is the most numerous: the anonymous crowd of simple craters—from a half-mile-wide to about nine miles. These are steep-sided bowls. You wouldn't want to try to hike out of one (the inner crater walls can slope at about thirty degrees to forty-five degrees while the outer slopes are about half that) nor would you want to spend a lot of time looking at them. They're—well, they're bowls. Pores made gargantuan in a lens. It's the sheer torrent of them. What they lack individually they compensate for in numbers. They're *everywhere*.

Jules Verne captured some of this in his novels *From the Earth to the Moon* and *Around the Moon*:

> The sight of this desolate world did not fail to captivate [the astro-nauts] by its very strangeness . . . watching heights defile under their feet, piercing the cavities with their eyes, going down into the rifts, climbing the ramparts, sounding these mysterious holes, and lev-eling all cracks. But no trace of vegetation, no appearance of cities; nothing but stratification, beds of lava, overflowings polished like immense mirrors, reflecting the sun's rays with overpowering bril-liancy. Nothing belonging to the *living* world—everything to a dead world, where avalanches, rolling from the summits of the mountains, would disperse noiselessly at the bottom of an abyss, retaining the motion but wanting the sound. In any case it was the image of death, without its being possible even to say that life had ever existed there.

The surfeit of simple craters is the very background of this death.

The violent meteoritic genesis for the huge craters of the southern highlands—for nearly every lunar crater—was inconceivable to James Nasmyth and James

Carpenter, who became the next great selenographers. Not long ago, I bought a 1903 edition of their 1874 book *The Moon Considered as a Planet, A World, and a Satellite*. It's one of the classics of nineteenth-century selenography and considerably more affordable than Edmund Neison's, the cheapest edition of which at present lists for $700. Nasmyth's is a thorough introduction to lunar motions, arguments against a lunar atmosphere, records of superstitions, and more, but most especially it's a theory of the Moon's landforms explained by contraction (and occasional expansion) of a cooling crust and by eruptions from volcanos. The book was a standard text to explain craters via volcanism and, while its science is now obsolete, the contentions were as well-argued as they could have been. Certainly, their arguments are well-illustrated: the book contains photos of lunar surface models that to this day remain famous and visually striking.

James Nasmyth was a successful and talented inventor, engineer and industrialist, the son of Alexander Nasmyth, a well-known painter and close friend of the poet Robert Burns. His autobiography includes sections on a nut-cutting device and "the uniformity of screws." It's pretty charming. (Chapter XIII is titled "My Marriage—The Steam Hammer," which he invented.) Of his father's death at eighty, he writes, "His life was fast drawing to a close. He had borne the heat and burden of the day, and was about to be taken home, like a shock of corn in full season." A few days before his death, Alexander painted a picture of an old man and a dog crossing a bridge toward a cottage. He called the picture *Going Home*.

James was as talented and driven as his father, and, though a capable illustrator himself, was inclined to the mechanical. As a boy in his Edinburgh bedroom, he cast mirrors for telescopes, and they and their views would become his passion, especially after an early retirement. As an adult, living in Fireside, Patricroft, near a canal, he had a long reflector of ten-inch aperture, the same as mine, though he had made his. A drawing shows it set on a grassy lawn surrounded by trees and beside what appears to be a greenhouse. A semicircular walk leads to servant quarters and a comfortable though not palatial house. Urns dot a crescent yard. Standing by the telescope is a black-coated figure with a cane.

"Sometimes I got out of bed," he recalled, "in the clear, small hours of the morning and went down to the garden in my night-shirt. I would take the telescope in my arms and plant it in some suitable spot where I might get a peep at a special planet or star." Sometimes he had "to move it place to place . . . in order to get it clear of the trees and branches." Once, a boater on the canal mistook the white-shirted Nasmyth and his telescope for a ghost carrying a coffin.

He next made a twenty-inch instrument, with a geared drive to move it on a turntable, fastening an arrangement of mirrors so he could sit comfortably in an attached seat and look at the views at eye level. He began his Moon researches in 1842. "I made careful drawings with black and white chalk, on large sheets of gray-tinted paper, of such selected portions of the moon as embodied the most characteristic and instructive features of her wonderful surface," he wrote. Eventually, he would make plaster-of-Paris surface models, rendering its "awful grandeur," photograph them, and, with Royal Observatory astronomer James Carpenter, write their book. He notes, without a boasting tone, that his lunar work was well-received, even by the queen, with whom he once had an audience. Ever practical, he also suggested that the Moon would provide electrical power, via generators responding to the lunar pull on ocean tides.

"I was," he admitted, "quite full of the moon."

Surely the queen saw the photographs I am looking at now. The models were based not only on drawings but also on the estimates of slopes, measurements of crater rims and mountain elevations. They lit all the models from the same angle. The results still astonish: even in today's era of Moon-orbiting, high-definition cameras, the Nasmyth and Carpenter lunar model photos are strikingly realistic, almost preternaturally precise. Consider "Plate IX.—Triesnecker," with the illustrative rectangular photo representing an area about sixty miles wide and one hundred miles long. Nasmyth and Carpenter capture the insanely nubbly surface, the tortuous ridges, and the crater mostly in shadow, the central peak rising like a lit watchtower, casting a conical shadow against the portion of the crater sunlit and stepped. Nearby are the well-known Rimae Triesnecker, the Triesnecker rilles, five or six main channels, a half to one mile wide, and requiring clear, steady air for good views. Several other rilles parallel or angle away, like strings left by a bored cat. Other craters dot the region, and there are gradations of gray that make more subtle this high-contrast, almost-too-sharp simulacrum.

The next plate shows the Moon's most impressive trio of craters, Theophilus, Cyrillus, and Catharina, each successively older and overlapping like interlocking rings. They are massive, unmistakable, and stuffed with detail. It happens that they were visible along the terminator the night before I write this, so I compare that view to what I see on the page now, the 1903 book sitting on the kitchen table beside my coffee. The grays are too white in Nasmyth and Carpenter, the topography almost pointillist, the hyperreal of the miniature, attractive in its God's-eye view, handheld and mastered. There is a peculiar edginess to seeing the Moon through the telescope, but one that is also softer than these models convey. Their views are those from an Earth without an atmosphere. Using more of my guides, including a volume of the massive *Photographic Lunar Atlas for Moon Observers* by Kwok C. Pau, an online atlas I printed and spiral-bound, I test the verity of this model Moon. Nasmyth and Carpenter seem to exaggerate the width and walls of a kind of table land that links Cyrillus to Catharina, but, on the other hand, they show a gully I didn't notice last night running outward from Cyrillus past the rim of Theophilus.

Well, this is esoteric. This is useless diligence. It's also the taxonomic pleasure of the Moon. The generosity of possibility, discovery of a high pass or a tight meander. The shock of the new in landscapes I know and the collector's pleasure of comparisons.

Why those craters existed in the first place is, however, the main business of these two British writers, and in this they took—perhaps even largely established—the majority view: craters on the Moon were like the craters they knew on Earth, always made by and leftover from volcanos. (In the seventeenth century, scholar Robert Hooke found that circular impressions were created both by bubbles burst from boiling alabaster and flinging pellets into moist clay. Like others, Hooke could not conceive of large-scale impacts.) Even though the scales on the Moon are far vaster than those on Earth and, while Nasmyth and Carpenter acknowledged varied differences—such as the key fact that terrestrial volcanos rise *above* the surrounds and are not depressed into plains like the massive craters on the Moon—they assert volcanism nonetheless. They simply assumed volcanic forces, in part because of reduced gravity, would produce these massive craters though nothing like them exists on the Earth.

Their basic volcanic process starts with a vertical channel from the subsurface that forcefully sprays material outward, forming the crater rim while destabilizing the surface, allowing the crater floor to drop below the surrounding terrain. Then, with the last gasps of less powerful ejection, the central peaks are formed. In some cases, continued lava flows fill the crater to bury the central peak and might even reach the crater rim, like a saucer full of milk, like Archimedes. Sometimes additional vents form to create peaks and hillocks in different parts of the crater floor. Their sketches rather neatly illustrate their stepwise theory, and, I have to admit, it's pretty tidy as such.

Unable to explain the bright rays emanating from some craters—these are the lines of impact-ejected rocks—they suggest the rays are cracks filled with once-molten matter. They compare them to an artist brushing white paint across the Moon. Mountain ranges were, in their eyes, the result of "exudation . . . comparatively gentle oozing of lava . . . [like]a water fountain playing during a severe frost." The unlikelihood that this would produce mountains up to and exceeding twenty thousand feet doesn't strike them.

Still, they're not arrogant. They admit that really large features, the massive maria such as Mare Crisium on the east limb or the dark circular Grimaldi on the west, are "beyond our explanation." They ask for "considerable freedom to conjecture" and ultimately suggest those areas were formed by "some very concentrated sublunar force of an upheaving nature." They admit "difficulty in accounting for such a very local generation of a deep-seated force" and how it would conclude in "a raised ring at the limit of circular distance."

In one retrospectively ironic passage, they even admit the possibility of impacts. "Were it not for the flatness of its bottom, [the Alpine Valley] might set one speculating upon the probability of some extraneous body having rushed by the moon at an enormous velocity, gouging the surface tangentially . . . and cutting a channel through the impending mass of mountains." This is not how the valley was formed but their invocation of possible impacts only highlights the unthinkability of applying the phenomenon widely. It's no wonder that pro-impact astronomers were shunted aside, especially when some of them, like Franz von Paula Gruithuisen, were just outrageous.

Further, as lunar scientist Peter Schultz points out, the impression that Nasmyth and Carpenter give in their photos of lunar-surface models was more than aesthetic. It was self-reinforcing. As Schultz writes in a historical article about lunar-impact theories, "The artistic renderings of Nasmyth and Carpenter influenced more than just popular opinion. They also affected scientific perceptions about the processes shaping the Moon. The rugged terrains and deep craters could easily justify interpretations of volcanic processes."

There is also inexorability. Nasmyth and Carpenter spend a portion of the book showing how when materials cool, they wrinkle, deform, and split. Hence, the cracks of the Moon. Their visually striking examples are of a shriveled apple and a wrinkled hand—Nasmyth's, it turns out. Sometimes when I am setting up the telescope, I chance to look down at my image in the mirror and I see the magnified furrows of my neck. My father's body shriveled rapidly at the end, from age and starvation; he simply couldn't eat anymore. The cancer, like a mineral, blocked the path. His skin sagged like vestments.

At the cabin once, I showed Kathe the picture of Nasmyth's hand, and we put ours next to it. While I could not claim for myself, as Nasmyth and Carpenter do for the Moon its "transcendent antiquity," I took off my reading glasses, looked at my hand, and announced it was, certainly, more lunar than hers. We agree with Victor Hugo: "Forty is the old age of youth. Fifty is the youth of old age." After that—

There is nothing more personal than death. Watching my father choke on mashed potatoes, hearing his raven-rattle hacking, seeing the body retreat—a kind of heat death—I bore witness to helpless agony. Valiantly, my sister, our brother-in-law and his wife, and my stepmother were "there for him," as we say. I counseled, visited, had driven the moving truck to the eldercare apartment. For a few minutes, I had stood in their bare Plainfield, Indiana, home, feeling that impossible-to-name emotion—ultimate counterpoint to the sublime?—provoked by empty rooms. Like my mother, I used to dread death. My father mustered grace while dying. As I've grown older, far from dreading the end, I will be glad to put an end to the chatter.

A fervent reader of Schopenhauer, the composer Richard Wagner once wrote to Liszt: "I have . . . found a sedative which has finally helped me to sleep at night; it is the sincere and heartfelt yearning for death: total unconsciousness, complete annihilation, the end of all dreams—the only ultimate redemption." This assertion is Wagnerian in its bombast. Rather than yearning for death, I am grateful for its inevitability. I think of it when I cannot sleep, a sedative, and the morning all the brighter. When I am, as my father was, closer to the end, I'll embrace a sincere desire for the only ultimate redemption. Until then, pain makes its tactical advances—a brief, almost epiphanic searing in the leg, weeks bent over with infection. If I can't make it to the eyepiece, show me a map of the Moon.

If the Moon has given me a new lease on life, it has done so in part because it is honest about death.

Dinosaurs would have seen, had they looked, some bright impacts on the Moon—Tycho punched into the lunar surface some 108 million years ago, ejecta zinging down on Earth, presaging the asteroid impact 65 million years ago that was the final blow against the saurians. There is nothing personal about death.

My excursion into discredited volcanic theories of the formation of the lunar surface reveal something at stake bigger than explanations of one orbiting rock's features. It's that eon after eon has shown us the radical unimportance of our species and our planet.

We now take impacts from space for granted, but the dinosaur-killer wasn't discovered until the 1980s, and we are uniquely capable of ignoring our mortality. Climate change comes to mind. As responses to mortality, whatever the threat, both capitalism and environmentalism are neurotic. The former seeks to transcend death by fixing treasures in the future, the latter by fixing the present or some recovered past in amber. One long look at the Moon should tell us both are impossible. Every crater is a middle finger at our lazy paradigms.

Geology was borne from the realization that we could interpret the past from forces operating in the present. That small rocks fall from the sky—and, eventually, it was understood from space itself—was accepted only in the early nineteenth century as legitimate natural phenomenon and not the folk tales of ignorant peasants. However, no one had seen a massive rock striking the Earth nor could anyone imagine the numbers, sizes, and energies of those bodies that populated the primordial solar system. Geologists early on also wished to elevate the scientific status of their work by contrasting it with the catastrophist tales associated with the Bible, such as the great Flood. The theory that all surfaces—here or on the Moon—came about due to slow-acting processes such as uplift, regional or localized violence (volcanic eruptions here and there, floods here and there), and patient erosion was called uniformitarianism.

One of the first and one of the few to think scientifically about space impacts in the nineteenth century was the great American geologist and leader of the US Geological Survey Grove Karl Gilbert, who, in 1891, traveled to a giant hole in the ground in Arizona. Could a large meteorite have caused what even then was called by some Meteor Crater? With exception of the obstreperous mining engineer Daniel Barringer, who owned the crater, most thought it was formed by a subsurface steam explosion. Yet Gilbert had been right in his hypothesis; it had been an impact. He would not be alive when it was proven that an iron meteorite 160 feet wide struck the area about fifty thousand years ago, causing regional destruction.

A cautious thinker and assiduous field geologist, Gilbert found evidence for and against an impact formation. In part, he also misunderstood the physics of high-velocity impacts, as did Barringer, who wanted to mine what he was sure was a meteorite below the crater's floor. But the main body of the impactor had mostly vaporized. Because Gilbert could not prove decisively the impact origin of the Meteor Crater, he ultimately sided with conventional opinions of the day.

A year later, Gilbert was looking at more holes in the ground, this time through an eyepiece of the twenty-six-inch reflector at the US Naval Observatory in Washington. In the fall of 1892, he studied the lunar surface. According to planetary scientist and historian Donald Wilhelms, one Congressman complained, "So useless has the Survey become that one of its most distinguished members has no better way to employ his time than to sit up all night gaping at the Moon." Gilbert not only observed, according to William Graves Hoyt, but immersed himself in selenographic literature and even fired bullets into surfaces in order to mimic high-speed impacts. The result was "The Moon's Face," a paper published in the out-of-the-way journal of the Philosophical Society of Washington. (Reprinted in a supplement to *Scientific American*, it did have a greater reach, but Gilbert's background as a geologist not an astronomer worked against widespread acceptance of his ideas.)

"The Moon's Face" was, if mistaken in some crucial particulars, a prescient piece of science. As Hoyt explains in a scholarly article called "G.K. Gilbert's Contributions to Selenography," the geologist noted an analogy to "crateriform structures formed by raindrops or pebbles falling on mud and projectiles fired into plastic targets, [and] he declared at the outset his belief 'that all features of the typical lunar crater and its variations may be explained as a result of impact.'" Gilbert argued that impact heat would melt material as ejecta and beneath the crater floor, even though he still didn't grasp that impactors largely would also be destroyed.

For Gilbert, not only were craters explainable by impact but even a massive maria—like Mare Imbrium, which covers an area not quite as big as Alaska—displayed the telltale signs of being the result of a collision. Gilbert spoke of the "sculpture" of many lunar features: troughs or furrows and parallel lines of hills or hummocks. Many of them converged at the center of Mare Imbrium. He wrote, "A collision of exceptional importance occurred in the Mare Imbrium . . . What must have been the violence of a collision whose scattered fragments, after a trajectory of more than a thousand miles, scored valleys comparable in magnitude with the Grand Canyon of the Colorado!" Even today lunar observers speak of the Imbrium sculpture, and I too trace those furrows back to the ground zero of that staggering event, which took place 3.8 billion years ago when a protoplanet at least 150 miles across struck the Moon.

The problem was circles: so many round craters. Gilbert didn't understand that objects traveling at cosmic speeds of say, thirty kilometers per second or 67,000 miles per hour, will form circular impacts except at the very lowest of angles. So he hypothesized that the Moon had formed as a result of moonlets accreting together to form our companion satellite, moonlets that themselves had orbited the ancient Earth like the rings of Saturn. Any remaining moonlets rained down on the new Moon, destabilizing its axis, so it always presented a more or less straight angle to the bombardment of remaining projectiles. Therefore, circular craters.

This was ingenious, if labored, and I have to think had Gilbert stayed with his lunar studies, he might have gotten at some of the complexities of impact dynamics. After all, the contemporary astronomy writer R. A. Proctor, who had spoken of the "plash of meteoritic rain," correctly asserted that rocks would be fluid enough after an impact to rebound, in most cases, to a circular form. Gilbert's paper was ignored. Proctor was forgotten.

Overturning a paradigm can take time. Don Wilhelms notes other figures in this sputtering effort. Nathaniel Southgate Shaler saw that high-velocity impactors would vaporize. Several other scientists, whose work was dismissed, when in the unlikely case it was read, demonstrated, as Proctor suggested, that craters would be circular except when the impacts took place at very low angles. Other geologists in the 1930s published obscure articles showing that a large impact would cause so much heat that rock would melt, wave outward, and rebound into a central peak in the middle of the crater: the very feature that most had taken to be evidence of a volcano.

Still others outlined the many differences between terrestrial volcanos and the so-called lunar ones, with Robert Dietz showing in 1946 that the craters on the tops of some central peaks were not openings for lava. *They were shadows.* He also, in Charles Wood's words, "tabulated morphological characteristics of lunar craters, none of which were consistent with known styles of volcanism," and Dietz explained that "the generally random distribution of craters was not consistent with volcanism, that crater rays were likely to be pulverized impact ejecta, that pervasive impact events would fracture the upper crust, that the high energy of impact would melt rocks . . . that the paucity of craters on the maria implied it [sic] formed late in lunar history." There was more, but Dietz was right. (Rays also trace lines of secondary craters from larger bits ejected by an impact; the brighter they are, the younger, as they have not had time to darken from radiation exposure.)

But Dietz went unnoticed because astronomers didn't read the *Journal of Geology.*

The Moon also was not a priority for most professional astronomers from the late nineteenth century on, at least until the Space Race. The volcanic origin of the Moon's surface remained largely accepted by obsessive amateurs, a few professionals, and pretty much ignored by the rest. Of those amateurs, some of whom had scientific training in other fields, Wilhelms notes one Josiah Edward Spurr, the author of four volumes of "tedious ramblings." Spurr argued, in Wilhelms's words, that "all lunar features were created by endogenic [internal] melting and fracturing triggered when the Moon was captured by the Earth." Spurr is famous for also systematizing the so-called lunar grid theory, which amateurs, especially in Britain, glommed onto with evangelical zeal. Map after map of these presumed lines of crustal weakness were invoked to explain the appearance of volcanic craters in lines. Theophilus, Cyrillus, and Catharina were often cited as the best example. They are right next to each other! In a line! This, it turns out, was an accident of random impacting.

The endurance of volcanic explanations for lunar features had, in the twentieth century, a good deal to do with the outsized personality of the British astronomy popularizer Sir Patrick Moore. A rotund Carl Sagan with a cooler accent, Moore wrote books, gave radio talks, and appeared on television. He, along with collaborator H. P. Wilkins, pooh-poohed the importance of meteorite impacts. With some good humor, they did at least dispense with notions that there were no craters on the Moon at all—that they were optical illusions caused by storms in the lunar air—or that the Moon's features had been carved by melting ice. Moore, Wilkins, V. A. Firsoff, and others carefully analyzed the Nasmyth and Carpenter approach and found it wanting. Rather than abandon it, however, they modified it, suggesting different processes for volcanic outflow. Not only, says Wood, did non-American scientists stick with the volcanic hypothesis; they simply assumed the craters originated at about the same time. As Moore and Wilkins wrote in their influential 1955 book *The Moon*, "The vast numbers of meteors necessary to produce the lunar craters and the arrangement of these craters do not suggest a meteoritic origin for the majority."

☾

Against the lava tide appeared, finally, Ralph Baldwin.

It was not until this businessman who had trained in astronomy and physics published a startling book in 1949, *The Face of the Moon*, that the impact theory began to take hold.

Baldwin—who had earned a doctorate at Michigan in 1937, the year before W. H. Pickering died—taught and worked at the Adler Planetarium in Chicago, conducted top-secret explosives work on the proximity fuse in World War II, then spent four decades running the family business, the Oliver Machinery Company. Don Wilhelms's 1993 book *To a Rocky Moon: A Geologist's History of Lunar Exploration* is dedicated to "the amazing Ralph Baldwin, who got so much right so early."

Ralph Baldwin could take in the big sweep, and to his eye the circular maria, especially Mare Imbrium with its curious curving mountain ranges on several sides, looked a lot like a crater, just as it had to Gilbert, of whose work he was unaware. A very, very big crater, one that required the collision of a colossal meteorite. Baldwin did more than guess. He quantified crater characteristics, and, in Wood's words, "plotted a log-log graph of crater diameter vs. crater depth. The result was a continuous curve linking together all these craters. Baldwin measured dimensions of volcanic craters [on Earth], finding that they display no such a correlation over many orders of magnitude." Influenced by his observation of wartime bombing craters, Baldwin was the first scientist to rigorously apply statistical methods to crater analysis instead of merely arguing from observational analogy. The inevitable conclusion was that craters of all sizes—including the giant maria—were formed by impact. (Just two years before Baldwin's book, a writer in the influential magazine *Sky & Telescope* wrote, "One seems to be in dreamland when he pictures the lunar heavens alive with flying blocks of stone.")

In prose that Gilbert would have loved, Baldwin writes of the accident that formed Mare Imbrium with nearly Biblical cadences:

> Downward the meteorite plummeted from the northeast, gradually gaining velocity. Probably it did not even glow from the effects of the nearly absent lunar atmosphere. Then it struck the surface and quickly disappeared beneath, leaving a small sharp hole to mark its passage. For only an instant, however, did the calm prevail, for then all hell broke loose, soundlessly, on a scale to shame the infernos dreamed of by little men. A great section of the crust, several hundreds of miles across, domed up, split rapidly and radially from the central point. Surface layers peeled back on themselves like the opening of a gigantic flower, followed quickly by a stamen of dust and fragments spreading rapidly in all directions.

The fragments gouged groves—Gilbert's Imbrium "sculpture"—while those landing closer helped to form the impact-uplifted, circular mountain ranges of

Imbrium, like the Alps and Apennines. Over time, the great basin would shift and crack and subside here and there, and lava would in fact flow, though later than Baldwin thought. The original mountain rings would largely be subsumed.

Speaking directly to the volcanists near the end of his book, Baldwin turns the analogy argument back against them:

> The existence of small but authenticated meteoritic craters on the earth allows a tie-in to be made between the craters on the moon and the smaller man-made explosions pits on the earth. The identification of the source of the lunar craters automatically follows from these correlations and their visible natures. The meteoritic impacts only are known to be capable of furnishing the requisite amounts of energy. Any nonmeteoritic hypothesis represents a fanciful extrapolation beyond anything known on earth.

Further, he said, there "is no volcanic neck funneling magmatic materials from basement reservoirs to the surface where they accumulate into a vast graceful mountain with a relatively tiny craterpit high in the peak." And there were arguments about the differing angles of repose for terrestrial volcanos and lunar slopes, the basic observation being that lunar central peaks are "craggy and irregular," unlike terrestrial volcanos.

Volcanism as the motive force for lunar cratering was dead.

"The vista opened up . . . is exhilarating in its magnificence," Baldwin writes in *The Face of the Moon*, "yet it also contains a disturbing factor." He understood that the impact origin of lunar craters meant it could happen again and not just on the Moon. "The explosion which formed the crater Tycho . . . left us an interesting object to study. A similar occurrence anywhere on the earth would be a horrifying thing." According to MIT researcher J. Scott Stuart, on average every six hundred thousand years, an asteroid one kilometer wide—more than a half mile—hits the Earth, which is enough to cause a global climate disaster. Only in the twenty-first century has our species become proactive in monitoring and potentially deflecting such threats from space.

If the argument between impacts and volcanism for lunar surface features—especially the maria and the craters—appears arcane, it's worth registering some other connotations of the debate: Impact theorists were edging closer to the reality of a solar system maturing through phases of extreme (to us) violence. By understanding that fact, one begins to comprehend how other solar systems form as well. The volcanists' inability to fathom the rocky chaos of the solar system seems also a legacy of the human need to confine cause to the visible present or the relatively recent past. Uniformitarians hadn't witnessed asteroid downpours. Science extends us beyond the moment, especially with instrumentation and computing, but our animal selves are, daily, in a world that seems relatively unchanging or, at least, predictable. That's Coventry Patmore's world.

The battle between volcanism and impacts would settle out in favor of the latter, helped by such figures as Eugene Shoemaker, who conducted research at Meteor Crater, on other craters formed by atomic blasts at the Nevada Test Site, and in the Ries Basin in Germany, where he discovered samples of a recently identified mineral that could only form under superintense pressures. Shoemaker helped seal the death sentence for volcanism as the major force in lunar features and, as I'd find, was a prime mover in the science of the Apollo missions.

And it wasn't until Apollo returned samples of something called "breccias" that lunar volcanism was placed in its proper context. Breccias are made of highly fragmented rocks, many containing high-pressure-shock minerals like those of terrestrial impact craters, that can only be produced by rocky objects hitting the Moon. During ABC's coverage of the Apollo 11 mission, scientist Robert Jastrow sought to demonstrate the impact hypothesis by lighting a firecracker in a pit of flour covered by black sand. When it ignited, the flour sent perfect white rays in all directions and a simple bowl crater remained. Had this been a real impact, there would have been breccias.

Just as the geocentrists had warped models to fit evidence, the lunar volcanists had done the same. The nearly unthinkable alternative, if the facts bear it out, is usually the right one. It's not that molten rock didn't matter on

the Moon. It did. But it is ironic that impact craters have experienced volcanism, albeit not the kind proposed by such figures as Nasmyth or Moore.
Impacts caused the craters, but lava has flowed on the Moon. Volcanism
was a consequence of crater formation, not a cause. By weakening the lunar
crust, especially on the nearside, where the crust appears to be thinner overall,
impacts gave lava permission. Thin lunar lava—compared frequently to syrup
or motor oil—later flooded the huge impact basins. They have spread their
drowsy gestures in the floor-fractured craters. Famously, on the far western
limb, lava filled the crater Wargentin to the brim and spilled over like an
untended coffee mug. Wargentin has in its center some low ridges that look
like a bird's foot. It's a weird crater, and I like staring at it.

This I can see through my telescope. I can also see, yes, volcanos on the
Moon. They are there. But they are not craters and they are not immediately
dramatic and they are not easy to find. And I love them.

The selenographers had pictured large, eruptive volcanos. The Moon's are shield
volcanos, in fact, slow-growing, oozing, composed of built-up layers a bit like
blood that makes a scab. Indeed, lunar volcanos are frequently described as
swellings or blisters, as though they are medical conditions. They are rarely seen
unless the terminator is near, with slanting sunlight catching their wide, demure
slopes. If big craters are loudmouths, volcanos are wallflowers. You can't see them
in shadow, and they disappear when the light is bright. They are crepuscular,
sitting like nesting birds on a dusk prairie. If large craters dizzy with their zillion
details of terrace, peak, and jumble, lunar volcanos steady with their three essential elements: base rising from the plains, slopes looking like gray note paper,
and, if you are lucky on a night of still air, their tiny summit craters, which,
if present and are visible, are little dots on an *i*. They are typically between
three hundred and a thousand feet high and between six and ten miles wide.

There have been explosions, but not akin to, say, Vesuvius. The Greek word
for "broken by fire" is *pyroclast*, and you can see the Aestuum Pyroclastics

and Bode Pyroclastics and the entire Aristarchus Plateau, which is deep with ash. From vents in the shifting lunar surface, pyroclastic eruptions sent up molten material, gasses, hotly formed glass beads, and ash in fire fountains. Such sprays of ash are laid down in delicate black strokes on portions of the lunar surface. Photos don't do justice to the brushed, painterly black lines of ash deposits between Copernicus and Ptolemaeus to the southeast. Part of their allure is that the black that one typically "sees" on the Moon is shadow, the obsidian dark of lunar night. But lit by lunar day are the black glissandos left from fire fountains. Their slender rarity is a kind of calm beside the busy textures of craters. That part of the Moon was once very busy with such eruptions, for in Alphonsus, hard beside Ptolemaeus, you can see the tiny holes where fire fountains spewed; they are called "dark-halo craters" and they look like cloaca. These craterlets were made famous by Ranger 9 in 1965 when that craft, the last of its kind, took photos of Alphonsus as it zoomed in for its crash. Millions of people watched the photos on television, the first live transmissions from the Moon.

The most famous lunar volcano or "dome" is Mons Rümker, which looks like it sounds: a long phlegmatic throat-clearing. Set in the northern reaches of Oceanus Procellarum, on the Moon's northwest quadrant, Mons Rümker is an exception to the usual small, low-dome shape of most lunar volcanos. It's big—some fifty miles wide—and composed of some thirty or so overlapping smaller domes, which gives it haggard look. This crusty mole of a plateau rises gradually to more than four thousand feet above the surrounds.

In December 2020, the Chinese landed a spacecraft near the site and returned samples to the Earth. They found a puzzle. The material was the youngest yet studied from the Moon, showing that lava was flowing just two billion years ago, about when multicellular life was arising on Earth. "Just" matters, because scientists cannot account for such young lava flows when the Moon's size tells them that it should have cooled off and died down before that. Subsurface radioactive materials giving off heat as they decay might have prompted this young volcanism, but the Chang'e 5 craft did not retrieve any.

Long-erased impacts might be a cause for late lava floods or even the gravitational pull of the then-much-closer Earth. One scientist told the press: "A
lot of people think we already know what's going on with the Moon, but it's
actually quite mysterious."

In part it's that air of mystery that returns me to the hunt for other lunar
domes. There's Mairan T, which, out of the norm, rises sharply at twenty-
five degrees and has a weird, fish-mouth crater on top. It's hard to see and
a pleasure when spotted. There are domes named for poor Franz von Paula
Gruithuisen. There are the Marius Hills, that far-west complex of about fifty
domes frozen like an army of hoodoos casting more shadows than one is used
to beyond crater walls. These and others have different compositions, histories
of faster or slower lava flows and, therefore, local implications for the Moon's
complicated history. Some domes are composed of basaltic lava, just like the
maria. Others, like the Gruithuisen domes, contain a lot of silica, something
that occurs on Earth—in the presence of plate tectonics and water. It's unclear
how silica-rich domes formed.

Mons Rümker is as easy to see—big and isolated in a mare—as the summit
pits of the Hortensius Domes are difficult. Six flat-topped domes west of
Copernicus and near their boring namesake crater, the holes at their tops seem
so tiny I could imagine, on the night I finally saw them, the sounds of squeaks
emerging from below. The view from the Tucson backyard was notable in its
smoothness and longevity. Just as Edmund Burke had put it, the astonishment
became a form of reverence. From normalcy to the sublime back to normalcy,
infused with reverence of fact. The air stayed so glassy I did not adjust my focus
once. When I was done, my wrinkled hands returned the eyepiece to its case,
the telescope to my study—pulling the metal springs off the rocker-box with
a twang, lifting the optical tube out, moving it, then moving the mount and
setting the scope back in—this, the closing ritual that had begun a couple of
hours before in reverse, I like a ghost carrying a coffin.

For a time, we couldn't conceive of a Moon smashed by giant rocks. For a time, I couldn't imagine a phone call in which a loved one announces he has cancer. Having seen volcanos mold landscapes, but not meteorites, we tended to the former's agency; having grown up with a father, I took his presence for granted.

In the same season I saw the summit pits of the Hortensius domes, I'd returned to Indiana to celebrate my stepmother's eighty-fourth birthday, the first without my dad, who had died the day after Christmas. I had booked tickets to see him but arrived too late. "It's for the best," Judie said on her birthday visit. "I wouldn't want you to remember him that way." She gestured at the room, as though some outline of his rictus that only she could see hovered in the air.

I brought home and stacked on a desk in the living room some of my father's photos and documents. There were pictures of a clear-skinned, tall, very skinny George Cokinos in naval whites and there were little travel booklets from Morocco and Lyon, ports of call for his aircraft carrier, the *Midway*, on which he served as a crew chief for a Douglas A-1 Skyraider. With a flinch of shame, I remembered the first thing I said when, while I was in graduate school, he called to tell me his mother had died: "She was crazy, right?" When my sister and I were children, Grandma Needler—a gigantic woman—would set us on the bed next to her and read Tennyson's "The Charge of the Light Brigade" and other poems. Only later did I learn she had abused her last husband. The picture of my father bent over a petri dish in college rescued me. He had studied chemistry, a discipline in which unrevealed order would be found by the sedulous. He looks at that petri dish with a studiousness I bring to the Moon.

We had almost parted for good, my father and I. One summer—when I was still with my first wife in Kansas—I had returned to Indiana to live with my mother and care for her after an aortal aneurysm. Her depression was unabated, as was her chain-smoking, and I didn't know how to endure her clawing panic attacks. Nor did I know how to convey this entrapment to my father, who expected me to drive the hour-plus to visit him. I didn't. All the distance, all the awkwardness, all our mutual failures erupted into angry letters

I've kept and never reread. I don't recall who called the truce. We learned to speak. Distances long kept collapsed.

The day before he died, my sister holding the phone to his ear, my father, unable to speak, heard me say, I hope, "We love you. It's okay. You can let go now. Go in peace."

Sometimes in his final weeks I spoke to him from my study in Tucson, where, on a metal bulletin board I have photocopies from books showing the geological ages of the Earth and the continents shifting over those eons. After saying good-bye, I would sometimes stare at a word like *pre-Cambrian* and listen to the silence. Perhaps our task, whether in mapping reality or living our lives, is to be less surprised.

CHAPTER EIGHT

Wide-Shining

The Moon was born at eleven A.M., eastern daylight time, on Sunday, July 20, 1969, in New York City.

A collapsing cloud of gas and dust heated up like a giant's twirling fireworks. Two black shapes began to spin, one a slightly larger yin to a smaller yang, each dark head jeweled with a primitive orange eye. Within three minutes, the early Moon was orbiting the early Earth.

"In the beginning God created the Heavens and the Earth," intoned a voice (male, of course). "And this earth was without form and void. And darkness was on the face of the deep. Some five billion years ago, whirling and condensing in that darkness, was a cloud of interstellar hydrogen, four hundred degrees below zero, eight million miles from end to end. This was our solar system, waiting to be born." The revisionist voice continued, describing "two protoplanets—the Earth and the Moon—now separate gaseous eddies mutually trapped in their gravitational pull, moving in tandem orbit around the sun and growing more dense." We are told of accreting dust and rocks, of chaos, of collisions, and, as the voice continued, the scene shifted to a cartoon-gray lunar surface, its craters filled with lava the color of childhood Tang. Explosions billowed curtains of debris as asteroids kept hitting, unforgiving as a swarm of playground bullies.

Six minutes passed, not six days, and the voice was CBS News reporter Charles Kuralt's, not God's. This vision of the birth of the solar system, the Earth, and the Moon was an animated film that began the broadcast-day coverage of the Apollo 11 landing.

It's a good story, and like so many theories about the origin of the Moon, it was both poetic and wrong.

It wasn't until after Apollo that scientists conjured up an idea that seemed to fit some rather strange facts, especially the startling chemical similarity between certain aspects of the Earth and the Moon, along with profound differences in their respective cores. Even into the twenty-first century, questions linger about the precise mechanics of the formation of our satellite, one that is, for reasons delineated below, unique in the solar system. To know the origin of the Moon, and here Kuralt was right, is "in part contemplating our own beginnings." The origin of the Moon is the origin of the conditions that led to our being here today, "gazing," as he said, "into that luminescent face."

How was it born? When? By what processes were contrived the distance, rotation, and orbit? These questions about the Moon were necessarily part of the larger mystery of the solar system itself, whose origin for a long time seemed special, suggesting that our Sun with its retinue of planets and moons was so rare as to be unique. Ours might be the only solar system in the galaxy, perhaps the only one in the universe. In retrospect, we know this is not so, and it is hard to resist pointing out that the notion of our solar system being solitary was another attempt to highlight some human importance in an indifferent cosmos.

In support of the rarity argument, some hypothesized that the solar system had coalesced out of gas, dust, and debris from the nearly statistically impossible collision of two stars, our Sun and some vagrant one flying by like a lost pilot. While the solar system did coalesce from a cloud of gas and dust—this was an idea, with variants, that reached back to, among others, Immanuel Kant in the eighteenth century—it did so not from a rare stellar collision. Rather it—and we—arose from old, red giant stars shedding materials into space and from some commonplace stellar explosions: supernovae that blasted out the materials of their dying stars, materials that gravity later brought together once more to birth a new star and its worlds.

The origin of our Moon within this maelstrom remained a mystery until the Apollo astronauts brought rocks from its surface and until two pairs of scientists independently theorized an impact so prodigious it makes the dinosaur-killing event of sixty-five million years ago seem like a hailstone falling in the desert. Between the nineteenth century and the late twentieth century, scientists struggled with lunar-birth theories to account for the Moon's presence—a large, relatively close body—and therefore singular among the inner planets.

For more than one hundred years, three ideas contended: The Moon had spun off a rapidly rotating, molten early Earth, like bits of clay flung from a mass turning on a potter's wheel. This was the "fission" theory. Or the Moon—or a few small objects that would become the Moon—wandered close enough to our world to be captured by it, the logically named "capture" theory. Or the Earth and Moon were born together—the "co-accretion" theory—in which these worlds are the conjoined twins of the ancient solar system, as in Charles Kuralt's creation myth. All three would be in their ways attractive and unsatisfactory. It would take Apollo, some big thinking, and modern computing power to generate a hypothesis that nearly all believe is a better story.

Cold wind whipped the unstaked tent and poles as I struggled to attach the fabric shell to the metal skeleton. Beach sand stung my legs and eyes. There was a storm off the California coast south of what was then Vandenberg Air Force Base, where, I hoped, I would see my first-ever rocket launch the next day.

"Yah, bro, you don't even need to drive into town," explained a man in cargo shorts. He wore a t-shirt so stained that whatever logo it bore looked like cosmic debris. "You can see them launch from here." He looked to a headland while I repeatedly failed to snap, loop, and pull taut the tent, which bucked in the wind like a bad argument.

"Pretty cool," he added, eyeing me so I didn't know if he meant rockets or my struggle.

The rocket, alas, didn't launch, first because of high winds—"No shit," I would later mutter to myself—then because of technical problems. So I drove home disappointed, looking forward to the day when I'd see a Falcon 9 or Delta Heavy or New Glenn send a satellite or human crew to space. But the storm changed course as I left the then-sunny coast, with heavy rains lashing the California interior, churning canyon sides into mudslides, crushing homes, and killing children in the night. Dismayed and unnerved, I was safe enough in my pickup, my escape capsule, navigating detours and hauling through ribbons of water streaming across back roads I never wanted to see again.

Tires gripping mud, wipers slashing across the windshield, I abstracted, I began to ponder watery violence and space: there were those who thought the Moon had come from there, back in what is now the Pacific Ocean, spun from the early molten Earth like a blobby discus. Where the Earth had given way, aqueous vapors later gathered to become the ocean named for peace. Where the remains of the Earth went self-molded into Moon. The image sustained me on the edgy drive.

For a century, the Pacific Moon origin story was the most popular hypothesis for lunar birth. A 1936 children's educational radio program put it this way, opening with, as the script said, "'Weird mysterioso' fanfare." The "Friendly Guide" asks the listeners:

Have you heard that the Moon once occupied the space now filled by the Pacific Ocean?

Once upon a time—a billion or so years ago—when the earth was still young—a remarkable romance developed between the Earth and the Sun—according to some of our ablest scientists . . .
In those days the earth was a spirited maiden who danced about the princely Sun—was charmed by him—yielded to his attraction, and became his bride . . . The Sun's attraction raised great tides upon the

earth's surface . . . the huge crest of a bulge broke away with such momentum that it could not return to the body of mother Earth.

The Moon gave way, released into orbit.

A girl responds, "How exciting!"

And it was. The "fission theory" had a flair missing from its competitors. It had imagery both colossal and poetic: our companion Moon born from a whirligig molten Earth, a merry-go-round with one too many horses, sending the surplus steed into orbit, the original dervish whirling in compliance with angular momentum. It brought space down to Earth. You could stand on a beach and imagine the ultimate molten Moonrise.

A son of Charles Darwin, a law student then, finally, a scientist, and Plumian Professor of Astronomy and Experimental Philosophy at Cambridge University, George Howard Darwin wrote *the* book on tides, *The Tides and Kindred Phenomenon in the Solar System*. The origin of his origin story about the Moon indeed involved tides—first of water then of lava. In the nineteenth century, astronomers realized that the Moon's orbital period was slowing down over time and that the Earth's rotation was too. (The slowing of the Earth's rotation, argue some biologists, helped cyanobacteria release more oxygen as more hours were added to the day; that oxygen helped multicellular life develop.) Science historian Stephen G. Brush writes, "The physical cause [of slowing rotation] was identified as dissipation by lunar tides in the Earth's oceans." In other words, the Moon's tugging on terrestrial waters was a kind of friction interfering with the Earth keeping a constant speedy rate of rotation. As Peter Wlasuk adds, however, it was a two-way street. The Moon bulged out toward the Earth and eventually these frictions led to the Moon obtaining "a rotation rate equal to its orbital period," which is why we only see the nearside facing us.

That's not all. "Darwin pointed out," Brush writes, "that since the angular momentum [or rotational energy] of the Earth-Moon system is conserved . . . [it] must be transferred to the Moon. As a result the Moon's orbit is gradually receding from the Earth; conversely it must have been closer in the past."

Applying sophisticated mathematics to the problem, Darwin concluded that sometime in prior ages, the Moon orbited the Earth every five-and-a-half hours and was, he estimated, once only six thousand miles distant. As the Earth cooled, it contracted and initially spun even faster.

If the Sun's gravity amplified the effects of a spinning Earth—pulling on its tides of molten rock in addition to the work of the Earth's own centrifugal force—then the conclusion was, that before this tightly bound pairing, the Moon had been part of the Earth itself. The Moon or bits of it in a molten state would have been hurled out to orbit, Darwin calculated, beyond the so-called Roche limit. (Inside that boundary, named for Édouard Roche, instabilities would have torn the Moon apart.) Period illustrations show such a genesis: the Earth as a sphere, then an ellipse, something like an egg with an umbilical cord extending from the molten Earth only to break off and become a blob, the Moon.

Or, as Darwin put it, his father Charles surely in mind, the process was like "the protrusion of a filament of protoplasm from a mass of living matter." The work that led to his lunar theory thrilled him. He wrote to his father in 1876, "I'm bursting with delight . . ." George Howard might also have had in mind the science poems of his great-grandfather, Erasmus, who wrote odes to and heroic couplets about plant sex, steam-powered water pumps, looms, balloons, canal building, and the possibility of inducing rainfall with electricity, among various marvels. Erasmus also sketched a rocket motor in the late eighteenth century.

Erasmus Darwin also wrote about the Earth birthing the Moon, perhaps a singular case of poetry preceding scientific theory as family history:

> . . . through the troubled air
> Roar'd the fierce din of elemental war;
> When rose the continents, and Sunk the main,
> And Earth's huge sphere exploding burst in twain.
> . . . when from her wounded side,
> Where now the South-Sea heaves its waste of tide,

Rose on swift wheels the Moon's refulgent car,
Circling the solar orb, a sister-star,
Dimpled with vales, with shining hills emboss'd,
And roll'd round Earth her airless realms of frost.

Actually, it wasn't any of the Darwins but geologist Osmond Fisher who popularized the lunar birthplace as the Pacific, though it's a nice touch that Erasmus articulated the idea.

There were complications. One involved the then-understood ages of the Earth and how long it would have taken for the Moon to reach its current orbital location. But the discovery of radioactive decay meant the Earth was far older than Victorians thought. Though Darwin came up with an age of the Earth that was too young, there was more than enough time for the Moon to reach its current distance from us, approximately 250,000 miles away.

More than 160 years after Erasmus Darwin died, researcher Donald Wise was noting the objections to this hypothesis while offering a modified version of it himself. It was a version that had been rejected from several scientific journals. At a 1964 conference, Wise, a geologist who would be an active member of NASA's lunar science team, began by saying, "Reasons why the Moon could not have originated from the Earth have been accumulating for half a century." These doubts largely involved whether there was enough energy to spin the Moon out of the molten early Earth and, if so, how far the bulge or umbilical cord would have extended . . . far enough to break off, after all?

Wise tried to answer these concerns by invoking the formation of the Earth's iron core, suggesting that it, too, would have deformed under high rotation rates, giving additional impulse to the molten rocks above, thus kicking the Moon into space. His paper is careful, however, noting both advantages to his hypothesis and the problems it faced. The specific gravity of the Earth's mantle is about the same as the Moon, which fits the fission model nicely. It explains broad similarities in composition. It explains the Moon's lesser density. Wise also invoked something called "tektites," little blebs of molten glass found on the Earth and then thought, incorrectly, by some to have originated on the Moon. If they had, their similarity to terrestrial chemistry is explained by the fission hypothesis. (Tektites actually are hot glassy raindrops that form and fall when large impactors hit the Earth.) The fission theory seemed to explain other mysteries, Wise argued, including why the Moon was so heavily bombarded early in its history: there would have been many bits of Mother Earth smacking down on the Baby Moon.

Nonetheless, Wise took stock of the problems. Scientists did not know enough about "viscous frictions of separation," they ignored variable densities in protoplanetary bodies, they did not understand enough about the early Earth's rotation, and so on. The modified fission model, Wise admitted, transgressed "our present energy assumptions" about all those things. George Howard Darwin himself knew of some of these complications.

Meanwhile the competing theories—co-accretion and cold capture—had their advocates. In co-accretion, the Earth and Moon formed close to each other in the proto–solar system. They formed hot then cooled down. In the cold-capture scenario, the Moon was formed by bits of cold rocklets and rocks—never melted—then wandered in to be grabbed by Earth's gravity well. This latter argument was championed by the commanding, Nobel Prize–winning chemist Harold Urey. He envisioned the Moon as totem of early solar system history, unchanged from core to crust, a veritable stepping stone toward understanding every bit and kibble of our nearly five-billion-year-old cosmic home.

Urey also had a sense of humor. As Donald Wilhelms remarks, Urey joked that reality simply demonstrated the impossibility of the Moon's existing at all. Because both hypotheses had their problems. Co-accretion had difficulty accounting for the energies of the orbit and rotations of the Earth and Moon, as well as the fact that the Earth had a big iron core and the Moon a very little iron core. If they formed together, why the huge difference? And "intact capture," as Wilhelms explains, "from somewhere else in the Solar System . . . is essentially eliminated by the unlikely coincidence of close approach trajectory and low relative velocity it would require." It was just too lucky for an object the size of the Moon to meander close enough to the Earth.

Throughout the Apollo years, sociologist Ian Mitroff asked forty-two lunar researchers about their support for these differing origins. In his words, "No theory enjoyed a statistical significant improvement in its average rating."

Unlike his nemesis Gerard Kuiper, who believed the Moon had been molten, Urey insisted that the Moon had been more or less cold, like a gigantic stony meteorite or no-bake pie. But when the robotic Surveyor 5 probe returned results from a chemical analysis in 1967, Urey was disappointed. The alpha backscatter experiment released particles from the radioactive decay of curium onto the lunar surface; by analyzing how much energy the particles had when they were reflected back to the instruments, scientists could generally discern the surface composition.

Lunar researcher John O'Keefe was driving Urey from Los Angeles International Airport just as this data were being analyzed. Wilhelms writes that "Urey commented glumly, 'It's basalt, isn't it?'" O'Keefe admitted that it was. Basalt is formerly molten rock. The cold-Moon hypothesis had been struck a serious blow.

Changing the subject, Urey told the excited O'Keefe as they careened through LA traffic, "'You are a very poor driver.'" He might have said to each of the Moon-birth scenarios, "You're a very poor hypothesis."

The word *origin* is interesting. It carries denotations of "rising" and "to set in motion, elevate," which latter word, in the Greek roots, gives us *oros* or "mountain." The origin of *origin* is set in the sky, that famous backdrop for the big mountain Moon. It's all of a piece. In 1681, Thomas Burnet in *The Sacred Theory of the Earth* suggested that this planet and the Moon had come into being naturally, despite the "sacred" in the book's title. As historian Michael Rawson says, Burnet's "influential work . . . used the earth's geology to explain the moon's and argued that similar processes had formed them both." Burnet was wrong on the particulars but right on the bigger point.

The particulars would come much clearer in the Tucson desert, where, decades before I arrived, Gerard Kuiper had happily left the cold and politics of Midwestern academia to assemble a team at the then-backwater University of Arizona. He and they would go on to make momentous contributions to the study of the solar system.

When Kuiper arrived to found the Lunar and Planetary Laboratory, he was one of just a handful of scientists interested in the solar system and, in particular, the Moon. He gathered graduate students and fellow researchers, many on the outskirts of astronomy. So slender was this community that when Kuiper circulated a call to help with a lunar atlas, only one person responded, Ewen Whitaker, who would go on to become one of the foremost mappers of the Moon. As the Apollo era got underway, LPL would be one of the few institutions at its scientific forefront. (Charles Wood relates that other envious University of Arizona researchers sometimes called the Lunar and Planetary Lab "the Dollar and Monetary Lab.")

One member of LPL was then-graduate student William K. Hartmann, the man who helped me decipher the photograph of Copernicus that had been hailed, in 1966, as "the picture of the century." Using an observatory in the nearby Catalina Mountains, LPL gathered photo after photo of the lunar surface in order to make what would become for its time the finest Moon atlas published. Photography had finally surpassed drawing in capturing lunar detail.

To aid in the process, Hartmann projected photos onto a blank hemisphere about three feet wide. (It's still owned by the university.) He did so in a dark corridor that allowed him to walk around the hemisphere then photograph the limbs as though overhead. Where this work took place, a Quonset hut—Building T6—is long gone, but I walk on its former location not infrequently when I visit the campus science library. One projected image of the Moon's western limb revealed to Hartmann an area just past the black lacquer of Grimaldi crater.

The Moon's "edges" are tantalizing. They reveal curvature, they reveal ruggedness—sunlit mountain slopes rising against the blackness of space as though a crescendo could present a serrated profile. Lunar edges reveal sun-touched tips at the southern polar region. There, a point of light is visible around day seventeen of a lunation. It's a mountain called Leibnitz Beta, and the elevation change from the floor of the nearby Shoemaker Crater to its tip is more than ten kilometers, according to the NASA lunar south pole atlas. Ten kilometers is some thirty-three thousand feet, far above Mount Everest. That's one impressive point of light. Lunar mountains are glorious even if you can't really see them.

Observers had previously glimpsed a vast range of mountains on the southwestern limb and a thread of solid lava, a trickle of mare. Then, when Hartmann projected this view onto the white hemisphere in the LPL Quonset hut, the photo made clear that this was a massive and well-preserved giant impact basin. It would be dubbed Mare Orientale, the Eastern Sea. (The name has stuck despite astronomers eventually adopting the usual terrestrial conventions of direction on the Moon: left is west, right is east, north is up, south is down.) Mare Orientale features not one mountain range but three, all uplifted due to the impact: the outermost Cordillera, the outer Rooks, and the inner Rooks. Hartmann had discovered one monstrous impact basin. (Or rediscovered, as it was mentioned in various, forgotten fragments in the observing literature.)

"Look at these pictures," he enthused to Kuiper in the professor's well-kept office. It was a winter day in Tucson in the early sixties. "There's clearly some kind of radial and concentric symmetry."

He published his findings in Kuiper's in-house Lunar and Planetary Laboratory journal, which kept it from wide circulation. Unlike other professors then and now, Kuiper granted Hartmann—a first-year grad student—primary authorship on the paper. (Kuiper wasn't always so gentle with his charges. Hartmann recalls how "we used to have jokes about Kuiper flying into a tizzy over something and saying, 'Call T6, call T6,' because a bunch of us graduate students over there were either about to be chewed out or he needed us to do something.") Those beginning to study the Moon in Flagstaff, with the US Geological Survey, loved the paper. Harold Urey hated it.

With a good libration, Mare Orientale is, as I have seen myself, an intriguing corrugation of mountainous arcs set between the dark, featherlike curve of lava called Lacus Veris, the Lake of Spring, and beyond the last mountain range, a glancing suggestion of the mare itself, the bull's-eye of Orientale. If, as I once dreamed, Mare Orientale appeared on the nearside, especially near the center of the disk, it would be as though an eye were watching us. Imagine the gods conjured from such a lunar stare-down! Eventually imaged from orbit, Mare Orientale would be revealed as the youngest (at about 3.85 billion years old) and largest of the Moon's many giant impact basins, what are, in essence, of course, just truly ridiculously large craters. Orientale is twenty-four thousand square miles, and the asteroid that caused the impact was an imposing forty miles wide.

Hartmann, who has recalled the magic allure of a daytime crescent Moon—emblematic of his love of the place and our possible future on it—was far from done. He was reading obscure Russian literature that took the hypothesis of huge impacts more seriously than scientists elsewhere. In 1962, Hartmann predicted that there was another giant impact basin at the Moon's south pole. A few years later, orbiting Soviet and US craft confirmed the existence of the South Pole–Aitken Basin where I can see the top of Leibnitz Beta.

So big impacts created big basins, molten after the initial collision, only to solidify then later experience thin but overwhelming lava flows that broke through the crust. The sheen of maria basalts we see today is not the original impact melt but subsequent flooding.

After the role of giant impacts was understood—Charles Wood calls basin-impacting the most basic topographic force on the Moon—it was not an illogical leap to think there might have been an even bigger impact than ones recorded on the lunar surface. There might have been an impact that occurred before the Moon was even born. An impact that birthed the Moon itself.

By the early 1970s, Hartmann and fellow former University of Arizona class-mate Don Davis had combined intuition with computer models to show a very large object could have formed separately in the zone where the Earth was congealing in the early solar system. Hartmann and Davis believed this second object wasn't a quiet co-accretion of the Moon. It was an impactor, the size, ironically, of the Moon itself, that hit the satellite-free Earth, creating a huge messy mix and thus forming the Moon we've come to love and know. Another scientist, Alastair G.W. Cameron, was the first person to ask Hartmann about the hypothesis after the latter gave a talk at a Cornell University gathering in 1974. It was the idea's debut. Hartmann expected trouble. But Cameron and his collaborator William Ward were on board, with one caveat: they too had concluded the Moon had formed from a giant impact with the Earth, but the wandering planet that engaged in such creative destruction was, they thought, the size of Mars. It would become known as Theia, the Titan mother of the mythological moon, Diana or Selene. Theia means "wide-shining," and its light would be cast on lunar questions to come. (The giant-impact idea had a longer pedigree than most know, a version of it being floated as early as 1946 by Reginald Daly.)

In the words of science writer Dana Mackenzie, the giant-impact hypothesis "solved the problem of why Earth has a large iron core and the Moon has almost none: in a giant impact, the iron in the impactor would have penetrated Earth's surface and stayed there." (The Earth's iron core contributes about thirty percent of the planet's total mass; the Moon has a core of only a few percent of its total mass.) "It explained the lack of volatile elements on the Moon: they would have boiled off in the explosion . . . It explained why the kinds of rock found on the Moon were similar to those in Earth's mantle: because a lot of the material in the Moon actually came from Earth." It also explained why the Earth-Moon system has the rotational and orbital energy that it does and why our Moon is so relatively massive in comparison to the Earth, so massive it can be considered a double-planet system, unlike anything else in the solar system.

Thus was born the so-called canonical giant-impact hypothesis, and when lunar scientists met just over a decade later in Kona, Hawaii, the verdict became clear. In an irony that at least some attendees must have noted, the Kona conference took place on an island in the middle of the Pacific, from whence the Formerly Darwinian Moon was once thought to have come.

The fission theory died at Kona. So did co-accretion and cold-capture. (The University of Arizona's Tim Swindle attended as a graduate student, running projectors to pay his way: "Little did I know that it would be the most momentous scientific conference I'd ever attend." Years later, Swindle would go on to become director of the Lunar and Planetary Laboratory.

Kona was a rarity in such meetings. Presented with evidence from the Apollo samples—on which, more in a moment—and the logic and math of the giant-impact model, nearly everyone at Kona conceded: this was how the Moon was born. It seemed. Mostly. Researcher John Wood presented a chart at Kona on the last day of the meeting with the pros and cons of each lunar-origin theory. While there were gaps in the giant-impact hypothesis, such as whether the initial collision was really that plausible and whether it would have been violent enough, the model still emerged as the strongest contender. It didn't hurt that by the time of Kona, geologists were beginning to recognize the impact event that had killed off the dinosaurs. As the decades continued, the fact of massive space impacts would become manifest, especially when Comet Shoemaker-Levy 9 plowed into Jupiter for all to see in 1994, including me, a budding stargazer, with my C90 telescope by an Indiana roadside. I showed my then-ailing mother. She was not as impressed as I was by the black impact dot on the striped disk of the gas giant.

As studies of the hypothesis continued into the twenty-first century, the story seemed to be set in stone: Theia hit the Earth at a forty-five-degree angle at a relatively low velocity, sending a disk into space, poor in iron, that congealed to make the Moon.

Critical to the giant-impact hypothesis and largely ignored in the 2019 fiftieth anniversary hoopla over the Apollo 11 landing was the program's most important scientific legacy: its rocks. They are very, very old. They indicated that the Moon formed early, likely some sixty million years into the solar system's history, a time of rocky helter-skelter destruction, with planets, asteroids, and everything in-between skittering away and into each other.

There had been a hurried decision at the end of the historic first Moon walk. From the moment Neil Armstrong scooped nine soil samples into a container—an "afterthought," one NASA official says—lunar science would be forever changed. Armstrong dropped the soil container among the rocks to keep them stable during the return flight, and he did so right at the conclusion of the brief exploration he conducted with Buzz Aldrin.

Planetary scientist John Wood and his team were one of the groups that received the priceless material, which, it would turn out, included that special something called anorthosite (this mineral would become a star in the Apollo 15 expedition). Wood writes vividly of holding two vials of those Apollo 11 last-minute lunar fines one September day in 1969. He was so nervous about handling the samples—he'd just picked them up from NASA—that he "borrowed a needle and thread from one of the secretaries . . . and sewed both small vials into one pocket of my sports jacket for the trip home." On the Eastern Airlines flight, Wood continues, he engaged in "much high-spirited conversation" with other lunar scientists and, forgetting himself, "tore off my jacket and stuffed it into an overhead bin." He only barely remembered to retrieve the jacket, and, had he left it there, his team's understanding of the importance of lunar anorthosite might largely have been someone else's scientific legacy, like Joseph Smith at the University of Chicago, who would have the same idea. Wood found eighty-four grains of anorthosite. Total. But he knew they matched results from the distant Surveyor 7 landing craft.

Anorthosite in two far-flung locales meant, Wood argued, that the Moon had once been at least partially molten—a state one would expect if it had been blown off the early Earth in a hot, violent collision. Perhaps it had been almost entirely molten. The early Moon was covered in a "magma ocean," as Wood put it, in which relatively light minerals like anorthosite rose in their brilliant white splendor to the surface, where a few billion years later, humans would collect some.

Anorthosite, a gray rock of plagioclase feldspar minerals, is the spumescent remnant of the magma ocean. When I see the bright lunar highlands, I'm seeing the hardened evidence of anorthosite rising from molten depths, crystals light enough to float, a 4.5-billion-year-old liquid-rock ocean whose heat seeped to space, anorthosite locking into place, wave crests and tides, cooling into solidity that awaited meteorite, asteroid and comet, scores of them that would slam into this former ocean to gouge and crater and push up mountains.

Now there is tantalizing evidence that beneath the Pacific and West Africa there are blobs that are the remains of Theia. Denser than the rest of Earth's mantle, these regions could be more than six hundred miles tall.

The Moon did not birth from the Pacific Ocean but its birth may be hidden beneath it, after all. And another ocean, that of magma, was part of the Moon's evolution.

When I look at the lunar highlands, I'm looking at its memory.

As Apollo samples from all the missions were subjected to more precise study, however, there emerged what became known as "the isotopic crisis." It was puzzling that the Earth and Moon were so different in *abundance* of materials and yet those materials were uncannily similar in makeup. If Theia had formed elsewhere in the solar system, as models suggested, then Theia—not the Earth—would have contributed most of the material that made the Moon. If so, the Moon and Earth should have different isotopes of elements like oxygen. Other bodies in the solar system have differing isotopes—where the same element has different numbers of neutrons—so why should the Earth and the Moon be so alike?

Researchers were also vexed by the amounts of hafnium-182 and its radioactive decay product, tungsten-182. The former likes to hang out with silicate rocks in a planet's upper layers, the latter with iron near or at the center. Why would the Earth and the Moon have such similar amounts of each when Earth and Theia would have formed their cores and upper layers at different times—*not* the precise nine million years it takes for hafnium to birth tungsten? In other words, it's almost impossible to imagine Earth and Theia following the same timeline for their formations of core and upper layers. Something else must have happened.

There are mornings when the waning Moon hangs like a sly, inverted wink in the blue western sky, as if chiding us for taking so long to understand how it got there.

It took more computers.

Modeling impacts takes computing power, and only recently have scientists been able to develop and test some wild scenarios based on the giant-impact hypothesis that can account for the isotopic crisis.

One promising scenario is called the "synestia," in which Theia's impact was so hot-tempered that Theia vaporized as did some of the Earth. In this model, developed by Simon J. Lock and Sarah T. Stewart, Theia doesn't hit Earth in glancing fashion but plows right into it. The resulting storm of gas and melt congealed not into in a new blobby Earth and a new blobby Moon but into a torus (think gaseous doughnut). This newly hypothesized in-between state—neither ring nor disk nor planet—swirled and mixed materials so thoroughly that, when the Moon formed out of the shrinking outer cloud, the isotopic composition would match that of the Earth. Heavy iron gravitated toward the proto-Earth's center. Meanwhile, the volatile gasses, such as potassium and nitrogen, remained in the gassy cloud, thus accounting for the relative rarity of such elements on today's Moon. As well, the impact and synestia model explains why the Moon is five degrees off the plane of the solar system. Theia's crash pushed the Earth over, the Moon then formed off the ecliptic plane, and the Earth eventually righted itself to the axial tilt we have today. (Which is fortunate, because without that twenty-three-degree tilt we would not have seasons, weather, or, possibly, life as we know it.)

After a century, a blink of the eye, the molten Moon hung above a still-forming Earth yet entrained in the heart of the synestia. The researchers named their "synestia"-formation model after "syn" for synergy and Hestia, the Greek goddess of the hearth.

As University of Arizona planetary scientist Jeff Andrews-Hanna explains, however, there are still other variations on the giant-impact theme. Theia may have hit the proto-Earth not once, but twice. One collision scenario has the Moon forming within *hours*. The Moon might have formed from the collision between two half-Earth-sized planets in the same origin zone, thus accounting for the isotopic match though we don't know how likely such impacts are. In a fission-theory variation, a "small impact into a fast-spinning Earth" could have sent more Earthlike material into orbit to create the Moon. Darwin's revenge? Or multiple impacts might have hit the early Earth to create enough lunar-formation material. In yet another possibility, as Japanese researchers suggested in 2019, the Earth itself was covered in a magma ocean when it was impacted, thus explaining the similar isotopes. Meanwhile, Nicolas Dauphas, among others, suggests there isn't an isotopic crisis after all—if the Earth and Theia formed out of a zone similar to that which formed the enstatite chondrites, a type of meteoroid, all of which have similar chemical origins. And an analysis by Erick Cano and his team found that the oxygen isotopes on the Earth and those on the Moon are, they say, different enough for an impact model to necessitate only partial mixing.

More modeling, new lunar samples, and images from developing solar systems in which a synestia or other planetary impact/formation structures may be detected will be key in refining this story. All this work will also help clarify not just the question of the Moon's origin but how commonplace such an event might be, for if a Moon like ours is needed for life as we know it, the implications of just how ordinary—or extraordinary—the Moon's formation is could tell us more about another question: Are we alone in the universe? Finding other moons that stabilize the axial tilts of their parent planets—thus producing seasons and, possibly, tides—will go a long way toward answering our concerns about cosmic loneliness.

The book of the Moon's origins does not yet have a final draft.

It was 4.6 billion years ago. Just a century after Theia's collision with the early Earth, the Moon had formed, and what would become our satellite loomed huge in the sky. At least George Darwin was right about that. In the old natural history museum in Salt Lake City, at the University of Utah,

I remember being captivated by a mural showing the massive Moon in the Earth's Hadean sky. It was at least fifteen times closer than now, and a day on Earth was a fast-clipped five to six hours long. (These days the Moon's closest point, or perigee, is 225,727 miles and its farthest, or apogee, is 251,995, and the distance is growing as the Moon moves off about 1.5 inches a year.)

Both worlds were hellscapes. Each roiled in varied states of melt. The combined heat of the violent early Sun, the early Earth, and the early Moon may even have given the Moon an atmosphere, reports a 2017 *Science News* article, albeit an atmosphere made of vaporized metals. Meanwhile, the Earth was a kind of infant Venus with nearly thirty times (or more) atmospheric pressure than we experience now, a choking carbon dioxide atmosphere and surface temperatures of 450 degrees Fahrenheit. Despite this oven heat, the pressure was high enough for Earth to retain surface waters.

Ages began. On the Earth the Hadean would give way to the Archaean and the later, eventual, first tiny swishings of life in warm little ponds (or deep in the oceans) and the rise of oxygen. On the Moon, the eons we've assigned unfurled with a violent, anonymous pageantry, pulses of impacts and spasms of lava, molding the Moon's major surface features, forging its chemistry, and structuring an interior we still do not fully understand.

In the pre-Nectarian, which begins with the Moon's formation, debris concussed the molten world, the lunar magma ocean plashing with blazing impacts in the metal air. What does an asteroid sound like as it streaks through a sky of gaseous metal? After the Moon lost its atmosphere, the collisions were soundless but no less stormy.

As it cooled over hundreds of millions of years, anorthosite rose like cork in water to harden into the bright terrae. When I look at Mare Nectaris, what is now one of the smaller basins on the nearside, I'm seeing the end of this age. About 3.9 billion years ago an impact formed that basin, shocking upward a circular mountain range that is now the partial arc of Rupes Altai and another arc, the low and battered Monte Pyrenaeus. Nectaris ejecta scoured the eastern limb to create the familiar, massive Vallis Rheita and the less obvious Vallis

Snellius, two targets telescopists seek out early in a lunation. British selenographer Edmund Neison was right when he called the Altai Scarp and Rheita Valley "great . . . frowning cliffs."

The collision that formed the Nectaris basin is the start of the vehement infall of major impacts that tore the Moon's semisoft crust into many of the major basins, from Crisium in the east to Humorum in the west. This Nectarian Era lasted only fifty million years, some still say, but others, according to Swindle "now think Nectaris is much older . . . The Nectarian Era is still brief on geological timescales, but might be as much as two hundred or three hundred million years. The problem is figuring out whether the rocks you're dating actually were formed or [were] last heated in the Nectaris impact." The ages may not be precisely set—yet. What is clear is the size of the lunar basins, which are between two and ten times larger than the ninety-three-mile-wide Chicxulub crater off Mexico's Yucatan Peninsula, the impact that made the dinosaurs extinct.

The next fifty million years, the era of the Lower Imbrium, saw the final two giant impact basins blasted out—Imbrium on the nearside and Orientale on the far. Some of the Imbrium impact's low-angle flying ejecta chiseled out those furrows radial from the cataclysm, the furrows of the Imbrium sculpture typified, say, by gouges near the crater Ukert. They alternate with lines of hills that the ejecta didn't touch.

Lunar scientist Barbara Cohen says that, while researchers have a good grasp on cratering physics, "That understanding starts to break down for multi-ring basins [like Orientale]. These large basins are much bigger than complex craters and so depend more sensitively on starting assumptions about target materials, thermal state, angle of impact," and more.

Lunar volcanism got underway just as impacts lessened, according to Charles Wood. In the Upper Imbrium, for several more millions of years, syrupy lavas oozed through shattered crust to spread their dark veneer across wide swaths of the Moon: the maria were born, Galileo's ancient dark spots. Domes extruded upward to become, mostly, the low shield volcanos I delight in finding.

During the quieter Eratosthenian Era, from 3.2 billion years ago to 1.1 billion years ago, the age of basin formation ended, some lava still flowed to create maria, and impacts—of course, impacts—still occurred. The namesake is the crater Eratosthenes, which formed at the beginning of this age. Complex craters from this time now lack ray systems and can appear a bit less sharp, with more slumping of rim and wall and terrace and more rounding of any remaining central peaks. Perhaps ironically Eratosthenes appears less so. Apart from the lack of rays, it might pass for a crater of the Copernican Era, the last and current such lunar age, whose namesake isn't far from Eratosthenes. Most obviously, Copernicus has splashed its white rays over much of the western half of the nearside. Craters from this age are less over-cratered so their rims are more intact.

Slowly, I begin to associate these features with their ages. The pell-mell hummocks of the southern floor of Copernicus, a place merely eight hundred million years old. This, on Earth, was the age of the Precambrian, a time of ravening terrestrial impacts and multicellular life, then only about a hundred million years old. I might swing the scope back to the black lake of Plato, whose giant rim-dislocated block, Plato Zeta, has, at high magnifications, striations, a detail so useless to human affairs that the first time I notice this I laugh out loud. Plato formed 3.8 billion years ago, in the Nectarian or Imbrium, depending on who is counting.

It is fitting that the ages of the Moon are named for impacts because impacting, more than anything else, has been the driving force of the lunar landscape. Craters cover the Moon like a scan of fossil migraines. From micrometeorite pits like pores in the skin to impact basins that can swallow whole nations, the Moon testifies to the ubiquity of chance cosmic violence.

Not only did massive impacts form the famous basins, but the impact energy shock-waved through the lunar crust to uplift the Moon's mountain ranges—which, at a glance, one can see as arcuate chains around the maria. Headlands ring all of Mare Crisium, whose northwestern section contains a couple of major, attractive, meadowlike passes between massifs. Montes

Alpes and Montes Apenninus border parts of Mare Imbrium, as do other less impressive ranges, which, from above, look like the rim of a broken cup. The basin-encircling ranges can be steep: the mountains around Mare Orientale have some slopes more than forty degrees, though a number may be the result of later grabens as opposed to the impact itself.

It's the Apennines I love most—for their brazen complexity, sheer scale, and ease of viewing. They are the nearside's most impressive mountain chain, hulking up in long strides from the east and dropping down at more than thirty degrees, sometimes forty degrees, in the west. Lacking forests, the Apennine ecosystem is rock and crater and passes gouged by Imbrium sculpture, draws one could hike in order to reach its three thousand mountains, including, Mons Huygens, at eighteen thousand feet. That would be a view, down the steep west face or back toward the way one came, the hummocks of the eastern Apennine back, where ponds of lava have gathered like high-mountain lakes. The entire range spans nearly four hundred miles. When the Moon is high in First Quarter, Montes Apennines tosses itself around Mare Imbrium's southeastern reaches like a squall.

Not all the impact-basin mountains look like this. Lavas have flooded some ranges so that only isolated peaks stand aloof. Part of the submerged inner ring of Imbrium—Mons Pico and Mons Piton—tower each at some eight thousand feet looking like the white sails of ships sinking in the Sea of Rains. Other ranges, like the Haemus, which borders the southern edge of the Sea of Serenity, are so worn that, for now at least, I barely give them a look.

For all the telescopic clarity of the lunar surface, the Moon has plenty still hidden.

Jeff Andrews-Hanna explains that we don't understand how deep the magma ocean was. We haven't dated the ages of the lunar basins precisely. And scientists disagree about the rates of bombardment of the lunar surface—and throughout the solar system—over time. Understanding these mysteries will help us understand how the conditions for life to evolve can change. Doing so requires more exploration, especially with samples brought to labs on Earth.

We don't understand why some elements, like thorium, predominate on one side of the Moon and not the other. Ditto for the existence of KREEP: *K* for potassium, *REE* for rare-Earth elements, *P* for phosphorus. Some of these elements are not easily comingled, and their formation history has challenged scientists to develop complex explanations for their lunar coexistence. Even the assumption that the area in which KREEP predominates—Oceanus Procellarum—is an impact basin may be wrong. It might be the result of what's been called an internal "magma plumbing system." In either case, KREEP produces heat from radioactive decay, and how much has been or is being produced remains unclear. Apollo monitoring equipment is too far from the area and might even have been set up imprecisely. The source of KREEP may be somewhere beneath the area where Apollo 15 landed, near Hadley Rille.

We don't know why the crust is thicker on the far side, which has very little maria. Was there a second impact of low velocity, which spread thick debris across the far side? Did the early magma ocean slop more anorthosite to the far side? Or was there some other mechanism?

Of the thousands of Moon quakes that record mostly meteorite impacts and gravity stress from the Earth, what about other tremors that researchers don't yet know how to classify? How to account for lobate scarps, thrust fault scarps that have formed only yesterday in geological terms, about a hundred million years ago? What of the some seventy irregular mare patches, all on the nearside maria? These are also recent, having formed volcanically within the past tens of millions of years. Scarps and IMPs are exotic objects of study but likely places to avoid staying at for extended durations. Human study of them will be dangerous if the exploration site is active. Is it possible that IMPs might still form today? So how warm does the Moon remain? Models have suggested the Moon should have been too cool for volcanic activity two billion years ago, but the Chinese far-side probe Chang'e-5 has found volcanic samples of that age. How is this possible? Some rocks have lower melting points, and the

Moon's interior may have had more heat-producing radioactively decaying materials; some of the lower-melting-point rocks may have been driven down into contact with those decaying materials. But Chang'e-5 did not find any unusually high concentrations of radioactive materials. Even wrinkle ridges have formed in the recent lunar past, which points to a legacy of catalyzing activity in faults formed from the 4.3-billion-year-old South Pole–Aitken Basin impact. Researchers have even found relatively fresh boulders on wrinkle ridges that may have been uplifted as magma dikes. Long-ago impacts may have an afterlife. The Moon quietly twitches.

Consider too that the tops of central peaks likely have material excavated from the mantle, samples of which we do not definitively have despite the Apollo expeditions and robotic Soviet probes. China's Chang'e-4 rover *may* have found mantle samples on the far side but exploration to find this elusive grail is necessary. Under the Moon's landscape there are questions we cannot answer until we return. Beneath the crust—some thirty or more miles deep—the lunar mantle extends about eight hundred miles and may include partially molten rock near the Moon's core, which itself may also be partially molten. The core is probably a mixture of rock and iron, though lacking enough of the latter in a molten state to generate a magnetic field. The global composition of these hidden structures remains unclear.

It isn't just about the Moon. As a NASA report called *Advancing the Science of the Moon* put it, our companion is "the cornerstone of planetary science." We'll understand more about volcanism on other worlds, such as Venus, by sampling lunar basalts. We'll understand more about impacts on the Moon and Mars by establishing seismic stations around the Moon and sampling impact melts. By digging down into areas that contain pristine ancient regolith—paleoregolith—we'll have an unprecedented window into the earliest times of our solar system.

Studying and sampling the mysterious white lunar swirls, like Reiner Gamma, will enable us to project how the solar wind and cosmic rays affect the surfaces of worlds with little or no atmospheres, with implications for how we can protect human settlers. Lunar swirls have no surface relief at all and drape over the terrain like gauze. Reiner Gamma looks like a faded paisley print or tadpole or "ill-defined white spots of doubtful nature," in the words of nineteenth-century selenographer Thomas Elger. It's been theorized that impacts on the opposite sides of lunar swirls altered subsurface materials and created "mascons" or "mass concentrations" of magnetically charged material. Local magnetic fields (on a Moon largely without magnetism) arced into space to create lines of force that have kept the solar wind from darkening the surface below. But Reiner Gamma isn't opposite a major impact, so how did its mascon come to be? No one knows. Lunar swirls are wraiths, and Reiner Gamma haunts the western reaches waiting for its visitors, the first of which will be a tiny rover called Lunar Vertex.

And it's possible that the Moon gathered up oxygen isotopes from the Earth's atmosphere over billions of years as it passed through our planet's trailing magnetic field. The history of our atmosphere may be archived in nearside hematite on the Moon, according to Shuai Li and others. Going to the Moon is going back in time, not just there, but here, on the Earth.

As to the Moon's atmosphere—what little there is: before the next astronauts arrive, researchers hope to study the extremely thin "exosphere" while it is still in a pristine state.

There's also the tantalizing possibility that scattered across the Moon are samples of other worlds, including Earth and Venus, in the form of meteorites.

And we've only barely started to explore the far side.

These matters, at least some of them, may seem arcane, but the fact that they exist—so far into our scientific relationship with the Moon—should both humble and excite. The nearest world in space has had its surface photographed and mapped, its mineral composition probed and delineated, its origins and history studied and laid out in a timeline. The Moon still refuses to yield everything.

Once, I saw a rock collected beside the lunar Apennines by Apollo 15's Dave Scott. Against an exhibit wall in Philadelphia, the rock was merely gray, until I saw it constelled with demure, almost fugitive, sparkles, and it was part of the Moon, even captive in a vitrine, because it had been scooped,

examined, bagged, carried by humans who had been there, so it too had grown
talismanic, a secular fragment—olivine basalt—of my journey to the Moon,
my endeavor to rendezvous memory with eschatology. Elsewhere, more than
once, I've touched a polished section of a Moon rock: as cold and slick as a
kitchen counter on a winter morning.

From these fragments to the varied levels of potassium in KREEP, from the
aluminum-rich, battered highlands to the iron-rich, magnesium-laced maria,
we travel routes and byways of connection and understanding. Embedded in the
lunar regolith is solar-wind deposited hydrogen and nitrogen, nestled in with
fragments of carbon-containing meteorites. Embedded in lunar silicates, an
abundance of oxygen—elements for survival.

But there's something else up there. The Moon has shocked us with a discovery that few could have imagined. To consider why, drink a glass of water
under Moonlight.

☻

For all the talk of magma oceans and giant impacts—all the rocky gossip
of the early solar system—there's another part of the tale, one that emerged
after the Apollo missions. At the time of those flights, it was mostly assumed
that the Moon was waterless, though as late as 1968, three UCLA researchers
wrote an article in *Science* suggesting sinuous rilles were the beds of former
"lunar rivers," with surface ice that kept water liquid beneath, at times freed
by impacting meteorites. They were incorrect. Yet too was the assumption of
a desiccated Moon.

New analysis techniques have detected traces of water in the Apollo samples. Scientists have found water molecules all across the lunar surface. And
both government agencies and corporations are interested in the vast swathes
of subsurface water ice in permanently shadowed craters at the lunar poles.

NASA had the foresight to lock away its precious lunar rocks, knowing that
future geochemical analysis techniques would allow for more precise measurements, even new findings. That policy was prescient, as researcher Alberto Saal

and others will attest. Scientists have found water inside Apollo's magmatic rocks. This water comes from, NASA researcher Barbara Cohen tells me, the lunar interior and is a "vanishingly small amount . . . parts per billion make it hard to measure and even harder to drink." The findings remind us that assumptions about the Moon can change when new techniques are developed to analyze previous materials and findings. And the tiny Apollo-sample droplets tucked inside rock are just part of the surprising story about the changing vicissitudes of the Moon's long association with water.

During flybys of the Moon in 1999 and 2008, respectively, NASA's Cassini and Deep Impact spacecraft used the Moon as a reflectance spectroscopy standard, a well-known source with which to calibrate their onboard instruments. When the spectrometers picked up absorption bands from water, both spacecraft teams dismissed the bands as the spacecraft outgassing Florida water into the field of view of the spectrometer! Everyone knew the lunar surface was dry. But in November 2008, India's Chandrayaan 1 satellite arrived in lunar orbit, observing the terrain below with a suite of ten experiments. Two of those, including the NASA-funded Moon Mineralogy Mapper or M3, also detected the spectrum of water molecules on the lunar surface in a pattern that varied with latitude, temperature, and surface freshness. That kind of variation couldn't be explained by spacecraft outgassing. Then a probe released by the orbiter persistently detected water vapor before it slammed into the Moon.

The three spacecraft published their results in a landmark issue of *Science*, showing that the lunar surface is literally covered in water molecules, created by the interaction of the protons of the solar wind—made of hydrogen—slamming into the oxygen molecules of the crust.

The same processes must be happening on other airless bodies as well, creating a small but ubiquitous source of water in the solar system. Unfortunately, the energy of the solar rays, cosmic radiation, and micrometeorite impact are also responsible for destroying those same surface molecules. Water on airless surface like the Moon's has a surprisingly convoluted cycle among surface, subsurface, and space—including preciously thin exospheres like the one that surrounds the Moon like an insinuation. There are depositions of water molecules, underground persistence of the same then micrometeorites freeing them to brief existence in the sky. The latter was observed by the LADEE spacecraft—Lunar Atmosphere and Dust Environment Explorer, observations that Mehdi Benna studied, finding, as well, neon molecules in the Moon's exosphere.

In 2020, scientists working with NASA's Stratospheric Observatory for Infrared Astronomy—a modified Boeing 747 commercial jetliner—found the unambiguous signature of water on the scorching sunlit surface of the Moon. The detection occurred at the massive Clavius crater, one of my old friends. Lead author Casey Honniball, then a postdoctoral fellow at NASA's Goddard Space Flight Center, said that between one hundred and four hundred parts per million of water were found at Clavius, or about the same as a twelve-ounce glass of water per cubic meter of lunar regolith.

"To be clear, this is not puddles of water, but instead water molecules that are so spread apart that they do not form ice or liquid water," she said. Researchers think the water at Clavius is trapped within glass beads in the soil that form during micrometeorite impacts. "These glass beads are about the size of a pencil tip and protect the water from the harsh lunar environment," she said. The discovery suggests an even greater stability and distribution of water molecules on the Moon's surface.

But the most promising place to find large stores of water—albeit frozen—is near the lunar poles.

In 1952, Harold Urey noted that the Moon's axis is offset from the vertical by only a degree-and-a-half. He speculated, therefore, that permanently shadowed regions (PSRs) existed at the lunar poles, places where, if you were standing at the bottom of a large crater, the Sun would go around the horizon, hidden from your view, year-round, for billions of years—just as the tops of peaks there would be lit more or less continuously. Any water molecules and other volatile compounds that found their way into these areas might be frozen and preserved. In 1961, Caltech scientists Kenneth Watson, Bruce Murray, and Harrison Brown extended that idea, suggesting that "water is actually far more stable on the lunar surface [than other volatiles] because of its extremely low vapor pressure at low temperatures and that it may well be present in appreciable quantities in shaded areas in the form of ice."

They were right. In 1994, the Clementine orbiter confirmed that PSRs do exist on the Moon, pinging some with radio waves. The signals came back strong enough to indicate that among the polar rocks were likely deposits of water ice, which reflects radio energy more coherently than rock. Four years later, an orbiter called Lunar Prospector found more evidence, detecting neutrons emitted from the subsurface that strongly suggested water molecules.

In 2009, a NASA team deliberately crashed a Centaur rocket body into Cabeus, a large crater situated close to the Moon's south pole where the signature from Lunar Prospector indicated the highest hydrogen abundance. Hydrogen, of course, is famously part of H_2O. The Lunar Crater Observation and Sensing Satellite, or LCROSS, trailed the empty rocket stage in order to observe the ensuing plume of debris, which rose some ten miles high. LCROSS kept recording data until it too slammed into the crater's floor. During the Centaur's impact, NASA's Lunar Reconnaissance Orbiter recorded a momentary flash and kept an eye on the temporary cloud of material. Its findings electrified the planetary-science community: Water ice and other volatiles were clearly detected in material blasted out from Cabeus.

Instruments aboard the Lunar Reconnaissance Orbiter, which has studied and photographed the Moon since 2009, have found that the PSRs are colder than Pluto's surface at a chilly minus four hundred degrees Fahrenheit.

Nearly ten years after LRO and LCROSS were launched, Shuai Li of the University of Hawaii and colleagues combed more carefully through the M3 data and found more spectroscopic evidence of light reflecting off water ice in many PSR locations surrounding the lunar poles. While less than four percent of the total sunless surface exhibits the spectral features of water ice, here and there up to thirty percent ice by mass might be mixed with the dusty regolith.

"Nobody believed there would be useful information in data collected from PSRs," Li told me. After all, how can you get reflectance spectra in a place without light? But Li decided to look after Apollo 15 astronaut Dave Scott told him the Moon's rough topography can scatter light into shadows.

Studying data with such weak signals was arduous, but the researchers found what they were looking for.

When Murray made his observation that PSRs would be thermodynamically stable regions for volatiles, it was natural to posit that billions of years of accumulation could have been happening from bodies such as comets, from volcanic gas leaked from the lunar interior and water formed by the interaction of solar wind with lunar silicate rocks. Water ice bonds tightly to bits of lunar dust and rock a few centimeters below the surface. Micrometeorite impacts release the water into the exosphere, and the sunlight's warmth and the solar wind can also send the molecules hopping about the surface or launch them into the sky. Some molecules must migrate randomly to PSRs, where they can last for eons.

Ariel Deutsch, then at Brown University, and her colleagues extended the analysis of Li's team by estimating the ages of craters near the lunar south pole. Most of the permanently shadowed region lies within twenty large craters at least 3.1 billion years old that contain patchy deposits of surface-exposed ice. Puzzlingly, some craters that should have ice deposits do not, possibly having baked during a past reorientation of the Moon's axis, according to an analysis by Matt Siegler. This would have happened at least two billion years ago. So the ice would be very old.

Another mystery is why only about a quarter of a hundred younger craters, all relatively small, show detectable exposures of water ice on their hidden floors. This ice is perhaps less ancient, having been delivered relatively recently by micrometeorites (instead of larger impactors in the distant past) and by the solar wind stirring up water molecules elsewhere on the surface. Or perhaps their ice originated in some of the large ancient craters nearby when recent impacts sent the ice elsewhere to be recondensed.

Some researchers suggest that the ice in both older and younger craters should be roughly the same age, because much-older ice itself would have been destroyed over time by impacts and other causes. No one knows for sure.

Asks Deutsch: "What are the origins of the ice? What processes are responsible for the delivery, modification, and destruction of ice, and what are the relative strengths of these processes? What is the physical form of the ice (thickness, abundance, texture, etc.)? How does ice vary with depth [and] on lateral length scales?"

There may even be more icy craters than scientists have suspected. With other researchers, Lior Rubanenko of Stanford has evaluated the depth and slope of thousands of lunar craters and found that the ones near the south pole are up to fifty meters shallower than equivalent sites closer to the equator. The likely culprit? Water ice filling the bottoms. Strangely, the shallowing trend does not hold with north-polar-region craters. Multiple missions from several nations—landers, orbiters, rovers—have been planned for the 2020s and beyond to scout for ice, determine its characteristics, and map its vertical and horizontal distribution. The polar craters where these probes will work—and, soon enough, where astronauts will walk—will be "unbelievably cold" and a "very strange, never-seen-before region" that may have very little relevance to the Moon we think we know, says Daniel Andrews, the manager of one such mission, a rover called VIPER, which will drill in the region to learn what it can of the Moon's icy world.

All told, Rubanenko's team suggests that there might be upward of a hundred billion metric tons of lunar water ice—not quite the volume of Lake Tahoe.

"On Earth, we've learned a tremendous [amount] about geologic processes and materials in the course of exploring for resources," says Kevin Cannon of the Colorado School of Mines. "I think scientists should welcome economic prospecting and development of these ices, because it will result in them getting more data, samples, and new findings than they ever could have hoped for with scientific exploration alone. And given the estimates that are thrown around (trillions of metric tons of ice), there's no concern it's going to be gone any time soon." (While this may be true, I'd learn that concerns regarding just how we gain access to this water ice in the near- and long-term is the subject of intense ethical and policy conversations.)

Though Pickering's blizzards in Theophilus remain a fiction, it's hard not to imagine how pleased he'd be with the discovery of icy polar craters on the Moon. And he'd be even more intrigued by a speculation that *liquid* water might have existed on the lunar surface, ages ago. After I learned of the pre-Apollo debates over lunar water and life, a study came out that speculated about the conditions that might have led to a very early lunar sea, one in which life might have evolved, only to be cut short by the Moon's small size, which kept it from sustaining either a protective atmosphere or a long-lived magnetic field.

Once upon a time, were there literal maria on the Moon?

Actual seas, *water*, waves, and wave crests, there upon the surface of the primal Moon? Magma hissing on a watery beach, steam disappearing into whatever air that was? Oceans of water on the Moon? It might have been.

The originating magma ocean would have outgassed water, leading to a short-lived dense atmosphere. And, according to a 2018 study by Dirk Schulze-Makuch and Ian Crawford, it's possible that the very, very early Moon then had a surface ocean of water that was *more than three thousand feet deep*. They admit that "this would be a very optimistic estimate . . . but it illustrates how much water might potentially have been available."

Another possibility is that about 3.5 billion years ago—the Nectaris impact basin would have been visible from the Earth by then—outgassing from lava flows might have built up an atmosphere with enough pressure for surface waters to be maintained. But if this lava-flow outgassing alone were responsible for putative lunar rivers, lakes, or seas—and *not* from the magma ocean—then the lunar water volume would have been far less. Such a later watery Moon would have had a surface layer of only about three millimeters—just .1 inches. Less an ocean and more a sheen. Enough, though, to reflect Earthshine, sunshine, starlight, the bursting colonnades of erupting fire fountains.

A 2016 study contends that the both the Moon and the Earth had a lot of water "from the start." More recent research suggests that Theia's origins were far out in the early solar system, and when it smashed into the proto-Earth, the impactor deposited tremendous amounts of water to what would become the new Earth and Moon.

Schulze-Makuch and Crawford wrote about early lunar water and atmospheres in an article that also was interested in life itself. The piece appeared in the journal *Astrobiology* with a provocative title: "Was There an Early Habitability Window for Earth's Moon?" The piece is highly speculative. Deliciously so.

Building out their hypothesis from research showing that the Moon "is not as dry as previously thought," including evidence for more water deep within, the scientists say that "associating our Moon with habitability seems outrageous," but the lines of evidence for water and an atmosphere ages ago means they can "speculatively identify two possible windows for lunar habitability . . . immediately following the accretion of the Moon and some hundreds of millions of years later following outgassing associated with lunar volcanic activity."

In short, where there is water, there is indeed the chance for life. Lunar liquid water might have occurred on the surface and beneath, threading through rock fractures formed from the shocks of impacts. Researchers have long argued for impact craters and fissures as the likely sites of microbial evolution.

"Whether life ever arose on the Moon," admit Schulze-Makuch and Crawford, "is of course highly speculative."

At least three obstacles present themselves to the rise of lunar microbes, quite apart from the uncertainty about the amount of water present. First is whether a dense-enough atmosphere ever accumulated. A 2017 study led by Debra Needham points to a span of seventy million years in which, due to lava-flow outgassing, a perfectly respectable atmosphere could have taken hold. Before that, as the lunar magma ocean crystallized, an even thicker blanket of gasses could have swathed the Moon. (Crawford, now, however, is convinced by other work arguing for the unlikelihood of any lunar atmosphere at all. As well, if it existed, any surface evidence for lunar liquid water is long since gone.)

Second, magnetic fields protect surfaces from dangerous solar and cosmic radiation. It's unclear whether the Moon had a magnetic field of any consequence. It's possible, as some have hypothesized, that the early Earth and

Moon had a kind of overlapping field. Whatever the Moon was generating on its own wound down, since the lunar core is small and its rotation eventually gave up the ghost. Work by University of Rochester scientist John Tarduno suggests the Moon's magnetic field, if it had one, was active from about 4.5 billion years ago—when it formed—then was a wisp of its former self within five hundred million years. What is clear is that a lot of magnetized Apollo samples are not relics of an early magnetic field, but rather were magnetized when impactors struck the Moon's surface.

Finally, how quickly can nonliving chemistry transition—or flip—itself into life? We don't know. If, as some argue, life can arise from its abiotic elements relatively quickly—within a few million years—then the Moon's chances of having hosted life are better than we might have thought a few years ago. Recent work on the evolution of life on Earth suggests it may have originated as early as three hundred million years after the planet's formation. Schulze-Makuch and Crawford also admit that "we do not know whether there were any intrinsic organic compounds on the Moon," though they surely landed there from asteroid impacts. Moony microbes are a far cry from tall, rational Selenites or W. H. Pickering's garden of Eratosthenes. But finding traces of early lunar life would be an astonishing development, suggesting life is commonplace in the universe.

Searching for evidence of such possible lunar life will be extraordinarily difficult and most scientists would say fruitless. But it could be done. First, drill down to the paleoregolith to look for signs of water or chemical biosignatures or even fossils. Second, drill down to locate and study lunar samples for evidence of clays, which can only form in the presence of water. Finally, use "simulation chambers" to produce conditions akin to that of the early Moon to see what transpires.

Crawford tells me "the problem will be knowing where to look." The other problem is just how hard it will be to drill that deep to obtain samples. It's never been done. "My own view is that these kinds of investigations, and many others, may not be practical until a significant scientific infrastructure has been developed on the Moon," he adds.

Meanwhile, Schulze-Makuch has been planning to run simulations of early lunar conditions—simulations put on hold when the COVID-19 pandemic hit—and, while he agrees with Crawford that we need to do the hard work on the Moon itself, such lab experiments can illuminate possibilities and parameters.

"Personally, I see the Moon as most exciting as a possible repository for ancient life on Earth (e.g., in the south polar ices)," he tells me, "but I also believe that for some time—even if it was only geologically brief—a thin atmosphere might have been possible, and active life as well—but it was probably spatially very constrained and quite short in duration, nothing like a global microbial biosphere we had on Earth billions of years ago. As Ian pointed out, even if it existed, it will be very hard to find. But I still find the topic extremely exciting."

So do I. Now I don't have to time-travel to Pickering's fictional living Moon. I can imagine lunar paleontologists working where my eye scans from Earth. The sound of passing traffic becomes instead the sound of hissing water.

☽

When I look at the Moon I see a scumble of violence and change that I register as terrain and that my mind knows is time. Inside me, and on my bookshelves, is an archive that puts the Moon's history in slow-motion. From the agèd impacts beneath the lava plains—the gray vellum of Imbrium, Nubium, Australe, Nectaris—to the relatively recent white-rayed craters—Messier, Proclus, Kepler—I have a growing understanding of the seemingly changeless Moon as an artifact, in this recent era, of change, if extraordinarily slow by human standards. Now I can even imagine a vanished ocean and putative microbes.

More and more, this understanding coexists in me, wordless if that makes sense, as I scroll across the visual precision of the Moon's surface, a precision at once adrenal and palliative. I wish I could explain this to Coventry Patmore or my now-gone father.

I could explain how sweet it felt one Tucson evening when I finally spied that little blister of a lunar dome in the southern end of the massive, variegated crater Petavius, how challenging this had been, because to see such a subtle feature one needs the terminator near, so the Sun casts its low light and because Petavius is one of several large craters on the waxing early crescent, so the Moon is usually too low to afford clear and steady views. So to see the volcano in Petavius was a thrill, the result of diligent looking and just-so circumstances and to know the volcano in Petavius is to know something of the story of how we have interpreted the surface of the Moon.

There was a time when that shield volcano wasn't there. Now it is. But the change I see is in the past, silent films of lunar violence whose coda appears complete. Apart from ever-shifting light on the lunar surface, what else moves?

What if I were to see changes now? Not Pickering's insect hordes or leafy tendrils. No, "just" a flash of light on a lava plain or a ruddy wisp of excited gasses sighing from a rille. A cloud. A mist. Some have said they've seen these and some still do. Some believed and some still do that the Moon is not so dead after all and that visual observers like me might detect some minute event. It's unlikely but alluring, and this quest has a strange history.

Two of the believers, I've learned, were pioneering scientists in the early days of the Space Age, including a NASA astronomer who briefed astronauts before their journeys and who collected thousands of reports of what has come to be called "transient lunar phenomena." Like these "TLP" themselves, Winifred Sawtell Cameron and Barbara Middlehurst have largely been ignored. But their stories are one of perseverance in the face of doubt and that era's endemic sexism. Ever-controversial, TLP have gotten some renewed attention in the twenty-first century. As they do, it seems only right that two of its pioneers receive the same. Whether certain kinds of TLP are finally proven to exist, the fascination they've held for centuries—and the forgotten endeavors of Middlehurst and Cameron—are inextricably part of the Moon's personal and scientific saga.

CHAPTER NINE

Transient

"Transient": lasting only for a short time; impermanent . . .

Let a single leaf fall from a tree on Earth and there will have been a greater change than may occur on the Moon in a hundred autumns.
— **Virgilio Brenna, *The Moon***

Inside the dome of the National Observatory of Athens—located on the Hill of the Nymphs across from the Parthenon—Julius Schmidt must have rubbed his eyes beside the ocular of the six-inch refractor. It was October 16, 1866, and the night was warm. Though the Greek climate is mild, the leaves of chestnuts and plane trees, of beech and oak, would be turning and falling soon enough. Cooler air would arrive in four days and drop temperatures into the brisk forties. For now, the evening was mellow and aglow. Looking at the Sea of Serenity, Schmidt, the respected observatory director and lunar observer, could not believe what he saw or, rather, what he did not. Though the Moon was changeless, Linné was missing. On the far reaches of Mare Serenitatis, a crater had disappeared.

In centuries past, others believed they had seen delicate eruptions, wispy clouds and dollops of vanishing color on the Moon. In the latter decades of the nineteenth century, however, such phenomena were déclassé. The 1837 *Der Mond* had largely established the Moon's changelessness. Schmidt, described by historians as visual astronomy's most dedicated practitioner, was therefore seeing something impossible. The crater was gone. In its place was a mere white fleck. But Schmidt's reputation preceded his startling public announcement, thus calming doubters and galvanizing new and various reports. Lunar observing enjoyed a revival after Linné "disappeared." And the legacy of the case of the "missing" crater would resonate for more than a century and a half, as astronomers dedicated searches to transient lunar phenomena: actual, real-time, perceptible changes on the lunar surface.

Some of those astronomers would embrace the absurd—W. H. Pickering's insect migrations in Eratosthenes is the most egregious example—while others would soberly record observations of apparent volcanic activity or "red events," outgassing, actual meteorite strikes, and more. Among those who took transient lunar phenomena (TLP) seriously were Barbara Middlehurst and Winifred Cameron, at the University of Arizona and at NASA, respectively. From the 1950s onward, they pioneered studies in the rancorous field and have largely been forgotten. Their stories illustrate the vagaries of lunar science and the sexism of the era. They're also fascinating, even inspiring.

The complexity and controversy of TLP revolves around several factors: the assumption of a dull Moon entirely in equipoise, the reliability and reputation of the observer, the terrestrial and lunar conditions during observation, the kind and quality of instruments being used and the often one-off nature of the alleged events. Still, over the centuries—from the sixteenth to the twenty-first—lunar observers have claimed to have seen what astronomy author Fred Price has called "mysterious happenings on the Moon." Many, if not most, TLP reports are simply mistaken, and understanding why is part of the work of one British scientist's current research. Yet some reports are worth a deeper look, and modern efforts by professional and amateur astronomers are constraining the validity of TLP while pointing toward a case for actual and ongoing processes on the Moon, processes that we need to plumb if we are to return.

A mile-and-a-quarter wide, Linné dots the drab Kansas of western Mare Serenitatis between two far more intriguing features: the pancake-shaped, low volcanic Valentine Dome and the torn-up Sulpicius Gallus Rilles. When the terminator is in this area, the morning Sun creeps toward dramatic mountains ringing the next sea over, Imbrium. The first time I paid any attention to Linné it was a midsummer night. I was waiting impatiently for sunrise over the magnificent Apennines, which, after a day or two, might again reveal the entire tricky length of Hadley Rille. Linné meant nothing to me. It was a ho-hum black period surrounded by ineffectual white ejecta that spans, I'd learn, about six miles.

Before Linné disappeared, it had been recorded by some selenographers but not by all, and what they saw was not consistent. Riccioli drew it on his 1651 map as a deep crater of moderate diameter. Schröter in 1788 called it "a very small, round brilliant spot, containing a somewhat uncertain depression." Others saw a crater either 4.5 miles wide or almost double that. In the white field, some saw a cone, some a pit, some saw a combination of the two. There were those who insisted that the change from crater to patch or cone or pit was the result of lava that filled and refashioned Linné. Reliably weird Pickering believed the changes in Linné were the action of hoar frost. Ten years after Schmidt's announcement, Edmund Neison spent seven pages reviewing the Linné affair in *The Moon and the Condition and Configurations of its Surface*. Neison suggested, as so many would, that inconsistent reports for features all over the Moon implied that some were recently formed, like the craterlets in Plato.

Those tiny pips push the limits of one's optics and demand steadfast clarity and steadiness of air. I always enjoy testing my ten-inch reflector against the small fry of Plato. I typically see only the three main craterlets but one of those is hard next to another, and there are more, extremely small and as elusive as Sahara raindrops. They've been there a long time. When I first started looking at the Moon and glassing the black flats of Plato, it would not have occurred to me that early observers might take the sudden appearance of a tiny crater to herald its birth. What could come and go on that inside-out crypt of a world?

Historical reports of transient lunar phenomena are complicated by the language of certainty that so often accompanies them. Though they cautioned observers against credulity, both Patrick Moore and V. A. Firsoff made many of their own untenable claims. So did H. P. Wilkins. All were British. That colossally important twentieth-century popularizer of astronomy, especially of the Moon, Moore remained a champion of lunar volcanism. TLP seemed to give credence to the volcanic theory of craters, and the volcanic theory of crater origins provided an explanation for TLP: Tautology is appealing, especially with a British accent.

In the case of Linné, the truth was mundane. The records of such selenographers as W. G. Lohrmann and Johann Mädler showed that they observed Linné only under high Sun when the craterlet appears to disappear, leaving only the white surrounds. The object was one of Mädler's reference points for mapping and not anything he observed with care. Like others, he didn't see a crater under the glare of lunar noon but assumed one was there. The Moon was past First Quarter, a little over a week old, on October 16, 1866, so the illumination was high not glancing, and Schmidt's prior records of Linné—from the 1840s—are not those of a seasoned astronomer, write William Sheehan and Thomas Dobbins. He was just wrong. Yet at least as late as 1959, Firsoff, in his poetic, pedantic book, *Strange World of the Moon*, accepted that the crater's apparent alterations represented "at least one well authenticated case of considerable volcanic upheaval on the Moon." Firsoff said that Linné's apparent alterations could not be the result of "mere periodical changes of illumination." Yet they were.

Part of the explanation for this and other lunar misconstrual might involve the law of simultaneous contrast, developed by the chemist Michel-Eugène Chevreul in 1835. He found that we perceive a color differently when it's next to another color, and that darks appear darker and brights appear brighter when they are adjacent.

As to Schmidt, it's possible, Fred Price speculates, that he spun the Linné observation to boost sales of his forthcoming lunar map, destined to be nearly eighty inches wide and dotted with thousands of craters. Disappointingly, publication did not occur for another decade. While the sensation did not put money in the astronomer's pocket, it did put eyes to the eyepieces. The Moon was back, and observers combed the records as diligently as they tracked the terminator to find evidence of the Moon's alleged coy behavior.

☯

In the "Region of Calippus," a "starlike appearance on dark [Earthshine] side" was seen from Worms, Germany, on November 26, 1540. On March 5, 1587, "A sterre is sene in the bodie of the mone . . . wherat many men merueiled, and not without cause, for it stod directly betwene the pointes of her hornes." Hevelius sees a "red hill" upon the Aristarchus Plateau in 1650. "Several New Englanders" see a "bright starlike point" on the Earthshine portion of the Moon on November 26, 1668. Thus begins the "Chronological Catalog of Reported Lunar Events" in NASA Technical Report TR R-277, which was published July 1968—the year of Apollo 7 and Apollo 8—and featured four coauthors, including the lead researcher Barbara M. Middlehurst, then with the Lunar and Planetary Laboratory at the University of Arizona. For her interest in TLP, she would clash with her boss, Gerard Kuiper, and put her job, reputation, and well-being on the line.

The chronology lists 579 reports, organized by date and time; feature or location, duration; description; observer; and reference. It deliberately omits topographical changes such as the alleged and mistaken Linné affair as too controversial. It reads like a found poem.

Plato: a "reddish streak on crater floor."

Plato: "a track of ruddy light, like a beam . . ."

Plato: "Yellow streak of light across crater floor while crater was in darkness."

Aristarchus draws the eye again and again: "red spot," "nebulous bright spot of light," "extraordinarily bright," "extraordinarily bright, like [a] star." Three times in 1790 a "small, hazy spot of light." On May 1, 1824, a "blinking light" near Aristarchus.

William Herschel believes he sees "at least 150 small, round, bright, red luminous points" during a lunar eclipse, and two years later Johann Schröter sees "vapors resembling a mountain" in, naturally, Mare Vaporum. There's a "black moving haze or cloud" over Mare Crisium in 1826. There's Franz von Paula Gruithuisen—he of the lunar city—watching a "smoky-gray mist" in 1839 over dark Grimaldi, that black eye of the western limb. And there's a young Julius Schmidt watching "a bluish glimmering patch of light not quite within the night side of the moon," near the isolated peak Mons Pico on April 25, 1844.

TLP reports have continued regularly for the past 150 years, with many areas recurring as alleged hot spots, Alphonsus and Aristarchus in particular. The sightings usually involve flashes, color changes, cloudy appearances or increases in luminosity. Occasionally, observers have claimed to see lightning or even moving lights. As the decades unfolded, the story of TLP turned from credulity to ridicule then to a measured assessment of viewing conditions, telescope flaws, human psychology, and the physical properties of the Moon.

The Middlehurst catalog still attracts attention, though critics argue it's so lax it's useless.

It is an unfortunate fact that, as the British amateur Gerald North writes in *Observing the Moon: The Modern Astronomer's Guide,* "The subject also attracted more than its fair share of cranks. These people tended to go to their telescopes and see all manner of coloured effects, plumes of smoke erupting from craters, etc. . . . There are still some today for whom the Moon is a fairground of amazing phenomena. Not only do these people bring the derision of many astronomers on the whole subject, their 'reports' pollute the data."

Organized campaigns using special equipment would be mustered, articles written and argued over, careers derailed or diverted, and the serene Moon of our night skies would once again divide the scientific community into camps that depended in part on a belief in or a disdain for volcanism in lunar history. It was more than a scientific mystery. It was matter of being right.

❦

Among the best-known historical accounts are three from the 1950s and early 1960s.

On November 15, 1953, Leon H. Stuart of Tulsa took a photograph of the Moon showing a meteorite impact in a mare flat between two chunky craters, Schröter and Pallas. The photograph of the eight-second-long flash shows a brilliant white dot between Sinus Medii and Sinus Aestuum, where its dark ash deposits are strewn like wind-felled rushes. The physician had just happened to attach a camera to his telescope when the flash occurred, according to a 2003 National Public Radio story. A photograph shows Stuart in warm clothes, a plaid scarf, and a beret next to a large reflector.

Decades later, reviewing data from the Clementine lunar probe, NASA researcher Bonnie Buratti and Pomona College's Lane Johnson found a mile-wide crater, apparently confirming Stuart's assertion, which had long been dismissed. Others found that the flash and the putative crater didn't match and that such a long-lasting flash indicated a bigger cratering event. Alas, the original image was lost during testing at the Lunar and Planetary Laboratory in 1967.

It is, however, no longer controversial that meteorites hit the Moon. NASA has a Meteoroid Environment Office that deploys two fourteen-inch scopes and video cameras. Between 2005 to 2018, this setup counted 435 meteorite strikes on the Moon. According to NASA, half of the meteorites are random while half come from meteor showers. Other monitoring occurs via the Moon Impacts Detection and Analysis System, and amateur astronomers are involved as well; the Association of Lunar and Planetary Observers has a zealous core of "lunatics." In 2015 observers videoed a candidate lunar impact plume catching sunlight. This form of transient lunar phenomenon is as real as it gets.

Science would not prove so straightforward with other cases.

In 1956, astronomer Dinsmore Alter, of Los Angeles's Griffith Observatory, thought he saw some kind of haze on the floor of Alphonsus, a venerably large crater located where the southern highlands push up into the Moon's midsection. It sits between the flat gray crater Ptolemaeus and the more rugged Arzachel. About seventy miles wide, Alphonsus features a low ridge, small center peak, floor fracturing and three distinct dark spots arrayed around its inner edge. We now know these are dark-halo craters, once thought to have been volcanic but now considered to be impact-strewn dark material from beneath the surface. They look like messy drill holes. Working with the sixty-inch reflector at Mount Wilson, Alter recorded obscuring detail in filtered photographs suggestive, to him, of an outgassing event that scattered light. More famously, and inspired by Alter, a controversial Soviet astronomer named N.A. Kozyrev used the fifty-inch reflector of the Crimean Observatory to photograph and take spectroscopic measurements of a red color atop the central peak in Alphonsus. The Kozyrev spectrographs would be met with extreme skepticism then acceptance by some of his doubters, but the case remains controversial, if not dubious, because the central peak is most definitely *not* a volcano.

Alter speculated that "for some reason the blue-violet photographs lose more detail [on one side] side of Alphonsus than they do in the floor of Arzachel. This is not true of the infra-red ones . . . There is a temptation to interpret these results immediately as being due to a thin atmosphere, either temporary or permanent, over the floor of Alphonsus. The theoretical difficulties inherent in such a hypothesis are, however, strong enough to forbid whole-hearted acceptance of it." Kozyrev suggested that a spewing of below-ground gasses that penetrated the surface—an outgassing—had been responsible. Rather being genuine evidence of gasses, Alter's findings have been ascribed to the way the telescope's filters scattered light and have not withstood the test of time.

As to Kozyrev's own Alphonsus spectra, the story is complicated. He began observations on Monday, November 3, 1958, a year to the day that the Soviet Union had sent the dog Laika into orbit. That night a spectrograph (the instrument reportedly was a good one) showed the crater's central peak dimming due to what Kozyrev interpreted as a thin exuding of carbon gas. The area changed in brightness as the apparent outgassing occurred—a "process of . . . considerable magnitude [that might have led to] noticeable changes of the crater," he wrote to Alter. On a dead Moon, this was surprising. Nearly a year later, on October 23, 1959, Kozyrev said new spectrographic observations once more suggested there were gasses escaping from the peak and he appealed to volcanism as the Moon's driving force.

His first claims landed like a bombshell—in part because they came via a TASS news agency wire report not a journal article. Newspapers around the world ran the story. Astronomer Ernst J. Opik called Kozyrev "trustworthy" in the *New York Times* and speculated that a buried meteor might be responsible for the heated gasses (an absurd proposition). As historian Ronald E. Doel documents, astronomers beyond the Iron Curtain were deeply frustrated not to have more details let alone the spectrograms themselves. Careful phone calls were made, and researcher Zdeněk Kopal learned from a different Soviet astronomer that the spectra—a series of vertical lines showing different elements—contained evidence of carbon molecules. The findings were written up by Kozyrev for *Sky & Telescope*, the popular astronomy magazine, where the editors noted that observations of the crater were not conclusive.

Gerard Kuiper was unimpressed. He'd been unconvinced by Dinsmore Alter's earlier photographs and dismissive of Alter's support of Kozyrev. To the fastidious Kuiper, it was like a feedback loop of mediocrity, inconclusion leading to inconclusion. Still, he did what a scientist is supposed to do: attempt to replicate the results, but he could not discern the presence of carbon when he used the eighty-two-inch reflector at the McDonald Observatory in Texas. As Doel points out, Kuiper's doubts were bound up in his distrust of Soviet science, which itself was bound up in Communist Party politics and technically more backward than in the West. Kozyrev himself was a cipher, and, it would be discovered, had struggled with imprisonment and mental illness. In the U.S, Kuiper also met a visiting Soviet astronomer who spoke poorly of Kozyrev.

As Kuiper tried to learn more, he set himself up as a kind of analyst of Soviet science, assisting the US government's evaluations of work emerging from that secretive society. (It was not an entirely new role: Kuiper had been part of a US mission to vet Nazi scientists as World War II ground to a halt.) Soviet researchers had come to no consensus about the Alphonsus findings, that much Kuiper learned. That did not stop him from writing a critique, Doel says, which "fairly bristled with contempt," contempt that "succeeded in stalling possible U.S. and Canadian plans to search for similar variable phenomena on the Moon."

Kuiper wanted to see the spectral plates but travel to the Soviet Union was rare and daunting. He'd been once for an astronomy meeting, lamenting a country that was poor, oppressive and isolated. Then came news of a conference set for Leningrad in 1960, a meeting that, Doel says, would be "the first major scientific conference devoted entirely to the Moon."

So on December 8, Kuiper was listening to Kozyrev, then to other astronomers who suggested there was an emission though not necessarily of carbon. They thought rather than a hot venting, the outgassing had been cold. Soon after, Kuiper held a magnifying glass over the plates. The emission lines were there. In a letter, Kuiper said there was "no doubt . . . that the spectra are genuine."

Yet Kuiper's former student Charles Wood says that the Alphonsus claims are "unbelievable." The peak is anorthosite, so "Alphonsus almost certainly did not erupt a half century ago," he says. Astronomers and historians Thomas Dobbins and William Sheehan chalk up the Kozyrev spectra to rickety guiding of the telescope rather than genuine spectral lines. Dobbins tells me that Kuiper was in fact more ambivalent about the plates and that his interpretation could vary with audience. There remains the possibility, however remote, that gasses had leaked from the crater.

The most spectacular TLP report in the twentieth century emerged three years after Kuiper hovered over the plates in Leningrad. A cartographer working for the US Air Force—then developing the best possible maps prior to the Lunar Orbiter missions—was at the eyepiece of the Lowell Observatory twenty-four-inch refractor on Mars Hill in Flagstaff, Arizona. It had been quite a journey for James Greenacre, who had been an archeologist then the manager of a classified effort to create maps for the Allied invasion of Europe. The Air Force Aeronautical Chart and Information Center took him on after the war, and the lunar cartography program there expressed interest.

First he had to detail his astronomy chops, in particular with the Moon, and he did so by noting that it "comes out at night" and "I wooed and won my wife under it." He got hired.

Greenacre did become a diligent and experienced visual lunar observer, making drawings at the eyepiece for the future maps. He knew the Moon's surface and how the Earth's atmosphere could alter its presentation and he knew the Clark refractor's tendency to fringe images with color. Chromatic aberration is a common problem with that kind of scope. No less a figure than the University of Arizona's Ewen Whitaker—one of the best in lunar mapping and observation—said Greenacre "had more hours of observing the Moon than any of us did." And, according to the then-director of the Lowell, Greenacre was no fan of transient lunar phenomenon.

On October 30, 1963, James Greenacre would begin to change his mind.

With his drawing supplies at the ready, he began observing when, twenty minutes into his run, he saw a red-orange color develop in two spots by Vallis Schröteri, the dramatic rille that snakes through the Aristarchus Plateau like an important, if meandering, thought. He observed a kind of "sparkle" and flow. Another observer who was with him, Edward Barr also saw a darkish orange tint.

Myself, I have seen no lights on this mesmerizing feature, though every time it comes creeping into view beside and above the wastes of eastern Oceanus Procellarum—the plateau's dawn heralded by light on the scattered peaks of the Harbinger Mountains—I look. I scan the chaos as though it were a gothic wilderness, rough volcanic remains bouldered and crinkled, sliced by smaller rilles and seeming as to the rest of the Moon as the twelve-tone scale is to "Claire de Lune." This diamond-shaped uplift is the strangest-looking place on the Moon, even presenting a different hue if one notices: a light yellow-green like poison. South of the plateau's most mountainous region are the large craters Herodotus, a Plato-copycat, and the terribly bright namesake crater Aristarchus, on whose slopes amateurs can detect "radial banding," swathes of darker landslides that for decades were incorrectly considered evidence of TLP. These I have seen, and Aristarchus remains a lure for its novel appearance and its place in lunar annals.

Greenacre and Barr—I imagine hushed, urgent tones and shaking hands—removed a filter used to tamp down spurious colors but the red-orange and sparkly appearance remained. The Moon was not yet high in the sky, the air was unsteady, and the observers were using high magnification, facts that critics would seize on. Still, they looked.

"I thought the motion seemed familiar somehow," Greenacre recounted in *Analog* magazine a year later. In a "slow drawl," according to the writer, Greenacre said, "Finally it occurred to me that it looked exactly like the sign in front of a supermarket in Flagstaff—dots of red and white lights are chasing each other across it." Greenacre was shocked, admitting that "my first thought was 'the Russians got there ahead of us.' Then I kept getting a sinking feeling—something must be wrong—it couldn't really be happening."

A third, pinkish patch then manifested itself just inside the south-western rim of the twenty-five-mile-wide Aristarchus crater. The first two spots—called R1 and R2—became "a light ruby-red," according to Robert O'Connell and Anthony Cook's thorough analysis, "but their density and sparkle were still sufficient to obscure underlying surface detail." Greenacre and Barr began to draw the widely separated locations of R1 and R2 and the third patch, R3. After about thirty minutes all three had disappeared. Greenacre narrated the event on a tape, audio that has since been mislaid. The night was not yet done with surprises for about three hours later there appeared a fourth patch, bluish, and this "B1" region grew for hours over the western side of Aristarchus crater.

The next day Lowell director John S. Hall wired a telegram to Harvard College Observatory to put the observations on the official record and he copied the telegram to *Sky & Telescope*. The *Arizona Daily Sun* ran a page-one article with a huge headline that read "Moon 'Eruptions' Seen Here," sub-titled, "Lunar Mappers Spot Possible Volcanoes," with quotes from Hall. He was mindful of the controversies that astronomer Percival Lowell had stirred with his earlier, dubious Martian canal claims at his namesake observatory yet Hall stood by his lunar cartographers. "These new observations are . . . well-documented," he said.

Greenacre, says the *Analog* article, was "dumbfounded." He described the observation to his daughter, Sabra Minkus, as being "like a red ball of fire." He recalled that "after it's all over, you feel stupid of course, to think of all things you could have been doing but didn't."

He'd have a chance to redeem that feeling.

The next lunation, under excellent seeing, on November 28, 1963, it happened again. Another "red event" at Aristarchus. Actually, three. This time four observers confirmed it. Ed Barr noticed something unusual thirty minutes into his mapping session—another pink light near where he and Greenacre had seen one the month before. He moved the field of view around, removed and returned a color filter, all to verify that atmospheric, filter or telescopic variables were not affecting the sighting. They weren't. Barr called over another lunar cartographer, who verified the pink-red light. A new glow developed, with an orange cast, on the crater rim. By then James Greenacre had reached the Lowell Observatory and looked for himself while Dr. John Hall saw the same reddish glow on the Aristarchus rim, a glow not visible in the smaller finder scope. Then two of the team saw another orange-red patch on the east side of Vallis Schröteri while the other two could not discern the change. It was estimated that one of the colored areas was twelve miles long and a mile-and-a-half wide. Photos were taken, and, crucially, calls were made to observers at another observatory—with a sixty-nine-inch reflector—but the crew there could not initiate spectral measurements in time.

Decades later, one of those astronomers at the sixty-nine-inch, Peter Boyce, would say that he and his partner could not "definitively rule out the presence of a red afterglow at the very limit of detection." He was hardly confident that they saw it and did not consider it a confirmation. Reports at the time, however, including the one published by the US Air Force, stated that Boyce had "within 15 minutes . . . called Dr. Hall and confirmed the sighting of the reddish-pink color on the southwest exterior rim." The next day "Boyce located the position of the color" on a map, which "agreed with the position observed and plotted by the observers at Lowell." As to the photos, the black-and-white film was not sensitive enough to record the colors.

As though following a pattern, later in the night a blue light grew on the western side of the crater in the same locale as the one from October. More such blue colors appeared, lasting for about an hour, during which the observing team double-checked atmospheric conditions, the effects of having a filter on or off and the possibility of the refractor causing false color. To the last, they believed they had seen something real on the Moon.

Greenacre wrote about the observations in *Sky & Telescope*'s December 1963 issue, as well as in formal papers for the *Annals of the New York Academy of Sciences* and the US Air Force report called "Lunar Color Phenomena" in May 1964. He wrote of the "astounding colors" he and others had seen on the first night. He also thought his use of the word "sparkle" might have been "ill-chosen" and described more formally the analogy to the supermarket sign. Hall wrote in one letter that "contention made here is that this coloration is not produced in the optics of the telescope or by the earth's atmosphere." Thomas Dobbins wonders if Hall, consciously or not, wanted the TLP report to further government interest for observatory support.

The effects of the Aristarchus red and blue events were electrifying. Amateur astronomers who believed in TLP felt vindicated. Some professionals were more willing to suggest the Moon was not quite so dead after all. NASA suggested that the Aristarchus region be a landing site for an Apollo mission. Along with the Kozyrev spectra of Alphonsus, the 1963 Aristarchus sightings catalyzed research and speculation. In the heyday of TLP activities, in the 1960s and 1970s, multiple reports would provide fodder to the credulous, the skeptical and the outright disdainful.

Part of that disdain was directly personally at Greenacre. There were rumors and accusations that he was a hard-drinking, publicity-seeking careerist, which amateur astronomer Bob O'Connell, based on his extensive research, categorically calls "FALSE!"

The crazies came out too: A priest in Massachusetts claimed the Lowell sightings were evidence of alien spaceships, thirty-one to be precise. Greenacre dismissed such nonsense, but the priest continued to harass him for years. James was "sort of in awe" over what he'd witnessed on the Moon, his son later said, but the letters from the priest made the cartographer mutter from time to time, "I wish I had never seen anything."

By the end of the twentieth century skeptics would argue that James Greenacre had in fact not seen anything, at least not anything on the Moon.

At the tapering end of the Spanish flu pandemic, Amos and Mildred Sawtell celebrated the birth of their daughter Winifred on December 3, 1918, in Oak Park, Illinois. She grew up healthy and stubborn. She hated school, preferred to play with younger kids and definitely did not want to go to college. Raised Presbyterian, Wini was interested in Greek mythology as early as fifth grade. Her daughter, Selene Green, tells me that "when she read about astronomy, she didn't believe in heaven. When she read about geology, she didn't believe in hell."

When Wini turned eleven, her parents gave her a telescope, and she never stopped looking up. Once, she'd been stunned to meet a woman astronomer on a school trip to the Field Museum; the astronomer discouraged her from entering the profession but a male scientist told her to try. In high school, she taught her classmates the constellations because, she once told a reporter, "The teachers didn't know, and I did." Her handmade "Astronomy Notebook" from an unknown school year begins with an outline titled "Supposed Importance of Earth" and continues with a poem that claims the Moon is "the most beautiful thing to me." Wini listed her address as 928 N. Harvey Ave., and continued: "Oak Park, Illinois, United States [of] America, North America, Western Hemisphere, Earth, solar system, Section A2 Galaxy G, Universe . . . Near Andromeda, Universe, Space." For her entire life, she kept a 1930s magazine clipping showing a telescope next to the labeled strata of fossils and an article titled "To the Moon by Rocket!" The first stop on her way to space? College, after all.

In 1940, the budding astronomer earned an education degree from Northern Illinois University (and later its Distinguished Alumnus Award), but for a time she kept working as a soda fountain girl and an elevator operator at a fancy Chicago department store. The *Chicago Sun* featured her in an article titled "Elevator Girl Hitches Wagon to the Stars: Ignores Job as Clothes Model: Blond Astronomer Lectures on the Moon." As part of a 1942 astronomy convention at the Congress Hotel, Wini lectured on the possibility of plant and insect life on the Moon. The journalist called her "scholarly" and quoted her as saying, "I [have] a vivid imagination and the immensity of the distances, spaces, and proportions [of space] appeal . . . to me." At NIU, she had added her name to a list of students who wanted to go the Moon.

While hosting astronomers from Indiana University, Wini got a ride home since she didn't have a car, and they had talked so late that public transit wasn't running. The professors were so impressed that they offered her an assistantship, and she went on to earn a master's degree in astronomy from IU in 1952. It was there she met her future husband, Robert, and his courtship was nearly as arduous as a trip to space. She wanted a career not a marriage. Bob proposed eight times. The successful one came after their planned trip to the American West turned into a lonely solo adventure for Wini, as Bob had suddenly taken a dream job at the US Naval Observatory in Washington, DC. She missed him so much she said yes.

From 1951 to 1958, Winifred Cameron also worked as an astronomer at the US Naval Observatory. Bob insisted that he would not take the job unless they were both granted positions, though Wini's salary was significantly less because she was a woman. She continued to study sunspots at the observatory even after their first daughter Selene was born.

Hopes of settling down were interrupted when Bob became part of a team that built a tracking station in South Africa to follow the course of Explorer 1, the first successful American satellite. He directed the development of the telescope and station, being rewarded with the first photo of Explorer 1 in space. In Johannesburg, the family lived in a pleasant white one-story cottage at the Orange Grove Hotel with a thatched roof set among palms and willows and surrounded by flagstones and planters. While there, the couple had their second daughter Sheri Carina, whose middle name is the constellation visible only in the southern skies.

There was darkness too: Bob was bipolar and while in South Africa, he became suicidal. Twice in his life he underwent shock therapy treatments.

"My mom was a very loving person," Selene says. "And she thought she could fix him."

$\circledast$

In 1959, the couple joined NASA, the same year the reclusive Soviet astronomer Nikolai Kozyrev would rev up lunar studies with his controversial Alphonsus spectra. Cameron was one of only three women astronomers at the agency, and she and Robert were the first astronomer couple at NASA. Winifred Cameron helped brief both John Glenn and Scott Carpenter before America's first two orbital flights, creating star charts, designing experiments and even developing some equipment for the cramped Mercury capsules. When they flew, Cameron worked as astronomer on duty in a backroom at Cape Canaveral and heard the excited transmissions from the astronauts about mysterious "fireflies" surrounding their vehicles.

"The Russians hadn't reported anything like this," she later remembered. "It was unexpected and a little alarming." Some speculated the particles might be alien life. Cameron got it right: The fireflies were particles of dust and frozen water vapor traveling alongside the capsule from which they had been shaken, inertia's little flock lit by the Sun and glowing against the black of space. Her last item with Mercury was helping astronaut Wally Schirra compile his observations from the flight of *Sigma 7*.

It wasn't all work. Cameron stayed with the Glenns in Houston and, in Selene's words, "bonded." The couple became friends with the artsy Scott Carpenter and his family. When I visited Selene in Florida to learn about her mother, an array of boxes and books and awards and papers spread on the dining room table, she found copies of a photograph of Carpenter smiling next to his daughter Kris, Selene, and Selene's sister, Sheri. The suburban colors have faded, those of the girls' shorts and tops of pink, seafoam, and yellow-green. Carpenter stands ramrod straight, hands in the pockets of his narrow brown trousers, a blue-and-white shirt tucked in, open at the collar, the epitome of casual dashing discipline as embodied in the early astronaut corps. Behind them, the ranch houses, lawns, and green trees of America, where the first launching pads were swing-sets.

It was through such verdant suburbs that the Cameron family would navigate to Maryland beach vacations at Ocean City's Stardust Motel. It was in the galaxies of living rooms and porches and kitchens that they would host fabulous parties—her mother "dressed to the nines," Selene says—for such luminaries as H. A. Rey, Arthur C. Clarke, Isaac Asimov and the lunar-science rivals Harold Urey and Gerard Kuiper. On those nights when Wini or Bob would bring home a telescope from the office, they'd show the girls the stars, the planets, and the surface of the Moon that Cameron knew so well, a Moon she would argue was not quiescent.

At NASA Goddard, the couple worked in Building 1, where Winifred Cameron had a navy-blue nameplate, an office with huge Moon photos, a glass-paned wooden bookcase, a Moon globe, and a chalkboard. She had calipers and, for a photographer, once put them up against some smaller craters by the large, lava-filled Archimedes. She helped compute landing zones for the Apollo missions, and Selene has NASA plaques given to her mother for work on Apollo 11. What Selene can't show me is the doctorate that her mother wanted. Winifred decided to support Bob and the children as he took evening courses to finish his PhD. If she resented this, she never let on.

"We've been lucky in our circumstances," she told a writer in 1968. "Bob helps with raising the children, with the housework and the cooking. At work, we're partners. Luckily, there aren't strict hours at the lab, so we trade off. One of us goes to work early while the other stays home with the girls until they go to school. Whoever goes to the lab early is home in time to meet them after school and fix supper." Perhaps that then-unconventional life helped keep Cameron's mind open to other new ways of thinking, like the possibility of transient lunar phenomenon.

Bob Cameron's mental health "was more of a challenge than sexism at NASA," Selene tells me. "I think they were proud of her." Indeed, her work was featured in a Goddard publication in 1966, and press coverage of her pioneering role and her TLP research would continue for years. Still. She had been banned from an Air Force eclipse-watching expedition in 1955 because no women were allowed, and a photo from a science conference shows Winifred Cameron in the front row, with just five other women, surrounded by a sea of more than one hundred black-suited men.

Cameron worked at Goddard Space Flight Center from 1960 to 1983 and she even brought her astronomical enthusiasm to her signature, putting a crescent Moon then a Sun over each *i* in the name. She needed that enthusiasm because controversies over TLP would not die down.

☽

To learn more about TLP, Winifred Cameron helped spearhead a network of specialized telescopes called Project Moon-Blink. Conceived by NASA's James B. Edson, the scopes consisted of a rotating wheel with red and blue filters. The effectiveness of the device was, said manufacturer Trident Engineering, due to its rapid shifting between filters—some two times a second—so that "when viewed with a suitable electro-optical device, changes in color . . . will appear as a 'blink' on the face of the image tube, thus drawing the eye to the spot." Just fourteen Moon-Blink devices were deployed, but in a 1967 peer-reviewed paper in *Icarus*, Cameron along with John J. Gilheany, affiliated with the Naval Academy and Trident, asserted that "color phenomena have been" seen on the surface of the Moon.

And in a little-known October 1966 report from Trident Engineering to NASA, Moon-Blink participants said they had seen and photographed "red colorations . . . on the lunar surface" and that "these colorations may persist for several hours." A sixteen-inch reflector detected a red coloration at the base of Alphonsus's central peak on October 27, 1964, and James Greenacre wrote to Gilheany that a "small crater within the breach . . . very close to the base of the peak" in Alphonsus "has everything in its favor as far as topography is concerned" if the event was an outgassing. Another "large-scale blink" in

Aristarchus on August 21, 1965, followed possible detection of luminescence around Kepler. The seeing that August night was said to be excellent. Another event lasted for some four hours on November 15, 1965, again on the Aristarchus Plateau. While the seeing was poor and the Moon low, more than one observer confirmed a possible blink. Two out of three observations made at other locations saw cloudiness or a red color on the plateau that night. The color pulsed so regularly for so long that the report's authors say this ruled out atmospheric effects, which are more random. Photographs were taken, though I do not know the location of the originals. The 1965 observations were detectable only with the Moon-Blink instrument.

It wasn't only detection that interested Cameron. She subjected data—from Moon-Blink, from the massive Barbara Middlehurst catalog and elsewhere—to statistical analysis, developing histograms and investigating whether TLP was related to such factors as the angle of sunlight on the Moon, solar activity, how far or near the Moon was to the Earth and where the Moon was in relation to the Earth's magnetic field. Convinced that the evidence demonstrated that TLP were internal to the Moon, Cameron found no significant relation among any of these let alone strong enough correlation to suggest cause-and-effect.

In 1971, she wrote in a NASA report that "even if the matter of short-lived brightenings and star-like points could be . . . explained by instrumental, atmospheric, and geometric effects, there are other phenomena that seem to represent genuine, abnormal situations . . . There are too many instances of obscurations or mists . . . while everything else around was very sharply visible." Both Cameron and Middlehurst found that most TLP sites were, in her words, "closely associated with the edges of the maria," a contention still not entirely accepted. In 1978, Cameron published her first iteration of a "Lunar Transient Phenomena Catalog," tackling—out of more than two thousand TLP reports, a few hundred observations in a hard-to-read grid that included far more information than the earlier Middlehurst compendium, such as where the Moon was in its orbit, solar activity, the telescope type, and sky conditions if known. She found nearly five hundred that "are probably intrinsically lunar." And she was less credulous than Barbara Middlehurst.

Cameron would back off from her previous acceptance of the many reports of blue and violet colorations at Aristarchus made by an assiduously unreliable amateur observer named J. C. Bartlett. Cameron noted that researcher L. E. Fitton had found that some weather patterns could cause both blue and red colors in lunar observations. Bartlett used very small telescopes and reported so many blue events at Aristarchus that it seems impossible that no one else saw them. (Cameron wondered if he was particularly sensitive to blue . . .) In any case, Bartlett's reports were many, mistaken and useless.

There were other credible TLP-detection efforts. In the 1960s observers at Northwestern University's Corralitos Observatory in New Mexico spent thousands of hours monitoring the Moon but detected *not one* transient event—or so it is often claimed. In the Cameron catalog, there are at least two reports of photos taken by the observatory that seemed to show bluing around Aristarchus. (The photos might be buried in the papers of directing astronomer J. Allen Hynek.) One entry says the observatory considered the effect atmospheric. The Corralitos twenty-four-inch reflector sent its images to an electronic monitor. The lack of any substantial Corralitos findings remains a major strike against TLP but that may not be the entire story. Says University of Aberystwyth physicist and current leading TLP scientist Anthony Cook: "Some of those observations were made hours after the original TLP reports, so they might have missed the events. Cameron, when she visited, noted that the refresh rate of their filter system was very slow . . . so she thought it was not reliable at detecting colours."

"It's a bit much," Winifred Cameron told a reporter in 2008, "to dismiss these reports as figments of the imagination. That's an insult to the people who've spent every night of their lives observing the Moon." Cook agrees: "What is interesting for the lunar science community is that a small number of [observations] remain unexplained."

If one path had opened for the tenacious Winifred Cameron, she might have been able to ground-truth transient lunar phenomenon. While she worked at NASA, she had her eyes on another plum job: astronaut. "They never even acknowledged my application."

While Cameron worked at NASA, fellow astronomer and TLP-cataloger Barbara Middlehurst was hired by Gerard Kuiper in the TLP-centric year of 1959. Kuiper took Middlehurst on primarily as an editor. Born in 1915 in Wales, Middlehurst trained in math and astronomy, earned an education at Cambridge and eventually became an astronomer at the University of St. Andrews. Of the three British employees Kuiper took on, Middlehurst was the only female and the only one with college degree. At the Lunar and Planetary Lab, she got the smallest office.

According to her friend Leonard Srnka, a geologist and physicist now affiliated with the University of California San Diego, Barbara was proud of her background as a Welsh girl who managed to get a small scholarship to Cambridge before it was officially admitting woman students. She did not receive a full diploma, however, given her "inferior" status.

"She told me many, many times about the kind of guff she got from people: 'What are women doing at night in observatories, we can't have that. It was very suspicious!'"

A fellow of the Royal Astronomical Society (and much later of the American Geophysical Union), Middlehurst published work in such major journals as *Science* and *Nature*, never vacationed and stayed under the intellectual sway of the British love of volcanism as the formative force on the Moon. A colleague wrote in an obituary, "There may be controversy over the reality of TLP . . . [but] there can be no doubt about Barbara's courage and determination in pursuing research in this field."

Kuiper, however, was no fan of her interest even after seeing the Alphonsus plates in Russia. He refused to publish Middlehurst's historical catalog in the in-house journal of his Lunar and Planetary Laboratory. When the catalog became a NASA Technical Report, the press paid attention, generating coverage in the *New York Times* and elsewhere. Kuiper—described as orderly, loyal, reserved and authoritarian—reportedly told Middlehurst, "I would suggest your list be very drastically reduced . . . after consultation with one or two competent observers, like [Ewen] Whitaker, and that a vigorous program of 'event' watching and analysis be started, not by amateurs but by professionals."

Middlehurst paid the price for the clash and resigned from her position. She would go on to work at *Encyclopedia Britannica* then at Houston's Lunar and Planetary Institute, where she had a visiting appointment. She left Arizona in 1968, the year her famous catalog was published.

"She and Kuiper had a falling out," Srnka tells me. "She claimed she did all the rat work for Kuiper. She was very diligent and very hard-working. And she could give a thirty-minute diatribe over coffee about various persons. I won't name names." He doesn't think Middlehurst was the victim of "real harassment but she did not feel treated fairly by various people." The now-deceased editor of *Sky & Telescope*, Leif Robinson, Srnka said, told some that there was another factor: that Middlehurst and Kuiper were for a time romantically involved.

What few letters between the two are filed in Kuiper's manuscripts at the University of Arizona, not surprisingly, give no hint of this. It's clear, though, that there was some overlap between the professional and the personal. At least once Middlehurst took care of one Kuiper's children while he and his wife were away, and she expressed affection for the couple in one letter and had a Christmas present to give them. Certainly the tone of Kuiper's critique is sharp. No documentation exists that I could find regarding her split with the lab, but at one point years before Kuiper had lobbied the university to make her associate director. After Middlehurst's departure, there was at least one congenial exchange of letters and a visit to Tucson.

To what extent Middlehurst's complaints about LPL and Kuiper were revisionist and inflected with personal disappointment is impossible to say, but her time in Houston was not entirely pleasant either, according to Srnka. "People would sort of shun her," he says. "People would roll their eyes. The view was . . . there was nothing on the Moon . . . nothing squirting out, no water. Some people thought it was Barbara's imagination."

"She did lots of good work. She was spirited, intelligent and brave. These women, this whole collection"—here Srnka speaks of female astronomers in the 1960s at the rise of popular feminism—"were a bunch of brave people who would do what they love doing, to hell with the establishment and fought fights through their careers. I was proud of her . . ."

Brown University lunar scientist Peter Schultz agrees. "I saw what was happening with Barbara Middlehurst or anyone else who touched the TLP phenomena . . . I was seeing people's reactions to it. This is not worth a career to get wrapped up [in]." Eventually, in the 1990s, Schultz himself began to make somewhat oblique references to changes on the lunar surface regardless of their putative visibility from earth. As early as 1972, using Lunar Orbiter imagery, Schultz was seeing evidence—now widely accepted—of recent activity on the Moon (geologically speaking that is). The work, he says, wasn't credited back to him when it was, in his words, "rediscovered" by the Lunar Reconnaissance Orbiter. By the time that NASA was taking up plans for possible lunar exploration again, during the 1990s and early 2000s, Schultz saw that the tide was beginning to turn.

So too did a Columbia University astrophysicist named Arlin Crotts, who came to TLPs late in his career, writing that "some scientists have studied TLP[s] . . . at their peril since the early 1970s." Such work, he says, "entered a strange and unfortunate state of 'pariah science' in which even its mention would bring derision and unreasoned skepticism." Crotts was once "admonished" by a prominent NASA official to "not pursue [TLP] because to do so would be a waste of the taxpayers' money." Peter Schultz says Crotts's doctoral advisors told him not to work on TLP. So Crotts also bided his time. Meanwhile, Middlehurst and Cameron led the charge.

❧

Project Moon Blink wasn't the only effort to systematically hunt for transient lunar phenomena. After Middlehurst and Kuiper parted company, she continued her research efforts with an endeavor called Project LION: The Lunar International Observers Network. LION's objective was simple: "To determine whether ground observations of lunar events can be confirmed by crew members of Apollo spacecraft during lunar orbit, and to obtain further evidence that may help determine the cause of lunar events."

Working with a Lockheed contractor, Middlehurst was the co-author on a LION report for Apollo 10, dated a month before the Apollo 11 mission. A missing link in the operation was that Apollo astronauts had not been trained to look for, spot and observe brief events on the Moon, a shortcoming that needs to be addressed as future lunar missions are planned.

The Smithsonian Institution's Center for Short-Lived Phenomena was the hub. Observers around the world would call the Center, which would then alert the LION desk in the Science Support Room at the Manned Spaceflight Center in Houston. For Apollo 10, the network had 46 observing points in fifteen states, 130 observing points in thirty-one countries, and something like 300 individuals looking at the Moon through everything from 2.5-inch refractors to a hundred-inch reflector. Fifty-four alerts were received by the Smithsonian, and nineteen "positive reports" were relayed to the Science Support Room. Aristarchus featured heavily in them. Despite the smooth functioning of the network, only two requests were sent from the back room to Mission Control. One was denied because the astronauts were asleep. The other resulted in Apollo 10 reporting "'no sign of activity in Aristarchus."

More would be heard from capsules above the Moon, however. A 1970 Project LION report—by now Winifred Cameron was monitoring the program—noted that the Apollo 11 crew observed that Aristarchus seemed brighter than expected. The event was rather more dramatic than the report suggests. German and New Zealand astronomers had noticed a glow at Aristarchus, and word was relayed to Houston's flight-control center.

"We've got an observation you can make if you have some time up there," Houston radioed to the Command Module Columbia. "There's been some lunar transient events reported in the vicinity of Aristarchus."

Command Module pilot Michael Collins responded quickly, "Hey, Houston, I'm looking north up toward Aristarchus now . . . there's an area that is considerably more illuminated than the surrounding area. It . . . seems to have a slight amount of fluorescence." Aldrin suggested it might be backscattered light—just normal light viewed at an angle. Arlin Crotts analyzed this possibility and found it "incorrect" but not everyone agrees. But it was news enough that the July 25, 1969, issue of *Time*—"Man on the Moon"—included mention of the sighting.

There would come more tantalizing hints. There were peer-reviewed findings of light over Aristarchus in 1967. That same year, Pluto discoverer Clyde Tombaugh and another observer saw a red event at Aristarchus with New Mexico State's twelve-inch Cassegrain reflector under what were described as "excellent" conditions. Tombaugh was convinced it was real. Crew members of both Apollo 16 and Apollo 17 saw bright flashes on the Moon, though whether they were meteorite strikes or cosmic rays causing bright spots in their optic nerves is not clear. In 1983 Torricelli B was observed by multiple competent observers to host transient glows. In a 2000 paper, French scientists reported brightening in the huge, eastern crater Langrenus, observations that Bob O'Connell says "should give even the most hard-core TLP skeptic pause—but it doesn't!" TLP doubter Thomas Dobbins admits he doesn't "know what to make of" the Langrenus report. The French team was observing the crater through polarized light. Someone looking through a telescope in ordinary light probably would not have seen a thing. The lead researcher told O'Connell that he thinks with a large enough telescope the brightness his team detected with polarized light in Langrenus could be visible. But a repeat illumination—when lighting conditions are the same as the original reported—was done several years after the observations and it suggested that one observation was reflected light not a dust or outgassing event. Two other observations from the Langrenus study have not been subjected to a repeat illumination.

There are more reports in the scientific literature. TLP have never fully gone away.

Middlehurst kept an eye on them until she died. The astronomer and WWII ambulance driver died of complications following a 1995 stroke. She was eighty years old. She went to the grave believing the Moon still had secrets.

Cameron agreed. Even in retirement, beginning in 1984, she continued to work on her catalog and analysis of TLP. (Wini seems to have been less affected by the scorn that Barbara Middlehurst endured.) She finished the study in 1994 with 2,254 reports, starting in 557 A.D. and including 645 that were "independently confirmed and/or permanently recorded" via photography, spectroscopy or other methods. Of these, Cameron wrote, 448 could not be attributed to the atmosphere or the telescope. She was still doubtful there is much external influence on TLP from the Earth's gravity, magnetic field, or the Sun's activity. It was Cameron who suggested a program of repeated observation to examine features under similar viewing conditions as pertained in the historical record. Anthony Cook would take this up with a vengeance.

Wini had been alone for years. Bob had died suddenly in 1972. A smoker of three to six packs a day, Bob's struggles included not only depression but drinking. Just seven years after leaving NASA, he died at age forty-eight. Upon retirement, Cameron settled at 200 Rojo Drive, Sedona, Arizona, where she'd ride her palomino Sundust from her two-acre red rock home that she called La Ranchita del Luna. She'd watch Nelson Eddy and Jeanette McDonald movies with friends or sit in front of a large brick fireplace reading Zane Grey. Above the blue patterned sofa was a black-and-white artwork showing a face-on spiral galaxy and the Sombrero Galaxy, a print that had appeared in the movie "Abbott and Costello Go to Mars." She kept a Moon globe on top of her yellow fridge and kept time with a clock that showed the Moon's phases and, inside her copy of a Moon atlas, she still had handwritten notes regarding transient lunar phenomenon.

As I handled them—scribbles from a forgotten history—I thought again about the word *transient*. In its roots, it means "across," a going over, a journey. Perhaps it is only in middle-age or later that one sees the crossings most clearly, for they have . . . accumulated. They are more evident than passing lights on the Moon. Wini Cameron and Barbara Middlehurst lived their lives, and, as scientists, they understood what H. G. Wells once wrote: That we are "just a film, just a thin zone of reflection halfway in the scale of size between . . . electrons and the stars." Yet that thin zone is thick with neglected epics.

When Cameron stepped outside the red brick house with her dog Ursa Minor, there was sky, lots of it, coming down to press against the juniper slopes of Courthouse Butte and lifting to the ecliptic and the zenith then down again to the juniper needles. She could walk to a small woodshed by a chain-link fence where she had improvised an observatory.

In her eighties, like Middlehurst, Cameron climbed a volcano, Mount Etna. She moved to Florida in 2010 and died of kidney failure on March 29, 2016. Friends remembered "Ms. Wini" as "gracious, elegant, a wonderful conversationalist and good listener" with "enthusiasm for life and sense of humor." Her daughter Sheri says, voice catching, "She was a great mother. I miss her."

As she faced her last days, Cameron could take professional comfort in the interests of other scientists, like Peter Schultz and Arlin Crotts and Anthony Cook, all of whom took up the study of transient lunar phenomenon despite a nearly devastating blow in 1999.

The Moon was six days past full the day she died. Night was falling on the Sea of Tranquility. The photos on the American Astronomical Society obituary page show Winifred smiling by a scientific adding machine between two pictures of the Moon, and there's one with a young Robert—handsome, dark-haired, in an overcoat—grinning at her, with her blonde hair cascading down a stylish coat. They are beside a telescope, that device which catches light from a distance and therefore from the past.

In 1999 two vociferous skeptics published an article in *Sky & Telescope* that profoundly affected perceptions of TLP among astronomers of all ranks. The well-known and accomplished duo of William Sheehan and Thomas Dobbins wrote a piece titled "The TLP Myth: A Brief for the Prosecution," in which they worked to take apart the major TLP cases of the 1950s and 1960s. Dinsmore Alter's filtered photographs of Alphonsus, they wrote, just showed generalized blurring in the entire area and could not be nailed down to any kind of gas or dust or eruption from rilles in Alphonsus. It was an effect of the filter, though Arlin Crotts would later disagree. It *is* difficult to detect anything suspect from the images.

N. A. Kozyrev's spectra of Alphonsus could be attributed more reasonably to "an artifact of faulty [telescope] guiding." Of course, the historian Ronald Doel would go on to note that the skeptical Gerard Kuiper was convinced by the spectra upon seeing the plates in person. "Again and again in letters to colleagues," writes Doel, "Kuiper declared that he 'had no doubt left that the spectra are genuine.'" But Kozyrev himself could not be trusted, Sheehan and Dobbins wrote, damaged as he was from years of Soviet imprisonment. Kozyrev had made other observational claims that were indeed suspect and proffered a theory of time in which, dependent on circumstances, could literally be thick or thin.

As to the Lowell Observatory 1963 Aristarchus sightings, Dobbins and Sheehan argued that the telescope's inherent color distortions as well as unsteady air were responsible. Years later, they again argued that Greenacre's "recourse to such a high magnification under poor seeing conditions certainly belies a lack of observing expertise." They stress that the Moon was relatively low in the sky during the first night of observations and that the seeing was truly unsteady. The image, Dobbins told me, was "really boiling." They speak from experience: "The authors have observed the Moon through the 24-inch Clark," and its proclivity to distorted colors casts serious doubts on the 1963 reports, they argue.

The 1999 article's influence was such that it took a Herculean effort by Bob O'Connell and Anthony Cook to argue against its assertions about the Lowell reports. They compiled a massive set of findings and documents, devoting an entire website to the controversial events.

Against the contention that the 1963 sightings were caused by a turbulent atmosphere working as a kind of prism, the Air Force report suggests otherwise: During the second observation, for example, the report says Edward Barr "immediately started to scan bright areas north and south of Aristarchus . . . to determine whether a similar color was noticeable in other places. If he had found this to be so, he would have simply attributed it to atmospheric dispersion. Like the October sighting, this was not the case as no color could be detected anywhere else." As well, the observers said that seeing improved from its initial poor quality on that fateful October night. The historical record also makes clear that the Lowell observers knew their instrument intimately and that they felt it was almost impossible that the observations were caused by refractor-induced colors. Greenacre understood the role of color filters and how they affected the Lowell refractor. After taking filters off during the observations, any lens-induced spurious color the filter was suppressing should have shown up elsewhere, but it did not. As well, there were no unusual weather patterns, such as an inversion, that could account for the colors.

Not long after the original sightings, Lowell director John Hall told *Sky & Telescope*, "Greenacre is a very cautious observer. He had long been skeptical of reported changes on the lunar surface . . . [Greenacre's supervisor] has stated that he could not recall that Greenacre had ever plotted a lunar surface feature not later confirmed by some other observer." Ewen Whitaker met Greenacre not long after the sightings and would say that he "struck all of us as . . . calm and level-headed . . . He had more hours observing the Moon than any of us did . . . and obviously knew all the phenomena caused by seeing and other atmosphere disturbances." Whitaker knew the Moon the way that the obsessive nineteenth-century selenographers did, so his endorsement is significant. Whitaker told O'Connell and Cook, "We concluded that the observations were genuine, but couldn't explain them."

What of the negative findings of the automated 1960s TLP survey at the Corralitos Observatory in New Mexico? Because Sheehan and Dobbins did not cite Winifred Cameron, who was still alive in 1999, they did not include her concern that human observers were fatigued by the set-up and that the sensitivity of the instrument may have been a factor in its failure to find TLP. The duo also focused on the most obsessive and least reliable of TLP reports, those of Maryland doctor James Bartlett, which, of course, Cameron had dismissed on account of weather patterns. Sheehan and Dobbins also failed to note that instrumental readings showed evidence of outgassing during Apollo missions and they ignored the Apollo 11 sighting at Aristarchus.

After reviewing the thick research that Cook and O'Connell had compiled on the Greenacre incidents, Dobbins argued again in 2013 that the
Lowell refractor and unsteady air had produced the colors. He and Sheehan
have pointed out that Aristarchus is prone to produce false colors because it
is so bright. And there was new evidence for the skeptics' case. A 2011 study
from highly experienced lunar observers found that during illuminations
similar to those in 1963 a red color was found near Aristarchus but one
caused by the atmosphere. Ironically, one of the authors of that study was
Jim Phillips, who had previously criticized Dobbins over the 1999 broad-
side. Charles Wood tells me that it's been found that under "rare conditions
when low angle solar rays illuminate crater floors where rims are low...[a]
temporary bright area" is seen. This could account for some TLP but not
the Aristarchus red events.

In an unpublished rebuttal, O'Connell suggests that the 2011 report has
shortcomings, including the use of a much-smaller telescope than the twenty-
four-inch Clark. O'Connell found the study misidentified locations for the
original TLP and did not find color at other areas where spurious colors should
have been visible. Given that the 1963 observers should have seen false colors
all over the Moon—if the telescope's optics were that insufficient—O'Connell
asks, "Why did the observers mistakenly interpret spurious colors which
only appeared at a few discrete locations as possible TLP just twice in eight
years and both times on the Aristarchus Plateau?" It's a good question.

Bob O'Connell still isn't sure what Greenacre saw. After the Lunar Reconnaissance Orbiter photographed the Aristarchus Plateau in detail, O'Connell
"intensely scrutinized the imagery." He saw "no evidence of recent geologic
disturbance at any of the sites reported by Greenacre and others. So I don't
think . . . the observations were due to lunar out-gassing/dust lofting." Inter-
estingly, skeptic Charles Wood wrote after his own review of LRO imagery
that he had found "various veneers of dark material but all contain numerous
small craters that attest to the deposits being considerably more than fifty
years old. If some physical event on the Moon did cause the TLP it left no
detectable deposit. So it was less likely to be a lava flow or ash eruption than

escape of gases." So this does leave open the possibility of outgassing. But Wood sounds a note of caution:

> Four different events scattered across tens of kilometers seems unlikely, although an ad hoc explanation can be invented. Additionally, the Lowell folks saw another TLP at Aristarchus at about the same phase during the next lunation, which is suspiciously consistent with the TLPs being visible only at certain observing conditions—i.e., an artifact of lighting. But someone will argue that a monthly recurrence might be expected if gases only can escape during monthly periods of maximum tidal stresses.

Cook says the lighting conditions were different by several degrees for the second set of events, and a 1965 article noted that the Greenacre sightings took place soon after a major solar flare.

O'Connell isn't arguing for a lunar origin for the Lowell reports anymore. "At this point, I am fully in agreement with Tom [Dobbins] that the observers did not observe something of lunar . . . origin. But what they saw? I have no idea." He admits, however, "I certainly could have missed something." O'Connell, who has become close friends with Dobbins despite their TLP disagreements, says the explanations that the telescope and atmosphere caused the sightings "all fall flat." Dobbins still disagrees.

O'Connell put intense effort into studying the strange events of the 1963 Aristarchus sightings, to the point that he collapsed from exhaustion after making the work public with Anthony Cook. While he is happy to have recuperated the reputation of James Greenacre, he tells me wistfully, "I am still haunted to this day as to what the Lowell Observers saw."

✿

Columbia University's Arlin Crotts had studied everything from quasars to galaxies when he became fascinated with lunar science, writing some long articles and culminating in the 2014 book *The New Moon: Water, Explora-tion, and Future Habitation,* a wide-ranging volume that captures every-thing from an early Adrienne Rich poem about lunar settlers to the politics of human spaceflight. The book appeared just one year before his untimely death. When Crotts turned his attention to transient lunar phenomena, the effect was nontrivial. Anthony Cook says of Crotts's works: "These have really reinvigorated TLP research and introduced new ideas and analysis." Crotts heaped praise on Barbara Middlehurst and Winifred Cameron, saying it was Cameron "who started me thinking about these issues."

He admitted that "TLPs have a bad reputation in select audiences," noting that their one-off nature is the great obstacle to studying them effectively. "Establishing their consistent behavior," he wrote, "is required to assess their applicability in understanding physical processes on the Moon." Crotts did not pretend to understand the precise subsurface mechanisms that release gasses like radon to sweep up and exhale from the surface, but the correla-tion between radon outgassing and TLP sites was, he contended, "stunning." Indeed, researchers writing in 2005 on the outgassing found by the Lunar Prospector probe noted that places where TLP have been commonly reported "were also . . . sites of radon release events."

Gasses, mixed with radon and/or argon, can vent from the Moon and send dust into the sky where it can become fluid-like and discharge static electricity. Dust clouds scatter light and create red and blue colors. The gasses themselves can emit light. In 1977, taking these factors into account, J. E. Geake and A. A. Mills set out the case that gas leaks can account for TLP. Lab experiments have also shown that lunar material can luminesce due to proton bombardment, which might explain some findings of seeing luminescence on the Moon, possibly due to solar flares, though the prob-ability of terrestrial visibility isn't high. Still, the Moon also does feature levitating dust. "Dust does become charged, and repels, levitating above the surface on the Moon," Cook tells me. "It moves from night to day on the terminator at sunrise and back again at sunset. This has been simu-lated in vacuum chambers in labs. It does similar things on shadow edges."

Understanding the presence of electrically charged dust on the Moon is another need before we can safely survive in some areas because lunar dust is sharp-edged, omnipresent, and toxic.

As to the most famous site, the Aristarchus Plateau, Crotts was emphatic. It's "responsible for undeniably objective transient anomalies associated with lunar outgassing." In fact, the surface material of the rilles there "either erases or cannot sustain cratering," which indicates geologically recent activity, though not all agree with this assertion. Aristarchus is in any case a special place, given that it's in the mare but the plateau is a bit like the lunar highlands of the south. Nonetheless, a report that used Clementine imagery purporting to find evidence of TLP in the volcanic valley that cuts through the Aristarchus Plateau was quickly retracted. There is still no widely accepted evidence of TLP at Aristarchus.

Are there factors contributing to increased outgassing or other triggers for alleged TLP? Could external and/or internal stresses play a role? The suggestion that outgassing is triggered by lunar perigee—when the Moon is closest to the Earth and thus subjected to more intense gravitational strain—dates to the Japanese astronomer S. Miyamoto in 1960. In 1977, Middlehurst was arguing for the link between Moonquakes triggering gas releases from subsurface channels and, far less likely, that such activity formed some craters. In 2014, Crotts said tidal stress and TLP correlation was not strong but did say that the correlation between "deep moonquakes" and TLP "is amazing." A 2020 study found correlations between moon-quakes, outgassing and TLP.

Although Crotts was secure enough in his profession and career by the time he turned his attention to TLP, he could be hesitant about the work. Bob O'Connell remembers Crotts as "kind, pleasant and friendly." But by then Crotts knew that O'Connell and TLP skeptic Tom Dobbins were good friends "and that made him suspicious of me despite the fact he knew I was working with" TLP researcher Anthony Cook. Crotts didn't even want O'Connell to take a picture of a poster—in a public setting—at a science conference. Of that presentation, O'Connell recalls "being underwhelmed . . . with a few photos of exceedingly small areas on the Moon (square meters of lunar surface) that he suggested were evidence of TLP. I felt he was trying to justify his position with very little hard evidence."

Crotts's last graduate student, the computational astrophysicist Cameron Hummels, remembers his mentor as being "always really in love with the Solar System." By the time Crotts was senior enough to tackle TLP, he and Hummels put together modeling that showed radon buildups could "burp" so hard that the dust goes "ballistic for minutes or an hour or so. This largely holds together for the time scales and size scales and . . . reported transient TLP."

The traditional lunar and planetary science community was mostly disinterested or resistant to Crotts's work, however. Hummels recalls, "We faced a lot of that." So much so that the papers on TLP were published in *The Astrophysical Journal* not the leading planetary science outlet, *Icarus*. "There's quite a break between the planetary science community and the traditional astrophysics community," Hummels explains.

As with Middlehurst and Cameron, Crotts "just went to the beat of his own drummer," Hummels tells me. It wasn't always easy. Crotts was eccentric and "a little bit socially awkward," but he also "came up with a lot of original ideas that I don't think can be said about the members of the scientific establishment. You can be at risk at rubbing people the wrong way. Sometimes you'll be wrong." Crotts's former undergraduate Robert Morehead said almost the same thing to me, calling his former mentor "a quirky personality that could sometimes rub people the wrong way." But as with Hummels, he praised Crotts's creativity and support. "He was really good to me [and] gave me good advice. I had no idea what I was doing!" Crotts attended Morehead's graduation in full regalia and even gave his student a graduation present. "He definitely was looking out for me."

And he has at least one enduring legacy: Columbia University's Arlin Crotts Radical Hypothesis Lecture Series.

Contemporary work has shown actual and recent tectonic activity on our neighbor, suggesting that its core is still churning and cooling. In *Nature*, in 2006, Peter Schultz, Matthew Staid, and Carlé Pieters wrote a short but significant article about lunar geologic change. They note that there are features on the Moon that show "patches of the lunar regolith" that were "recently removed" and "perhaps are still forming today." These are the irregular mare patches, and one is called Ina, and it looks all the world like a pumice or sponge you'd hang in your shower. Ina is located in Lacus Felicitatis, the Lake of Happiness, north of Mare Vaporum near the center of the nearside. It's the strange—and relatively fresh—terrain around the IMPs that caught the eye of Peter Schultz, who used Lunar Orbiter photos to discern the features back in the 1970s. Years later, in 2006, Schultz and his colleagues had spectral data to back up the newness of these regions.

"This should be a destination because it may be possible to monitor episodic releases of gas from the deep interior, Schultz says, who adds, however, that "even if looking, I suspect that an amateur or professional astronomer would not use the proper lighting or filters to capture the event…not to mention the patience to sit on one small feature."

Ina is not alone. There are similar patches elsewhere—including one near . . . Aristarchus. What could cause such a pumicelike appearance? Outgassing.

Then there are the lobate scarps, which are, as Schultz explains, "scarps caused by active uplift (or down dropping) on one side." He again used Lunar Orbiter imagery to discern these recently formed features and discussed them in his 1976 book *Moon Morphology*. They were re-confirmed by the Lunar Reconnaissance Orbiter. LRO in fact is discovering hundreds of these features, more than three thousand of them, according to Arizona State University researchers. These lobate scarps are so young that new ones could be forming today. Some are only a yard or so high, though most are about forty yards high and present a steep face. The formation process could also release gasses from beneath the surface. A 2019 study concluded that "the proximity of moonquakes to the young thrust faults together with evidence of regolith disturbance and boulder movements on and near the fault scarps strongly suggest the Moon is tectonically active."

More instrumentation—and human exploration—can help determine how these processes occur and whether and how they might manifest themselves as colors or obscurations. As Crotts argued, tapping into released lunar gasses will also give us a pristine picture of the Moon's interior.

Meanwhile, TLP skeptic Chuck Wood remains just that. He dismisses Crotts's statistics because "it is virtually impossible to correct for observer bias . . . I am ready to close the book on TLPs, let them rest in peace—RIP." Wood characterizes the legacy of historical TLP reports as "tens of thousands of observer hours spent looking for TLPs" all wasted while they "failed to discover impact basins...I bemoan our fascination with the improbable and unlikely." Tom Dobbins has even stronger words: "I regard Crotts as a charlatan. He repeatedly claimed to have imaged TLP . . . but never shared the data and was evasive when O'Connell gently prodded him for details at a conference." I have tried to track down the data from those observatories but to no avail. "I was rather incensed," Dobbins tells me, when Crotts dismissed his anti-TLP case because it had not appeared in a peer-reviewed journal. (That is a bit of an overstatement, I think, as Crotts presents arguments against those that Sheehan and Dobbins mounted in their important 1999 article.)

Dobbins also tells me that his reputation as "an arch skeptic about TLPs" has been blown out of proportion. "I am certainly not a visceral naysayer and regard many anomalous observations as fascinating and worthy of re-examination. Over the years my work has often involved such subject matter and has lent credibility to observations that were dismissed by skeptics." Dobbins's work on as-yet-unexplained flares on Mars is, for example, just one part of his important contributions to solar system studies in addition to his well-respected historical research. His perspectives justly carry significant weight.

Dobbins's own journey includes spending "a considerable amount of time into the 1990s looking for TLPs." A research chemist, Dobbins's fascination with the solar system began in childhood with an illustration in *The Golden Book of the Moon* showing George Darwin's fission-theory of the Moon's birth.

Living in light-polluted Cleveland and armed with a good telescope starting at age six, Dobbins soon owned a telescope-making manual by lunar denizens Patrick Moore and H.P. Wilkins. He was hooked. Before the advent of digital cameras and the web, Dobbins, like others, knew that visual observers could detect more detail on the Moon than professional instruments. "There was the notion," he says, "and I don't think in my case this was some inchoate idea . . . it was an explicit idea . . . that a visual observer with a decent scope was really at the cutting edge" of solar system astronomy.

That cutting edge did not include seeing TLPs. "I know vacuums and distillations," he says. "There is no physical phenomenon that I know of that would in the known lunar environment that would give rise to reddish glows." Does that mean outgassing doesn't happen? Not necessarily, he notes. The question is what, if anything, would be *visible* for visual observers on Earth. Today "lots of people are imaging the Moon with very decent telescopes and resolving sub-kilometer detail," Dobbins points out. As amateur equipment has grown increasingly sophisticated, with imagery rivaling that of the Lunar Orbiters, reports of TLP "all but disappear." Crotts wondered if the increase in imaging might mean that visual observations—which are sensitive to color—have fallen to the wayside. In any case, he also recognized where the burden of proof fell: "The onus of the argument must burden those who would convince us that TLPs are real."

On that, perhaps, all would agree.

Meanwhile, the arcane field of transient lunar phenomena will not rest in peace.

As Peter Schultz jokes, "The Moon is still passing gas, so to speak."

The vigil continues. Hakan Kayal, a professor of space technology at Julius-Maximilians-Universität Würzburg in Bavaria, Germany, has a small telescope looking for TLP and is developing software to detect events on the Moon, thus eliminating any expectation bias. He also hopes to utilize a larger telescope at some point. Events will be compared to European Space Agency telescope observations, and the ultimate hope is to use a similar system on a lunar satellite. As for now, he tells me, "We have no new insights on TLP."

Arlin Crotts had set up a similar facility in Chile and at Columbia, and his friend Paul Hickson, now at the University of British Columbia, says the Chile scope "autonomously recorded images of the entire Moon several times per minute." Robert Morehead, now a professor at Pennsylvania State University, helped set up and run the telescope on the roof at Columbia, and, while he was not part of data analysis, recalls that processing imagery was hard because of the constant change in lighting.

Bob O'Connell, as an amateur astronomer, has found that more professionals are "open to the possibility [that] TLP [are] real, objective lunar phenomena" though the area remains "treacherous." He credits his own interest in TLP to reading Arlin Crotts's work.

Indeed, there's yet another possible twist in the search for TLP causes. "One of my grad students and I actually did impact experiments at the Apollo-era NASA Ames Vertical Gun Range," Peter Schultz tells me. They "captured the nature of the flash, photometrically and spectrally. We found that the 'flash' even at laboratory scales (at much lower speed) was much longer than expected and could be explained by the cooling vapor plume." NASA's high-velocity projectile used on a comet during the Deep Impact missions showed the same kind of result. "In other words, some short-lived TLPs (less than 2-3 seconds) could be attributed to occasional large impacts, not gas release."

Led by the University of Aberystwyth's Anthony Cook, efforts are ongoing to clean up the massive number of historical TLP reports in order to find those that withstand initial scrutiny. The work is tedious and valuable. Cook and volunteers observe those repeat illuminations when the Moon and Earth and Sun are aligned just as they would have been for a prior TLP report. This is more than just a monthly matter because of the Moon's very complicated motions. Sometimes it takes years before a surface feature is placed for observation exactly as it had been once before.

"I just methodically go through repeat illumination observations sent in each month and use these to eliminate 'a few' past TLP," Cook explains to me. "Rather than take on an active TLP hunting role, which has generated 3,000 reports in the past, with very little analysis of these, we concentrate on attempting to disprove past reports by asking observers to visually describe/sketch/image features" when conditions are right. "Any images provided can be used to model atmospheric spectral dispersion, chromatic aberration [and] poor seeing conditions . . . to see if these tricked past observers into thinking they saw a TLP."

These repeat sightings eliminate a lot of noise from the data and have helped constrain what physical factors may or may not influence TLP events. Cook has found that less reliable TLPs spike in the northern hemisphere summer—possibly a function of atmospheric distortion, since the Moon is lower in the sky. Contrary to Cameron, who suggested there was more TLP at lunar sunrise or sunset, reports cluster in places that are experiencing lunar mid-day. Cook also suggests that the contention held by Crotts and others—that TLP cluster around sites along the edges of maria—does not hold but quickly adds, "To be honest its rather difficult to define where a mare boundary is."

Tony Cook is the stalwart in TLP studies today. Inspired by Patrick Moore, Cook fell under the sway of the Moon during the Apollo missions, got excited by looking for TLP, then "started taking a more pragmatic view. Though to be honest [I] had to give up much observing for a number of years till my degrees were done and dusted."

With an international Moon revival underway that includes Chinese projects and NASA's planned Artemis missions, the time seems right for lunar orbital missions to scan for outgassing and understand better their nature and what they reveal of the Moon's interior, especially the nature and possible visibility of electrically charged dust. Indeed, interest in TLP now seems keen among Chinese researchers. Barbara Middlehurst's contention that the Moon still has secrets remains true today, though they may not be what she thought. Yet perhaps some secrets will be revealed by today's work in the obscure field of transient lunar phenomenon and perhaps some will be revealed when astronauts walk, finally, on and near the strange Aristarchus Plateau where, decades ago, an Apollo mission might have landed and where, for centuries, humans have claimed to see mysterious colors on a world that is, like the humans who watch it, not so black-and-white.

CHAPTER TEN

A Meaning

a light beyond our lights, our lives—perhaps
a meaning to us
O, a meaning!
            —Archibald MacLeish, "Voyage to the Moon"

There are many moons we cast upon the one. The Romantic Moon of bygone sonnets, the melancholy Moon of nineteenth-century nocturnes, the symbolic Moon of love song after love song after love song. We watch the shadows of pine branches cast by the Moon, and Basho looks and Juliet sighs. Ralph Blakelock paints a golden disk in the cloudy dark, and Frank Sinatra sings. The Moon still lights prayers. It is for everyone rich with meanings. In the 1960s, before humans put boots into lunar dust, the poet May Swenson asked, "Dare we land upon a dream?"

We did. And we landed on a surface. We landed on dust that smells of burnt gunpowder. A real place, as real as the terrain Galileo inferred from light and dark more than four centuries ago. Poets and lovers and landscapists, however, need not worry. The Moon rises and sets, waxes and wanes, and we are free to allegorize and harmonize all we want.

In her poem "Landing on the Moon," Swenson wrote about Apollo's prelude—the automated Surveyor craft that touched down before the astronauts did: "A probe reached down / and struck some nerve in us." She wondered, "Can flesh rub with symbol?" Mute equipment and footprints answered the questions. A nerve was touched. Even now—more than a half century after the Apollo landings and only a short time, at this writing, before humans will land again—even now, as the Moons of song and painting and religion abide, we wonder: What did Apollo mean? What might the next steps, whether they are giant leaps or not?

In an obscure but engaging book called *Apollo 11*, published in 1969 by ABC's radio division, commentator Mort Crim wondered before the famous voyage, "What will the journey mean? . . . What will spaceage man [*sic*] think of himself . . . What will spaceage man demand...Will he [*sic*] settle for such achievements, forgetting that life is more than computers and rockets? No doubt man will continue to make great progress. But will man progress?" For Crim, the Saturn V was either "an embryo of triumph or tragedy." The same was true for Norman Mailer in his brilliant, often egotistical account of Apollo 11, *Of a Fire on the Moon*. Mailer cast the journey in stark terms of good or evil. He saw nuance only in character not in quest, as though they were somehow set against each other.

In Archibald MacLeish's masterfully ambiguous poem, "Voyage to the Moon," which was published on the front page of the *New York Times* to commemorate the July 1969 Apollo 11 lunar landing, the poet describes how the Moon has been a "wonder" and "unattainable" throughout history: "A longing past the reach of longing." Deploying Biblical language and cadences—"Three days and three nights, we journeyed / Steered by farthest stars . . ."—MacLeish then gives us the view from the Moon: an Earthrise, another meaning, another form of wonder. We had landed on the Moon and looked back upon ourselves.

As Apollo unfolded, public opinion was divided, more than we remember. Most creatives, unlike MacLeish, were skeptical. Anne Sexton compared the Moon landings to macho men barging into barracks. W. H. Auden thought it all "a phallic triumph." James Dickey wrote Apollo poems that appeared in *Life* magazine, and, by the end, he was as sour on it as Mailer. Many politicians, most engineers and a chorus of science-fiction writers were enthusiastic. The landings were akin to the lumbering first wriggle-steps of fish onto land, they thought. And President Nixon, who wanted to reflect in the glory of a program started by his dead rival, said upon meeting the Apollo 11 crew after splashdown: "This is the greatest week in the history of the world since the Creation." Ralph Abernathy likely disagreed. The reverend had led the Poor People's Campaign to Florida for the launch to protest using money on space instead of social justice. Perhaps Robert Hayden hit the right note in his poem called "Astronauts": "Why are we troubled? / What do we ask of these men? / What do we ask of ourselves?"

As a child, what I needed from the voyages to the Moon were things a child needed: escape, adventure, imaginary skill. What do I need from them now? What do they light in my life and other lives? What meanings, disappointments, what hopes? What have the Apollo voyages given us and, in that gift or its lack, what suggestions, if any, for tomorrow's trajectories as NASA's Artemis program gets underway and other nations plan to land there? I do know that my Apollo Moon is no longer a child's. I know too much about a place called the Mittelwerk for that to be the case.

During the fiftieth anniversary celebrations in 2019 of Apollo 11's landing, the historian Roger Launius tried to capture some firm views. In *Apollo's Legacy: Perspectives on the Moon Landings* he wrote that the dominant ways of seeing Apollo's meaning are as a triumph of national will and technological expertise or as a waste of money or even a hoax. Even if one agrees with Auden, it's clear Apollo was a triumph of national will and technological expertise. In eight years, the United States went from never having sent a human into space to landing humans on another world. The landings still inspire me. And the technological consequences do too—the first serious miniaturization of computers occurred due to volume constraints in the lunar landing module. Your cell phone has its ancestors on the Moon.

There is a deeper stratum. MacLeish and Hayden are pointing to it, somehow.

Launius avoids, however, as do many, in seeking the meaning of the Moon landings through another lens. It obsesses me: the Mittelwerk, the criminal origin of advanced rocketry, an origin I was unaware of as a child and only learned of in my thirties when I visited a museum in Kansas. Ever since, I've been haunted by Wernher von Braun and how we remember him, that handsome and brilliant and opportunistic engineer who built missiles for the Third Reich, then for America, including the massive Saturn V that sent humans to the Moon. We have largely failed to reckon with this meaning of Apollo. Yet we have to. Because we cannot build a humane future in space if we forget its inhumane origin.

❁

I've come one February with this uncertain cargo to see Florida's Cape Canaveral and the Kennedy Space Center, where, despite years of fascination with all things space, I have never traveled, where launch gantries, lightning towers, and assembly hangers form a sparse and cryptic skyline across the grass and water prairie, jutting glyphs that, while made legible by knowledge and distance, have constructed no prophecy our planet has embraced. The tropical sky is as hazy with salt and vapor as the gantries and hangars are hard-edged with machined precision, their logic wavering in a dream.

The Space Coast's road-fill-providing ditches, its overlit discount stores and mid-century resorts, whose battered facades have been redressed with corporate logos, and the strip clubs and the Jesus-themed coffee shops, the abandoned boat storage sheds and grubby lawns, gravel pull-outs for pickups spilling anglers, and cotton candy ground into the soles of work boots and flip-flops *Made in China*, all sprawl around me in some temperate and tacky elaboration that it almost seems a practical joke or a piece of performance art: This is the place where we say *ad astra*?

When I arrived in Cocoa Beach, passing parking lots hacked out of Florida's Max Ernst jungle, after roadside osprey and strip-mall egrets, after chunky boars rooting about drainage ponds rippled with vestigial alligators, after miles of stumpy palms and NASA's titanic Vehicle Assembly Building looming like a protocol, beyond the cruise ships docked at Port Canaveral, my room at the Sea Aire Motel smelled of sticky bleach and salt, like the cheap rooms of beach vacations on which my mother would take me and my sister. *Per aspera.*

Decades ago in Cocoa Beach, at George's Steak House, diners were directed to restrooms labeled "Astronets" and "Astronauts." There was the unironically named Celestial Trailer Court and the Astro-Dine Outer Space Eat-In, as well as motels named Sea Missile, Vanguard, and Polaris. The Moon Hut restaurant is closed, and the Satellite Motel with its giant globe-sign spikey with two faux satellites is gone. The Missile Town Shopping Center survives as a vision yet-to-built in a vintage photograph with its billboard advertising it is "Coming Soon." That retro Space Age is all gone.

I've brought along my first-edition copy of J. G. Ballard's *Memories of the Space Age*, in which the Cape is drained of time, in which the first exploratory forays lead to tragedy—strange ennui, corpses forever circling in metal capsules, a spreading plague of slowing moments, halting perceptions in which "nouns and verbs were separated by days" and "vowels were marked by the phases of the sun and moon." Our dream of space burns up upon reentry, ablation's fiery, entropic sigh. Ballard's future is the past returned, eons creeping up to choke time off altogether. The vision appeals to the catastrophist in me, the child who grew up at the tail end of duck-and-cover drills, and the adult who can imagine walking in reverie among the abandoned plazas of surrealist paintings.

Lightheaded with anticipation, I'm on a pilgrimage to time—mine, the Space Age's—and I walk along the edge of a future—New Space, the promised democratization of access to orbit and beyond made by, of all people, billionaires, and the global revival of lunar exploration, the missions that should return humans to the Moon in my lifetime. I'm a free-return trajectory, an odyssey come full circle: from my fondness for musty Funk & Wagnall's yearbooks in which silver-suited pilots rode rockets shedding ice in blurry photos—heroism as aesthetics not as motivation—to the fifty-something son whose chemist father is dead, whose childhood telescope sits by a desk, whose Moon is manifold, a teleidoscope made of facts and fancies and, lately, a recognition of mortality. I am almost weightless with zones of time, including this week's near-future, the dark beach from which I should witness a rocket launching a probe to study the Sun's gales of light. I have come to see that, to see a rocket fly, to erase the disappointment of the storm-cancelled launch I tried to witness in California. In the days ahead, I will stand before an ancient console of dials and toggle switches and see one knob labeled "decade selector." I'll move the decade selector back and forth. I always have.

Primeval sea to land. Neoteric land to sky. A wren chatters in grass beside the beach, pivots past the aerodynamic blades of agave piercing the foreground of this evening's Atlantic. The dream: sea to land to sky, hatched like a schematic egg in Room 8 of Cocoa Beach's old Sea Aire Motel, four tall, salt-crusted windows, two facing east to the ocean, the golden-brown-umber wooden panels a comfort to Wernher von Braun, who stayed here, in this very room, on his trips from the development facility in Huntsville, Alabama, to supervise work as NASA burgeoned.

Von Braun traveled with his pillow, like I do, and his secretary would have to call someone to fetch it after the absent-minded engineer invariably left it behind. All those bright, amoral dawns, hot coffee, the distance from World War II's V-2 tests at Peenemünde, where the oxygen plant was built to the proportions of a basilica, from that gray Baltic to Florida's green seas, from leering Hitler's V-2s to Cold War Redstones and Titans and soon, then, von Braun's Moon-bound Saturns, the orogeny of the rocketeer's time-drenched passion to cast our species into space, toward immortality, a dream that requisitioned pain and delivered it.

I am in his favorite room. Tonight I eat takeaway spaghetti at the white table, then walk a few paces to the empty beach where the Atlantic is a churning planet of molten agate and the Moon tosses its light like a reminder.

And to see yourself as a force of history is to be absolved from both pity and guilt.

—Robert Hughes

Origins blossom and recede like fire and smoke from an engine exhaust.

When does the Space Age begin—or begin to begin? With Icarus flying too close to the Sun? With Montgolfier and the first human flights—in hot-air balloons—over astonished crowds in France? With the Wright Brothers and those who followed with the rapid development of powered flight, higher and higher? Did the Space Age begin in the imagination of an obscure Russian teacher, Konstantin Tsiolkovsky, who foresaw rocket-propelled travel outside the atmosphere, and with the first crude mechanical rockets launched by the American Robert Goddard in the early twentieth century? Or did the Space Age begin in the imaginations of such writers as Jules Verne and H. G. Wells, each of whom sent humans to the Moon in their novels?

A closer source: the Nazi rocket launches during the last months of World War II, when Adolf Hitler ordered V-2 ballistic missiles, the first, to be launched against England and other targets. They were also the first to reach space. Those missiles were manufactured and assembled by concentration-camp labor with the direct knowledge and participation of Wernher von Braun, Arthur Rudolph and other members of the German research team, rocketeers that the US government captured in order to hasten American development of such vehicles and to keep them out of Soviet hands.

The V-2 slaves worked in an underground factory. The Space Age began in darkness.

I learned of this in the 1990s when I walked into the German Room, the first exhibition room of the Cosmosphere, an air and space museum in Hutchinson, Kansas, which was a day-trip from my then-home in Manhattan, Kansas. (In 2021, the Cosmosphere would renovate the exhibition, keeping intact information about the Germans and slave labor.) The German Room was a version of the Mittelwerk or "Central Works," the banal name for the factory where V-2s were manufactured before being deployed to launch. Hitler was in a frenzy to find some final way to turn the tide of impending defeat. V-2 attacks against London, Antwerp, and elsewhere killed nearly ten thousand people. They were not the only victims.

Elevated at stomach level was a platform on which a real V-2 rested, supine on artificial tracks, a simulacrum of the rails that ran through the ten-million-square-foot facility, which had been built beneath a mountain as protection from Allied bombers, whose raids already had damaged the test facility at Peenemünde. The effort was led by von Braun. When I stood with my hand hovering over rocket metal, I looked at a mannequin of a V-2 slave laborer, a George Segal-like man making missiles for his captors. There was a photo of one V-2 slave laborer, thin as Saturn's rings, a man whose hands may have made some part of this very missile, an original captured from the Nazis. About twenty thousand prisoners—mostly French and Eastern European— died of cold, starvation, executions, beatings, and overwork.

My hand landed on the metal, and I felt angry and stupid, betrayed by Walter Cronkite, by his enthusiasm that had been mine, by the explanations of orbits, the exclamations of attainment, the voices, the later, celebrated ghosts of American astronauts. Betrayed by transmission and ticker-tape and by von Braun, who led the development of NASA's Moon rocket a quarter of a century after World War II.

The museum had this quote on a display: "Pay no attention to the human cost," said Waffen-SS. General Hans Kammler, head of the forced labor for the V-2. "The work must go ahead and in the shortest time possible." Apologists for von Braun and his rocket team still stop there, putting the entire moral burden of this calamity on the Waffen-SS. It is, I have seen, a heinous revision of the facts, one that still has purchase in popular culture and even in places where NASA has visitor facilities.

Yet, as one museum label read:

> There is no doubt that von Braun and many of his rocket team were quite aware of the deaths occurring inside of the V-2 factories. From historical records, it appears that von Braun was so consumed by the concept of space flight that he was perfectly willing to shut his eyes to the genocide to realize his dreams . . . There is still active debate whether his actions would have classified him as a war criminal if his knowledge had not been so important to the Americans after the war.

ABOVE: The Copernicus crater as seen from Lunar Orbiter 2 in 1966. This oblique photo was hailed as the "picture of the century" and published on newspaper front pages around the world. The Lunar Orbiter Image Recovery Project later reprocessed the photo to the clarity seen here. BELOW: This is the Abri Blanchard bone that independent scholar Alexander Marshack said recorded lunar phases from 25,000 years ago. In many places, lunar time-keeping was a crucial middle ground between daily and seasonal variations. While Marshack's claims are not universally accepted, many artifacts from around the world demonstrate a sophisticated awareness of the Moon's phases and motions.

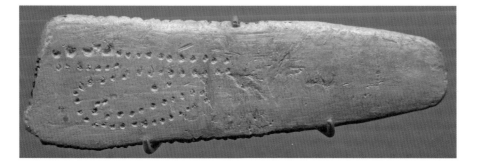

ABOVE: This Fremont culture petroglyph at Parowan Gap, Utah, is said by some archaeo-astronomers to represent "The Three Moons of Winter." Present-day Paiute say the rock imagery is the story of a famine. The narratives may not be mutually exclusive. LEFT: The title page for Henry Cornelius Agrippa's 16th book of magic, which includes a treasure trove of lunar folklore in the European tradition. It also includes a recipe for a ritual concoction with a cow's eye and dead frog to be performed under the Moon, which this author attempted.

HENRICI
COR· AGRIPPAE
ab Netthesheym,

DE OCCVLTA PHILO-
fophia LIB.III.

ITEM,

Spurius Liber de Cæremoniis Magicis,
qui Quartus Agrippæ habetur.

Quibus accefferunt,

Heptameron Petri de Abano.
Ratio compendiaria Magiæ naturalis, ex Plinio
defumpta.
Difputatio de Fafcinationibus.
Epiftola de Incantatione & Adiuratione, colliǥ
fufpenfione.
Iohannis Tritemij opufcula quædam huius argu-
menti.
Diuerfa diuinationum genera à quodam Anti-
quitatis ftudiofo collecta;

LVGDVNI, PER BE-
ringos fratres.

ABOVE: This painting shows Chang'e and her rabbit on the Moon, where she must live forever, according to a classic Chinese folktale. Her story is the inspiration for the Moon-focused Mid-Autumn Festival. BELOW: An Aztec depiction of a rabbit on the Moon from the Florentine Codex. Multiple cultures across the globe have seen animal shapes—and stories—on the nearside face of the Moon.

LEFT: Jan van Eyck's fifteenth-century painting, *The Crucifixion*, was Western culture's first realistic portrayal of the Moon's face. It is seen to the lower right of the sky behind the thief Gestas. BELOW: A 1636 portrait of Galileo by Justus Sustermans. Note the telescope in his right hand.

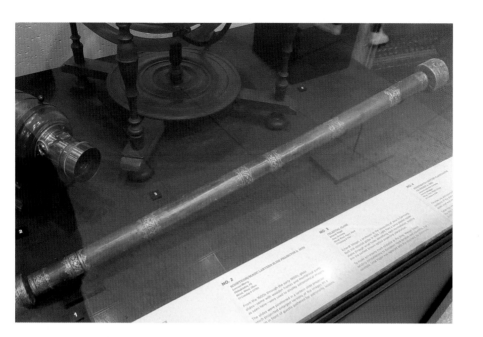

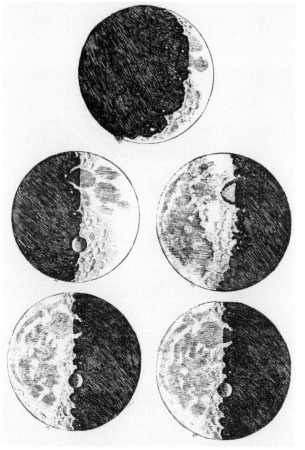

ABOVE: A replica in the Franklin Institute in Philadelphia of Galileo's telescope. BELOW: The revolutionary drawings of the magnified lunar surface made by Galileo as published in the 1610 book *The Starry Messenger*. The book revolutionized our conception of the Moon and was one of the origins of modern scientific thinking.

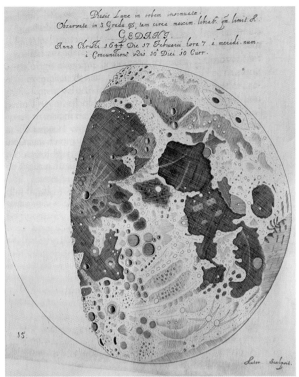

Moon maps of varied accuracy and quality—but almost always aesthetically pleasing—proliferated after the invention of the telescope. Here is a portion of a lunar phase map from the 1647 *Selenographia* of Hevelius, the first book devoted to the Moon.

Scenes from a lively Moon—as imagined by illustrator Benjamin Henry Day for an 1835 series of hoax newspaper articles that claimed Sir John Herschel had discovered unicorns, giant minerals, and bat-men on the Moon.

LEFT: American astronomer William Henry Pickering, the last great advocate for lunar life, who made multiple claims for having detected weather, vegetation, and insect migrations on the Moon well into the twentieth century.

RIGHT: One of William Pickering's illustrations that he claimed showed plant growth in a crater, what he called "the gardens of Eratosthenes."

A photograph of a photograph of a plaster-of-Paris model of the Moon, one of several extraordinary illustrations from James Nasmyth's and James Carpenter's 1903 edition of their 1874 book *The Moon Considered as a Planet, A World, and a Satellite.* The model shows the vast flat Plato crater and sharp shadows over its surface cast by the rim. The Alpine Valley runs through the lunar Alps. The valley is a graben, and in the middle, barely discernible is a rille that once ran with lava. The entire Alps range was uplifted into being like all such on the Moon—by a huge maria-creating impact. Another common lunar feature—wrinkle ridges—can be seen in the upper right in Mare Imbrium.

ABOVE: This Lunar Orbiter photograph of the crater Ukert, to the right, and Mare Vaporum in the background, clearly shows the raking texture of what nineteenth-century geologist Grove Karl Gilbert first called "Imbrium sculpture"—the after-effects of radial ejecta from the incredible impact that formed Mare Imbrium billions of years ago. BELOW: Tycho crater is one of the many stunning "complex" craters that can be seen with telescopes. This crater category, which includes Copernicus, features terraces, well-delineated rims, secondary crater chains, built-up material called a nimbus around the rim, and central mountains rising from the floor. They are formed by impact not, as so many believed until the Apollo era, by volcanism. Tycho is more than 50 miles wide, some three miles deep, with the central mountain peaks rising 5,000 feet high.

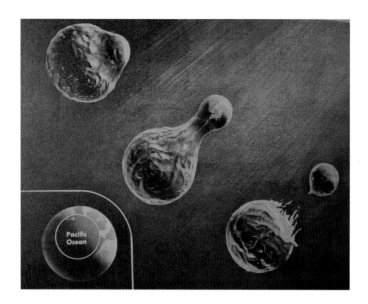

ABOVE: A vivid, vintage illustration of the "fission" theory of the Moon's birth, in which our companion was flung out of what is now the Pacific basin. Romantic as the idea was, it's been replaced: the Moon was formed when a Mars-sized primordial planet hit the early Earth. The Earth re-accreted, and the Moon was born. BELOW LEFT: Astronomer Barbara Middlehurst pioneered a more systematic approach to the study of "transient lunar phenomena" and clashed with her supervisor, the imposing planetary scientist Gerard Kuiper. BELOW RIGHT: One of the first women astronomers at NASA, Winifred Cameron took up the study of transient lunar phenomena and created the most comprehensive catalog of TLP. Her work inspired others, including renegade astrophysicist Arlin Crotts.

ABOVE LEFT: Rocket pioneer Wernher von Braun headed the development of modern ballistic missiles and launch vehicles for the United States, including the Apollo Moon program's Saturn V. His career began in Germany, where he willingly worked for the Third Reich. Here he is seen with Fritz Todt, who organized forced slave labor for the regime, a fact von Braun would have known. The lapel pin on von Braun's suit is the Nazi Party badge. ABOVE RIGHT: Inside the underground slave labor factory called the Mittelwerk, where von Braun's V-2 missile was built by such prisoners. Not seen are the slave laborers from the Dora-Mittelbau camp. Von Braun once calculated the number of slaves needed for maximizing production. He knew of the hellish conditions in the factory and the camp. The V-2 killed more than 2,500 victims in its attacks. Some 20,000 slave laborers died at Dora or at affiliated camps and facilities, like the Mittelwerk.

RIGHT: Wernher von Braun's recreated office with original furnishings at Huntsville's US Space and Rocket Center.

LEFT: Arthur Rudolph's glasses at the White Sands Missile Range Museum. Rudolph was a member of von Braun's team and the project manager for the Saturn V. He was a war criminal and choose to leave the United States instead of facing a trial and/or deportation.

ABOVE: For a long time, many at NASA opposed both still photography and live TV during missions. The famous Apollo 8 Earthrise image was one of the many stunning visuals that helped to silence such doubters. *Earthrise* is one of the most iconic and widely circulated images in history and helped to energize the growing environmental movement. BELOW: NASA concept art for Apollo often featured steep, jagged mountains—part of the visual legacy of space artist Chesley Bonestell. Professional and amateur astronomers had known for a long time, however, that the Moon's mountains were rounded not sharp. Defenders of this visually fictional Moon argue that the real Moon was too boring to inspire the lunar landings.

ABOVE: A more realistic, more rounded, softer-edged Moon, as depicted by French space artist Lucien Rudaux in a 1943 booklet. Rudaux's Moon was more scientifically accurate than the more popular, Alpine depictions as popularized by Chesley Bonestell. BELOW: The first truly scientific expedition to the Moon was Apollo 15, and its crew felt at home both on the lunar surface and in orbit. Among other places, Dave Scott and James Irwin explored the area around Hadley Rille, a mile-wide, 1,300-foot deep former lava channel.

ABOVE: The Lunar Module *Falcon* landed near both Hadley Rille and the Moon's most spectacular mountain range, the Apennines, which rise to a muscular 18,000 feet. Note the lunar rover tracks in the foreground, Apollo 15 being the first to use the vehicle. The mountains were uplifted in seconds by a giant impact that created Mare Imbrium.

The Moon rises behind NASA's new rocket for its human return to the lunar wilds—the Artemis program. The Space Launch System combines new and old technologies and has flown successfully, though critics question its cost.

ABOVE: NASA's human-rated deep-space capsule, Orion, reached the Moon on the uncrewed Artemis I mission in late 2022. BELOW: A view of the 60-inch telescope at Mount Wilson with Moonlight pouring through the slit in the dome.

ABOVE: The author with the Mount Wilson 60-inch telescope, one of the largest telescopes available to amateurs for visual observing. BELOW: Holding a smart phone to the eyepiece of this telescope, David Hasenauer captures a view of the rugged lunar landscape. The prominent gash is Vallis Alpes or the Alpine Valley, through which the author saw a grail—a .3 mile-wide rille or former lava channel, which had eluded him through his backyard 10-inch telescope for years.

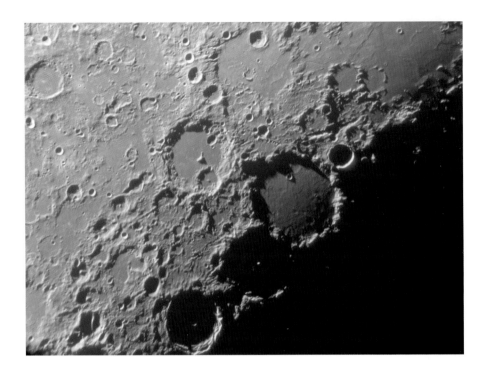

ABOVE: Sunrise on the trio of Ptolemaeus, Alphonsus, and Arzachel. It was from the crater Alphonsus that astronomer Dinsmore Alter thought he saw evidence—using photographic filters on the Mount Wilson 60-inch—of gases erupting out of the Moon. That assertion has not held up over time, but scientists continue to study a Moon that outgasses, quakes, and lofts up electrically charged dust—a definite hazard to astronauts. This trio of craters is a famous landmark for observers. BELOW: The stunning Malapert Massif in the lunar south pole region, one of the areas rovers and astronauts will explore, looking especially for water locked in the form of ice. Such ice preserves a record of the early solar system and can be used for drinking water, oxygen, and rocket fuel.

The German Room was rendered in metal, rough white concrete, girders and a fake office window, the glimpse into and out from the bureaucracy of National Socialism, the machinery of state and corporate coordination, of technocracy and torture. Perhaps in an office like that von Braun or his brother Magnus or their friend Arthur Rudolph, the production chief at the Mittelwerk and later the Saturn V manager, reviewed reports on changes that needed to be made to smooth out poor workmanship as the slave laborers built their many rockets.

Despite his youth—von Braun was only thirty-two in 1944—he had become an astonishing leader of his team, managing countless details to effect the development of the missile. In his adolescence, von Braun had built and launched his own rockets. Coming from an aristocratic conservative family, this interest in engineering was unusual but the brilliant boy would not be deterred once he found his passion. He dreamed of space at a time when few thought travel beyond the atmosphere was possible. There was a cultish feel to those true believers. Perhaps there still is. At the time of Apollo, von Braun wrote an essay that makes his sense of meaning clear; it was called, simply, "A Step Toward Immortality." In it, he suggested that spaceflight was part "God's plan for man [sic] to spread the spark of life to other heavenly bodies." He calls it a dream—one that can also redound benefits to Earth, like lessening social and ecological ills. To enable such a dream required something else with which von Braun was familiar: will.

In his twenties, he had joined a group of World War I veterans, mechanics, and a few charlatans to create an amateur rocket society, and their efforts attracted the interest of German officers. In short order, von Braun was working for the rearming military, and his rocket researches and his talent for industrial organization continued with the support, sometimes reluctant, sometimes enthusiastic, of the coming Hitler regime. He expressed no real qualms. In a 1950 article for the *Journal of the British Interplanetary Society*, von Braun wrote, "We felt no moral scruples about the possible future use of our brainchild. We were interested solely in exploring outer space. It was simply a question with us of how the golden cow could be milked most successfully."

The Cosmosphere's V-2 was painted black and white, the checked pattern so useful in test flights for easier tracking at altitude. I kept touching a fin that the inmates from the Dora-Mittelbau concentration camp made in the Mittelwerk. Cold smoothness and dust. I once shook the hand of author Primo Levi, the chemist who survived Auschwitz in part by remembering lines from Dante's *Inferno*. The sculpted prisoner's eyes looked up but would not have seen the stars or the Moon or Mars, all the destinations von Braun dreamed of and was convinced would be our destiny. The statue slave laborer, he was white as a cloud, would have seen only the rock roof of a tunnel and perhaps a crane on which the SS hung suspected saboteurs, wood blocks in their mouths as they dangled in the air. One of the prisoners under Rudolph's factory management, Jean Michel, wrote in his memoir of Dora-Mittelbau: "The weight of the machines was so great that the men, walking skeletons at the end of their strength, were often crushed to death beneath their burden." Levi was not at this factory, but when his sleeve moved, I saw the numbers his captors had tattooed on his skin.

Brought secretly to the United States, the Germans who designed the V-2 became the core of our new national missile efforts. War records were overlooked. As the German team eventually moved from army bases where they had been confined to more public facilities in Huntsville, Alabama, first to work for the army then for NASA, the publicity-conscious von Braun set about to sanitize the past and put our eyes on the future. He starred in Disney TV specials extolling spaceflight and wrote articles in popular magazines. In the 1950s, this all seemed a bit science-fictional, but his genial, teacherly ways and the threat of the Soviet Union began to convince Americans that he was on to something. Just as the country did not want to deal with the "forgotten war" in Korea, it did not want bad press for its spoils of the prior war, a group of truly brilliant engineers, scientists, and technicians who were now working seamlessly within the American military-industrial complex and the values of their adopted country. An early example of such public relations is Dieter Huzel's 1962 book *Peenemünde to Canaveral*, which describes the "mass production" of the V-2 as a "miracle" organized by the "genial" Albin Sawatzki, who was well-known for his brutality. Huzel described the "gigantic underground factory" as "spectacular." Von Braun and his Germans were celebrated as heroes as they led American rocket success after early failures by other teams.

As details of the Mittelwerk slowly began to become better-known, von Braun and his group defaulted to a position of moral helplessness. A month before the Moon landing in 1969, von Braun gave a deposition in an East German war crimes trial in which he admitted going to the Mittelwerk but left out the part about slave labor. According to friends and revisionist biographers Frederick Ordway and Ernst Stuhlinger, von Braun called the Mittelwerk "repulsive" and "the treatment of the prisoners . . . humiliating." Stuhlinger himself does invoke "German 'rocket slaves'" in the 1996 *Yearbook of German-American Studies*. However, he means German scientists captured by the Soviets not the prisoners of the Mittelwerk. Von Braun would say that the Mittelwerk was "hellish." Friends and co-workers, after von Braun's death, say that rocket pioneer protested: "My spontaneous reaction was to talk to one of the SS guards, only to be told . . . that I should mind my own business, or find myself in the same striped fatigues . . . I would never have believed that human beings could sink that low."

My first night in Cocoa Beach I thought of my loss of Space Age innocence in Kansas. After walking by the sea, thinking of how easily oceans become metaphors, I crawled into my bed at the Sea Aire Motel. I searched the grain of wood, assuming it was original and that Wernher would also have looked into its lines and swirls to help him sleep. I found the lines forming letters, this quote from Jean Baudrillard: "Forgetting the extermination is part of the extermination itself." Between that drive across the Kansas grasslands and my hearing the almost-furious Atlantic in von Braun's room, I had learned the story.

❦

The first spectacular public blow against the hagiography came in the 1980s when Arthur Rudolph was found to have advocated for the use of forced labor and supervised prisoners as well as abused them in the V-2 factory. A Department of Justice probe by the Nazi-hunting unit, the Office of Special Investigations, found that Rudolph had forced inmates to witness the hanging of suspected saboteurs, reviewed daily reports about which prisoners could work or were dead and asked the SS to provide more slave laborers, workers whose jobs he assigned.

Rudolph voluntarily renounced his US citizenship instead of face trial and returned to Germany in 1984. The former engineer agreed to never contest the findings though he complained in the press as did his stunned former colleagues, largely in Huntsville, where they denounced the charges and the OSI. In 1989, the case got attention again when Rudolph flew to Canada, apparently to try to enter the United States for the twentieth anniversary of Apollo 11. A Canadian hearing was convened to determine whether he could cross the border. He argued that he had been, in the words of the OSI, "shocked to learn" that the V-2 factory would use slave labor. But this was a complete fabrication. Historian Michael Neufeld supplied to the authorities German documents proving Rudolph's guilt. As well, in a sign of a true believer in Nazi ideology, Rudolph had joined the party quite early; one American official, upon screening Rudolph for his possible usefulness to the United States, wrote that he was "100% Nazi, dangerous type . . . Suggest internment." (As of this writing, the Wikipedia article for Rudolph has been sanitized to avoid mention of his war crimes.)

Linda Hunt's 1991 book *Secret Agenda* was a journalistic expose of the military effort called Project Paperclip to bring German scientists, engineers and technicians to American shores. Then, after extensive documentary research in Germany and interviews with the rocket team, Neufeld wrote *The Rocket and the Reich*, a 1995 book that the rocketeers considered blasphemous. In that book and his massive von Braun biography, published in 2007, Neufeld eschewed moralistic hyperbole in favor of the sheer weight of evidence. Conclusion: von Braun was opportunistic, if not amoral or worse, in his pursuit to build a rocket that could someday take humans to space. Defenders often point to the fact that he was briefly arrested by the Gestapo for talking a bit too much about space and not the rocket's combat necessity. (Late in the war, the ostensibly politically indifferent von Braun began to see the evil of the Third Reich, Neufeld writes.) Von Braun was quickly released, and the episode has no bearing on the use of slave labor. Von Braun, who had joined the SS as part of his ingratiation to the regime, of course knew the rocket would be used to bomb civilians, a fact that he later justified not because he was a Nazi (he was not a fanatic true believer) but because he was trying to defend his country. Also, he was engineer. He wanted things to work.

Not only did von Braun know of the conditions of the slave laborers, he visited several concentration camps and spent time in V-2 factory of the Mittelwerk. The only time he tried (apparently) to assist a slave laborer was because of his educated background and expertise: he needed him to be part of more technical production. Still, this involved von Braun in directly transferring prisoners—forced laborers—a violation that alone could have put him on trial at Nuremberg.

At least once, Neufeld found, von Braun *calculated* the number of slave laborer needed for part of the assembly process. Multiple popular accounts since this revelation have not included this fact. On this basis alone, as well, von Braun could have been charged and tried at the Nuremberg tribunals.

The atmosphere that could encourage educated and sophisticated men like von Braun to accede to a regime as ugly and brutal as the Third Reich's was an atmosphere of isolation, privilege and mechanical challenge, as historian Michael Peterson has found. Before Allied bombing forced the German project to move to the Mittelwerk, the rocket research had begun at the isolated Baltic town of Peenemünde. Peterson's *Missiles for the Fatherland: Peenemünde, National Socialism, and the V-2 Missile* shows the "Peenemünders" were willingly engaged in an exciting, shared technological enterprise. Peterson also shows how the German rocket team, from scientists and engineers to military officers and technicians—found collective identity and well-being in the work itself. They did so at a place that was built especially for them—and constructed for comfort and aesthetics—in a location that was full of natural beauty, from the beach

to the pine forests. There were nice houses, parks, gardens, sports facili- ＇
ties, hunting trips, sailing, nearby resorts, parties, and more. The technical
facilities were world-class, rivaling, in retrospect, only those devoted to the
Manhattan Project.

The secrecy that necessarily surrounded this "community of elites" was
far from an impediment. These men were not prisoners. Peterson shows
that the secrecy surrounding the project provided cohesion and a kind of
status; indeed, excepting of course the forced laborers at the test facility,
the Peenemünders enjoyed better housing, better pay, better food, and more
professional, social, and recreational opportunities than nearly anyone else in
war-torn Germany. The place was a technocratic paradise, and the relationship between the civilian team and the military was far more nuanced than
the former merely taking orders from the latter. Certainly the program's
army director, General Dornberger, exhorted everyone with patriotic calls
to duty but this was just one contribution to the strong bonds holding the
team together. Everyone had a say in making the project work and everyone
pulled their weight.

As the war went on, and the test area was bombed, the grim realities began
to sink in more readily, but the rocket team threw itself even more into refining
what had become the V-2 and developing more exotic weapons. While some of
the team—both civilians and military—were ardent Nazis, even those, who,
like von Braun, were conservative German nationalists, found common cause
in defending themselves, especially against the threat of Soviet troops and
communism. Nazism was, in Peterson's words, "an ideology that they could
live with." The German rocket team members were kept under surveillance,
but far from restricting their relative freedom in a dictatorship, it proffered
more mobility and access to privilege. One rocketeer called Peenemünde a
"paradise," and nearly all looked back on the development of the V-2 while
there as a kind of golden age. Arthur Rudolph called the time "fantastic."

It was this "culture of consent," as Peterson puts it, that fostered "a narrowed
technical and patriotic vision that consented to some of the worst crimes of
the Nazi regime." The myth that the SS forced slave labor onto the rocketeers

is just that, a myth. Rudoph brought slave labor to the Baltic paradise, and no one blinked an eye.

A sense of German victimization often held Nazis and non-Nazis together, and, as Peterson notes, the "everyday" racism of Germans then made it easier for the rocketeers to ignore the fates of others. V-2 managers at or visiting the Mittelwerk, including von Braun, had to ignore the sounds of slave laborers huffing in exhaustion, screaming in pain, collapsing after beatings or hanging in a row from a crane in the tunnels. "Quietly," wrote survivor Yves Béon of the various civilian engineers, "they indicate location points desired for machines, for junctions, for joints and fixing points for the electric and pneumatic air ducts." Civilian rocketeers in the factory also reported sabotage and slow work to the SS, knowing the slave laborers would be punished with beatings or execution. Some civilians themselves brutalized the workers. Though technically skilled slaves were treated better, there is little evidence of serious civilian help for the prisoners. Peterson says, "The fate of prisoners not on the factory assembly line was common knowledge."

Perhaps the most striking revelation in Peterson's history is, of all things, a strip of cartoons. A set of them were provided to dinner guests in late 1944 when Dornberger, von Braun, and another official received the Third Reich's Knight's Cross medal. Von Braun's caricature shows him waiting for a solar array to cook him breakfast. This seems apt: self-absorbed in his dream of the next technological innovation that could fill him. Others' drawings are more of the moment. They show Peenemünders supervising slave laborers hauling out a stalled private car or building a missile.

This history was largely known only to a few in the military and government, those with access to Project Paperclip, an effort to which there was some internal resistance, especially from some in the State Department. But, as historian Brian Crim explains, the need for scientists to combat the Soviets in a Cold War trumped all hesitations, moral or otherwise. The Soviets themselves had grabbed as much material and personnel from defeated Germany as well. In the backwash of the October 4, 1957, Soviet launch of Sputnik, the advocates for von Braun, his team, and the Paperclip effort seemed vindicated. The "Reds" had powerful rockets, ones that could lob more than satellites into the sky. They could lob nuclear weapons. Sputnik came as a shock to the American people, and the Space Race began in earnest.

So here, on my first morning in Florida, away from the visitor center crowds and current launch towers, is the American origin of the Space Age: the open blast door of the cinder blockhouse of Launch Complex 26. This was where the first successful American satellite, Explorer 1, was sent aloft on January 31, 1958, more than a month after a competing US effort called Vanguard rose all of one yard and exploded. The tiny satellite landed in the shrubs, its faithful beep-beep transmission prompting one wag to write, "Why doesn't somebody go out there, find it, and shoot it?" The whole fiasco— "Kaputnik"—was broadcast live.

While both Explorer and Vanguard were scientific instruments meant to contribute to the International Geophysical Year from 1957 to 1958, the military and geopolitical repercussions were plain. Although von Braun's V-2-derived Jupiter-C could have launched even before the Soviets, his project was, at the time, purely an army effort swarming with former Germans. Vanguard, though tied to the navy, was run by the civilian Jet Propulsion Laboratory. Dwight D. Eisenhower had no love for von Braun—a charming, single-minded egoist and a former enemy—so he picked the civilian Vanguard project for first launch attempt. After the disastrous failure, von Braun got the go-ahead. He was right to hope that it was the first step toward the Moon. It was.

On the day of the Explorer 1 launch, he stood beside the consoles in the cramped blockhouse. I position myself behind one, just where he had been, and I inhale decades-old must and sweat and smoke from Camels and Pall-Malls once heated by banks of antiquated computers and electronics and now baked into the air. Sun pours through the thick, angled windows, whose glass is layered and gas-pressurized and one-foot thick, transforming white Florida light to a protozoic green that washes over my hand as I move the loose black switch that launched the rocket from a tower hundreds of feet away.

The aqueous green spills across consoles, dials round and rectangular, knobs and toggles, Burroughs tape readers with their wheels of magnetic tape, cabals of color-coded wire and connectors, the brushed metal handles of a primitive computer with its hand-wired logic circuits of simple inverters, gates, differentiators, single-shot and flip flops—the circuits look like a tapestry—the data punch and tape comparators, thick electrical cords and plugs, speakers, grand polished microphones like those used by big band leaders, the red lights of the operating console, the green and orange and white lights bejeweling other panels, the ceiling-mounted video monitor, the building's thick pillars and acoustic tile, and the green spreads its gentle tide inside these two-feet thick walls, beneath the heavy concrete roof, lapping against overhead lights shaped like upside-down flying saucers and gleaming off the polished floor. Above the four main windows are four clocks: Eastern Standard, countdown, hold, Zulu.

Explorer 1 would do more than beep which is all Sputnik did. Using an experiment designed by the University of Iowa's James Van Allen, Explorer 1 discovered the radiation belts that swaddle the Earth, the so-called Van Allen belts.

Somehow—I don't know how—I had read copies of *Soviet Life*. The enemy fascinated me: black-and-white missile parades and smiling peasants on spring days like mine. Or, more accurately, perhaps, the potentially apocalyptic nature of our contest fascinated me. As a kid in the Civil Air Patrol, I avidly read a manual on nuclear, biological, and chemical warfare and knew what a single blinding flash in the sky meant for the fate of Indianapolis, Indiana. At the very least, it would make Indianapolis far less boring than it was. I would don a hazmat suit and map the empty hallways of my school, a vast new continent. Surely, had I visited Blockhouse 26 as a youngster, I would have felt a surge of nationalist fervor.

Today, the song of experience still allows a kind of pride—in the technological achievement—but also a wistful stoicism that the leap into space had been catalyzed by competition and fear and moral decrepitude. Still, sometimes to help myself fall asleep, I watch videos of jet fighters and Cold War bombers with the sound muted. They just fly in a silent sky. Weirdly, the technological sublime is my innocence and my disillusionment and my hope.

Outside the blockhouse I clamber up an old but vividly red transport crawler, accessible only because a chain-link fence has collapsed. A few feet above the ground on its wide platform I climb a stair to nowhere, the steps just stopping in midair like a piece of Dada architecture. I survey the preserved ruins like an explorer: Narrow palm trees each with one spraying crown. Chain-link fence and wide grassy stretches. A rocket on its side, a rocket standing vertical. A thickening layer of white clouds pressing stasis on the scene. When the decking clatters, I jump.

Thirty-six missions used this crawler and gantry, which was, I learn, derived from an oil rig, from deep earth to upper reaches, including the test flights of chimpanzees before the flights of American humans, those first two suborbital Mercury-Redstone missions of Alan Shepherd and Virgil "Gus" Grissom. Like a beached barge, the gantry looks like it's been hauled up for salvage on the grass-cracked concrete, all strut and brace, door and corrugate, paint-peel and metal mesh, some of which the nearby Air Force Space and Missile Museum has appropriated for souvenirs. Glued to a piece of white cardstock paper is a rusted metal X cut from the mesh. I will put a dollar donation in a jar as payment.

In the distance, to where this red metal thing would be moved with a live rocket, there stands, guy-lined against the wind, a full-sized replica of a Mercury-Redstone. Seen here on a wide concrete embayment, without the apparatus of the gantry, the Redstone with its cramped Mercury capsule looks almost more decorative than suborbital. Largely white, with black-and-white vertical stripes beneath the black capsule and topped with a red skeletal framework holding an emergency rocket to sheer the capsule away in case of an accident, the Mercury-Redstone is both a piercing tower—plenty phallic and nicely air-streamed—yet narrow, almost dainty, only ten feet longer than the trailer we lived in for a time after my parents split. Surrounding the rocket at a distance a yellow circle has been painted on the tire-tracked pavement.

Farther are the striped concrete barriers behind which fire-dousing pipe stands rust. Metal plates cover trenches dug into the concrete. Beside the rocket now, I touch the clunky fins and trace my hands over rivets. Somewhere close by, Alan Shepherd paused in his TV-silver spacesuit to crane his neck and look at the rocket, wondering just what in god's name he was doing. Famously, any nub of concern the hardened test pilot felt was never expressed. After peeing in his spacesuit after several countdown delays, he barked to the engineers to "fix your little problem and light this candle."

They did so on May 5, 1961. The launch was controlled from another, nearby bunkhouse, whose museum-like displays feature translucent human figures—engineers in ties and white shirts—pasted onto protective glass so one can see through them to the equipment they worked with, a ghostly merging of simulated skin and relic machine. As befitting a launch complex for creatures—chimps like Ham, humans like Grissom—there are panels labeled SPACECRAFT CHECKOUT, CAPSULE COMMUNICATION, AND ASTRONAUT BODY FUNCTION RECORDER, its output paper scrolling like a Chinese land-scape painting. From the firing console, one can look out through the algal, lacustrine light and see the replica Redstone.

After Shephard's suborbital toss, President John F. Kennedy set the goal that would become Apollo: land a man on the Moon, return him safely to the Earth, and do it within ten years. We had fifteen minutes of human spaceflight experience when he set the challenge. Embarrassed by the failed Bay of Pigs operation and Soviet space dominance, Kennedy needed a geopolitical win. He did not care about space per se. But he was convinced that the Moon landing project would leave the Soviets behind. He was right. Between Mercury and Apollo, the Gemini program taught NASA how to fly a more complex two-person craft, stay longer in space, do space walks, and rendezvous and dock, all steps needed for the Moon missions.

Out of clouds, the Sun becomes a circle in the sky again, and it all seems so small, this blockhouse out of which I walk into a clearing in the coastal jungle, a portal archived against time. Strung up and down the Cape are the launch complexes, abandoned or preserved, renovated or active, while the

Earth we warm expands its ice-fed seas, lapping up along the sands, slowly toward the patient forest, a vegetable predator of palmettos, swamp maples, live oaks, and moss against which the launchpads loom.

I had moved the decade selector.

❦

At home, through the eyepiece of my telescope, I can see the regions of the six Apollo landing sites. Of course, you can't see evidence of the landings—no landers, no flags—but I seek the Apollo regions often. From 250,000 miles away, at, say, three hundred magnification, I am brought closer, an illusory few hundred miles above the Moon. Though first, the Apollo 11 site in the Sea of Tranquility is the most boring—flat gray to my eye. The more demanding targets are the three nearby simple craters named Aldrin, Collins, and Armstrong, small enough to require steady air and high magnification. The history of Apollo 11 is more impressive than the landscape on which it sits.

Next, Apollo 12 landed southeast of the beautiful complex crater Lansberg in Mare Insularum, hard by a Surveyor probe, from which the crew took bits and pieces for scientists to examine on Earth. (Arizona's Ewen Whitaker was tasked with locating Surveyor to make Apollo 12's landing as precise as possible.) The sites became more interesting with Apollo 14, landing in the hills north of the degraded Fra Mauro crater. There's topography, and I match my eyepiece view with photos and memories of photos of the surface—I am in three places: the Tucson backyard, eyepiece orbit, the surface of the Moon. I am in three times: the past of a mission, the past of my childhood, the present of my looking. The mission that obsesses me, Apollo 15, landed by the sublime, massive Apennine Mountains and that gorge called Hadley Rille. Then the highlands were visited by Apollo 16—another hilly region, this time between the Cayley Plains and the worn Descartes Mountains.

Some fifty years ago, as I write this, the exploration ended with Apollo 17 by some beetling mountains in the valley of Taurus-Littrow, landmarks as impressive as those of the Apollo 15 site. Apollo 17 astronaut Harrison Schmitt, the only geologist to walk on the Moon, would later say to Congress in a moment of rare, possibly ghost-written eloquence, "I would like to tell you about a place I have seen in the solar system. This place is a valley on the Moon, now known as the Valley of Taurus-Littrow. Taurus-Littrow is a name not chosen with poetry in mind; but, as with many names, the mind's poetry is created by events. Events surrounding not only three days in the lives of three men, but also the close of an unparalleled era in human history." The words are engraved on a plaque at Kennedy Space Center's Apollo Visitor Center.

That unparalleled era may not have begun had it not been for more deaths, this time of three astronauts stuck in a test capsule on Pad 34. Launch complex 34, far more elaborate than the Explorer and Mercury-era sites, is approached by a long road that feels like an avenue to a sacred site, and one could call Pad 34 just that. I feel sick to my stomach when I see the abandoned complex for the first time. This is where the three Apollo 1 astronauts died in their Command Module on an unfueled Saturn rocket during a completely routine test called "plugs out." Plugs-out meant the capsule was either on internal power or simulated internal power; the test was meant to determine how well the capsule functioned under those conditions. It was January 27, 1967. The crew was comprised of commander Mercury and Gemini vet Gus Grissom, America's first spacewalker Ed White, and rookie Roger Chaffee.

One thing that wasn't working well was communications. At one point, Grissom complained, "How are we going to get to the Moon if we can't talk between three buildings?"

It was Grissom who would exclaim, "Hey! Fire!" Two seconds later, Chaffee yelled, "Got a fire in the cockpit." Then silence. What follows was garbled but has been interpreted as "we've got a bad fire. Let's get out. We're burning up . . ." Then an agonizing cry.

On the nearby blockhouse monitor, some saw Ed White, who was seated beside the hatch, reach up to try to release it. Gruesome as the fire was—melted spacesuits, third-degree burns—it was the toxic smoke and cardiac arrest that killed the three men. White had died trying to open the hatch. Grissom had removed his restraints and was on the capsule floor, probably trying to help White. Chaffee died in his seat, where he was supposed to maintain radio connection with ground personnel. It took five minutes for pad workers to get to the blistering capsule interior. At the end of the worst day of their lives, the workers stowed their laminated passes in a metal rack.

As I walk about Pad 34, I look beyond a strip of trees planted on a canted embankment to see an ocean flat enough to fold like a paper bird. I wander this six-hundred-foot concrete expanse laid atop a mound of grass that reminds me of a Mesoamerican ruin. Massive flame deflectors lurk on the concrete's edge as though the symbol for infinity were disassembling itself. Diamondbacks nest there. A long concrete wall extends beside the road and, from the pad itself, partly obscures the squat, round dome of the blockhouse. Scattered across the pad are knee- and shin-high concrete foundations—small squares and rectangles and a lonely, single wall, like the post-battle wall of an office building, black-stained and containing an empty doorway that composes views to the jungle or the ruins, depending on where you stand. There are postholes at the top of the wall, as for vigas, and a red and rusted metal frame around the empty door, a de Chirico-like architecture of abandonment worthy of a depopulated Roman Forum or the ghost town of the Nevada Test Site. Weird engraved stones are underfoot, with the words "A. P. Green" above "Clipper D. P." I have no idea what they mean. The concrete is beaten, crumbling into degrees of gray and white and black like the surface of rogue planet. Incised in the stained, tire-tracked concrete—long arcs of turning trucks recorded in more shades of gray and charcoal—is the narrow-gauge railroad that drew the service tower to the launch site.

The tower's base remains. It is surrounded by waist-high metal barricades that keep me from touching the thirty-foot-high pillars or standing under the concrete platform where Apollo 1's Saturn IB sat. Where its engines loomed there is a circular cutout. When it was there, the rocket and Apollo 1 capsule pointed up another 160 feet. Circuiting and circuiting the platform, craning my neck, I regard, like a sad collector, its patina of blast-black and mundane rust and storm-cloud gray, its holes and cracks and intestinal piping, the chaos of scratches and divots, its preform panels and metal plates and the faded stencils that say "Abandon in Place." A bit of concrete decking overhangs each side, rectilinear or angled crescent, and within one pillar there is a nook, an inset, a kind of niche against which leans a memorial wreath on a tripod. Above it, a kind of shelf and the business end of some outlet tube. Its cavity must attract birds who emerge from the dark to fly.

The fire halted Apollo flights but not the program. Investigations found blame everywhere. The contractor for the Command Module, North American Aviation, had allowed sloppy workmanship, including wires that frayed—one was considered the likely spark for the fire. NASA thoughtlessly filled the module with highly pressurized one hundred percent oxygen, a highly flammable environment. The hatch was a balky, inward-opening affair that, as the fire and gasses expanded, pressed tight. The capsule atmosphere was twenty-nine times the pressure outside. No one could have removed the hatch.

All these issues were fixed after months of delays. Hindsight seems to agree: had the Apollo 1 fire not occurred, the likelihood that space-faring missions would have led to death in space—or on the Moon—seems high. The four hundred thousand people who worked on the retooled program made it a success. Adjusted for inflation, the cost of Apollo was some $250 billion. By comparison, the twenty-year-long war on terror, in Afghanistan and the Middle East, cost the United States some $8 trillion.

When it's time to leave, I look back at the raised concrete platform of Pad 34, its pillars and beams exposed like the excavated undercroft of a church. Another meaning of Apollo: religious. A shared endeavor, sacrifice of prophets, a redemption. We reached the dream of the Moon. In his elegy "Abandon in Place," Ray Bradbury writes, "When will the harvesters return / To gather further wonders as a fuel / And let them burn?" When we memorialize the tragedy of Apollo, it's with Apollo 1 in mind. I force myself to listen to the Apollo 1 crew's screams for help.

As we remember Christ on the cross, we forget the two beside him who also died that day. It was the thief Dismas who accepted his fate and who, Christ said, would be welcome in the afterlife. It was the thief Gestas who had the temerity to say, "Are you not the messiah? Save yourself and us." I wonder how we would memorialize the V-2 slave laborers if we could hear them in the tunnels of the Mittelwerk. My sympathy is with Gestas, who sought the end of evil on a planet.

✳

While scholars have excavated the ugly history of the V-2 slave laborers and the story of Project Paperclip, popular culture has largely ignored the rock roof of the Mittelwerk. Apollo was a technological sublime, an engineering triumph, a national victory. If we can land on the Moon, why can't we . . . ?

Neither lunar science nor wartime history have mattered much. From the hagiographies of the 1950s, 1960s and 1970s and some truly indefensible programming in the 1990s on such channels as A&E (which aired a scandalously sanitized "biography" of von Braun), popular scrutiny has increased somewhat but the record remains decidedly mixed.

One example is Huntsville journalist Bob Ward's von Braun biography, published by the Naval Institute Press. The 2016 British-produced documentary, *The Saturn V Story*, available on Amazon, notes the V-2 as the ancestor of the Saturn V Moon rocket and says the Third Reich pursued the V-2 with "great costs," but there is no mention of slaves who built it nor those who died by its use. The first episode of the otherwise credible 2019 documentary *The Apollo Chronicles*, also available on Amazon, has one clip of a V-2 launch as a prelude to a discussion of Sputnik. An historian, with an unfortunate turn of phrase, says that Wernher von Braun and his fellows "ended up in the United States," when they were, in fact, brought here deliberately. There is no mention of the victims of the V-2, whether in cities or the Mittelwerk factory, nor discussion of complicity and war crimes. When von Braun is discussed, it's in terms of his genius and visionary dedication to space. Arthur Rudolph is not raised.

In 2019, the fiftieth anniversary of the Apollo 11 landing, PBS's excellent *Chasing the Moon* referenced von Braun and slave labor, but the Smithsonian Institution's series *Apollo's Moon Shot* did not, foregoing even any background on the secret importation of the Third Reich's rocket team. Two years earlier, the Smithsonian Channel's series *America's Secret Space Heroes* included an episode on the Saturn V. Of the German rocket team, all that is said is that they worked on V-2s in Germany, were led by the "charismatic" Werner von Braun, and were brought to the United States to work for the American government. He was "one of the most remarkable men I have ever known," one American engineer says of him. "He was a great leader, he was a very good engineer, and the team that he brought over here was absolutely outstanding."

NASA itself hardly has grappled with the past of its first visionary and those who worked with him. When I began my quest to understand the public history of the V-2 slaves and the German rocket team—and what it might mean for our renewed lunar ambitions—I turned, of course, to the web. Back then, more than two decades ago, I searched in vain to find any reference to the V-2 slaves on NASA's history and space-exploration pages. Agency biographies did not include Rudolph, who had been given medals for his work. And now? Arthur Rudolph is still missing.

A recent search on the front page of NASA's History Office website will yield Google-assisted hits, the first page of which directs readers primarily to NASA histories of the Saturn V with no mention of Rudolph's deportation case based on war crimes. One hit takes you to a PDF of the NASA-published book *Remembering the Space Age*, which contains Michael Neufeld's essay "Creating a Memory of the German Rocket Program for the Cold War," another of his scholarly and unstinting looks at the whole episode. One would need to be a dedicated internet searcher, however, to click on the link and track the information down within the book. *Remembering the Space Age* also includes historian Monique Laney's analysis, "Operation Paperclip in Huntsville, Alabama," a topic she would later expand into a book of her own. NASA's online *Biographies of Aerospace Officials* entry for Wernher von Braun says nothing about his wartime link to slave labor. As of this writing, that page was last updated in 2017.

But one NASA history link does say this: "Von Braun is well known as the leader of what has been called the 'rocket team,' which developed the V-2 ballistic missile for the Nazis during World War II. The V-2s were manufactured at a forced labor factory called Mittelwerk. Scholars are still reassessing his role in these controversial activities." The von Braun biographical page at the Marshall Space Flight Center website is much the same: "The V-2 assembly plant at the Mittelwerk, near the Mittelbau-Dora concentration camp, used slave labor, as did a number of other production sites. Von Braun was a member of the Nazi Party and an SS officer, yet was also arrested by the Gestapo in 1944 for careless remarks he made about the war and the rocket. His responsibility for the crimes connected to rocket production is controversial." Both statements avoid mention of von Braun's slave-laborers-needed calculation, and Arthur Rudolph is invisible. They also don't grapple with the ethics of Project Paperclip.

Not surprisingly, at the Kennedy Space Center itself, this past is also invisible. There is no exhibit, no sign, no brochure, no information at all about the V-2 slave laborers and their connection to von Braun or Rudolph.

At the Apollo Visitor Center, sectioned like a metal whale after the hunt and suspended from the ceiling like a trophy, is an honest-to-goodness, unused Saturn V. Though *it is* huge—nearly sixty feet taller than the Statue of Liberty—the impact of its scale is diluted by the horizontal posture, at rest and almost defeated in the air, sectioned like a carcass. The expansive power of the enterprise had served to preserve a small thing: a human body or two or three. The Astrovan, in which crews rode on their way to the launchpad, had chairs they sat in, like any RV tourist. The astronauts wore spacesuits, like miniature space craft against the hot and cold vacuum. They walked through human-sized doors. Their capsules were so small, the tip of the invisible pyramid. And taking the shape of a pyramid are the plaster casts of six human hands, three pairs, belonging to Aldrin, Collins, and Armstrong, to make their fitted space gloves. The casts, as though in prayer, each impression of a fingernail precise, each vein a landscape ridge, each pore a divot.

In all this the Moon is also absent, though you can stand on a photo projection of it on the floor and you can touch a polished slice, smooth as marble, and encounter "fun facts." Just as with Apollo itself, the program, not its deeper history nor its destination, is the focus.

Tides of tourists wash through the cavernous building, which echoes with the sounds of feet, inspirational music and barking parents, fellow pilgrims in florid cargo shorts holding cell phones like pennants. I look up. My spine crawls. What this testament to genius and single-mindedness deserves is to be set on its five big-ass F1 engine nozzles, standing straight up beside an iconic red gantry at the end of a narrow path in the jungle hundreds of yards away, forcing us to walk, one at a time, to silently confront this machine. It would tower on its base and taper whitely to its capsule and the red escape-rocket assembly, making our necks hurt like we were bird watchers, our eyes stinging as they strained to see in the salt-air Sun. In my imaginary exhibit, I place close-up photos of the lunar surface on one side of the walkway and pictures from Dora-Mittelbau, snapshots from the Mittelwerk, on the other.

☙

Later I would travel to Huntsville, Alabama, home of the Marshall Space Flight Center and the US Space and Rocket Center, an affiliate museum of the Smithsonian Institution. There I would crane my neck to see what I had imagined: a vertically mounted replica of a Saturn V. From my hotel room, I'd watch it with its blinking light atop its some 360 feet, standing beside a freeway. Spot-lit, that shell of a rocket would seem brilliant with promise.

Neither at its base nor in the museum exhibits would the V-2 slave laborers be mentioned, let alone the Rudolph case. I would find von Braun's midcentury desk and bookshelves and model rockets—a re-creation of his Marshall office—encased in a giant vitrine, like some glass-cubed time capsule. Only at the public library would I find a small sign that called out von Braun's wartime immorality, a sanctuary of historical accuracy in a city where the German rocket team is venerated nearly without question. Scholars even refer to "the Huntsville school" of space history.

As Monique Laney suggests in *German Rocketeers in the Heart of Dixie*, this loyalty is bound up in an "us versus them" attitude that wishes to forget the past or explain it away. This reaction is more complicated than it can appear. The German rocket team's work in Huntsville energized a sleepy town's economy. The very whiteness of the Germans led to bonds explicit and subtle with the dominant white culture in the Jim Crow south, though von Braun defenders are quick to point out, rightly, that he opposed segregation. As long as the children and grandchildren of the German rocket team remain in Huntsville, they will remember the team members as fathers and grandfathers, men whose daily gestures of love and child-rearing are hard to square with having worked for the Nazi regime. Listening to Margrit von Braun describe in a local TV interview how her father would take each of his children separately on trips to give them love and attention, how he would delight in boating and swimming with his family on Sundays at the nearby lake and how he had that talent of listening intently to the person right in front of him—listening to her say these things, I can imagine his presence, brilliance, and charm. Magrit did not respond to any of my questions of substance.

Another local TV clip introduced me to Heidi Collier, a retired teacher and US Army civilian specialist who has emerged as the leading denialist among the Huntsville space community. She has been chair of Huntsville's German-American Heritage Committee and is the daughter of one of the original engineers. "She's as American as anyone you'll ever meet," said WHNT News 19 host Steve Johnson in 2019. In a fawning interview, Johnson also made the point that Collier is now seen as "protector of the legacy" of the rocketeers.

When I tried to talk with her about this past, she was blunt.

> Christopher Cokinos—
> Yesterday I found your letter in the mailbox of my last residence. I'm sorry, but must decline your request. It should be pretty easy for you to figure out how the group of German rocket team descendants must feel about negative assessments made by outside historians. This doesn't require an interview.
> Thanks, Heidi Collier

By e-mail, I then asked her about an article that had appeared in the Red-stone Arsenal newsletter, one in which Collier told a reporter that the German rocket team members "were kept in a prisoner of war camp at Fort Bliss, Texas." While the Germans were restricted to the base, it was hardly a POW camp. The Germans eventually were allowed freedom of movement off base, including trips to El Paso. In the same article, she contradicts herself, saying, "Then they were confined in a community behind a fence at Fort Bliss, but, of course, this was for their own safety." As few people then knew of the team's location, this is largely a fabrication. The US government still had security concerns about the rocketeers, concerns that relaxed as the group proved themselves more-or-less committed anticommunists satisfied to be building rockets for the Americans.

Before that, Collier says, "They developed rockets for space and then were forced to build missiles for Nazi Germany." This is simply not true. It was exactly the reverse. The German rocket engineers may have wanted to build spaceships but they willingly dropped that vision in the short-term for military and Nazi support. No one forced Wernher von Braun to take up with the Wehrmacht in the 1930s. No one forced Arthur Rudolph, Kurt Debus, or Konrad Dannenberg to join the Nazi Party or the SA or the SS.

In another article for the Redstone newsletter, Collier again distorts the record, saying, "To the Germans, the questions about the use of prisoners of war for labor and the war itself make Peenemünde something that they want to forget, to bury. The German scientists at Peenemünde were civilians working on rockets and then there was the military. When work first began at Peenemünde it was all separate." Not so. The rocket team, military, and civilian employees worked at the behest of the entire state. Calling the Dora-Mittelbau slave laborers "prisoners of war" is inaccurate. German prisoner-of-war camps were reserved for captured Allied troops. Dora-Mittelbau was a concentration camp specifically utilized to work slave laborers to death.

"Collier has heard all the claims that von Braun and his team were aware of prisoner of war conditions," the article continues, "and the death of prisoners of war in connection with V-2 production. But the German scientists were, in their own way, also prisoners of war, being forced to work at the command of the German army, she said." Collier, like others, suggests the rocketeers were the victims: "They were taking orders from the military without any communication with the outside world and in a locked, isolated area. They were always looking over their shoulders . . . I'm grateful for the freedoms I was able to grow up with. These scientists didn't have that freedom. What happened to them was schickshal, a matter of fate."

When I asked her if the article correctly represented her views, she told me that it did.

At the Sea Aire, the night before I fly back to Arizona, I had my last chance to see a rocket rise into the sky for the first time. The launch of a European solar probe had been delayed. I'd even bought a ticket for a bleacher seat near the Apollo Visitor Center—as close to the launch as the public could get—but the venue had been cancelled due to prevailing winds: should the rocket explode, exhaust and debris would rain down on the audience. *But what a way to go.*

I had come to witness a rocket launch but also to confront the origins of that rocket launch, as though the countdown would go back to the 1940s. And the Moon was out each night, the Moon that has entranced me for years, observing it with a telescope, imagining myself on that pocked, rolling surface, wondering about the motives for such longing in myself and others. What was all this? What did it mean? What can it?

After dinner at a former Cocoa Beach jazz club that surely saw astronaut antics back in the day, I picked up a plastic chair, brought water and my phone, then walked onto a night beach empty of spectators but for a couple with whom I exchanged pleasantries. Near the surf line, I set up at a respectful distance and looked: the Full Moon was halfway to the top of the sky. The ocean seethed with lines of waves. The wind whipped sand around my ankles in pecking flurries visible in the glow. In the blocky condos and apartments fronting the beach I could see the blue light of televisions and the fluorescence of kitchens before bedtime. Trawlers dotted the horizon like notes toward an epiphany, and the Big Dipper pointed down to where we launch rockets.

A cool wind mixed up memory and moment, and I held the phone to my ear, listening to countdown—all was well, one system after another was "go"—and suddenly an orange light flared in my periphery—clearly, someone deciding to drive down the beach—until I realized the countdown broadcast had been on a time delay and I'd been looking at the wrong spotlit gantry miles off. I had missed the beginning. I cursed, then fell silent.

That warm golden light on the beach flared in a hemisphere as the rocket rose. I could not see the rocket itself. There was only an orange shield of licking flames bigger than a Full Moon. The flames were steady, ragged at their ends, surprising the beach in a glow from the world's most powerful streetlight or some uncanny campfire. The shine flashed darkly into low clouds for a moment. The Atlas V rose, invisible, quite studious. There was only its fire, fire in the sky.

It was silent because the sea was not, and the arc above the cloud described by golden flames was graceful and rational. This was beautiful velocity, the artful force of a plotted curve, and as the Atlas V moved over the Atlantic, it seemed as though its engine exhaust were aimed at me. I stood wordless by the Moonlit churn, glassing the rocket. Shock waves streamed off its skin like white sculptures made of cloud in the dark, like the translucent tendrils of some benthic speedster. (The contrail of exhaust had sheared into ragged clouds.) The main stage was cast away unseen, the action made manifest by new configurations of twirling orange flame and white wave how many miles up? In the binoculars, I saw how the exhaust was a pattern of partial curves, petaled orange nearer the rocket then petaled blue at the end of its trailing violence. For a few seconds I thought could see the rocket as a rocket, a different dark, a metal stretch, against the night it climbed to, and the machine blossomed a white corona of supersonic waves, a short-lived flower flicking toward the Sun, exquisite and very fast.

Smaller it became, an orange disc, like a fat and messy Jupiter hunting the ecliptic, like a satellite on fire. Then gone, and a gawky shadow landed by the surf line, a tall heron there to hunt the silver shallows, the waters breaking into bright-topped creases, the creases falling into dark valleys of moving water that suctioned sand and spit it forward. I walked slowly back to my room, wishing next time, whenever that would be, to have not only sight but sound—a roar from a Moon-bound mission. I danced around the sand in happiness. The rocket had curved beneath the Moon, and it felt like an embrace.

Then I was in von Braun's room again and I thought of Arthur Rudolph's glasses. I had seen them in another museum, in New Mexico, at the White Sands Missile Range, where the German rocket team first launched V-2s for America. I stopped cold when I saw the glasses in among the overstuffed and pell-mell rooms of the Range's historical museum, full, as it was, of mannequins of Union soldiers, impossibly large cameras for tracking rockets, innards of missiles, and a life-size mock-up of a fallout shelter. His tortoiseshell owl glasses are kept in a felt case. With them, he looked at calculations both for missile trajectories and prisoner reports. Here was his white slide rule. Here, his photo album of the transatlantic ship on which he and his mates traveled to New York. Here, a small wooden box of drafting tools—steel and bronze compasses—with his name inked on the inner lid: A.RUDOLPH. Unlike the invisibility of Rudolph and the slave laborers at the NASA visitor centers, the museum sign noted the use of such labor and Rudolph's role and case—"still controversial."

My innocent longing for a morally unsullied human leap to space has been gone a long time. But if we can leap upward and care deeply there, if we can stay put and care deeply here, why not? Why not both, staying and going? Earth and Moon. Remembering. Taking care. To do all that, we need to tell these stories.

Perhaps the question is no longer if should we condemn Wernher von Braun. Nor is it if we should venerate Wernher von Braun. Perhaps the question is whether we can forgive him. I believe we can. (It is too much, for me, to forgive the unrepentant Arthur Rudolph.) But such forgiveness will be more palatable if those who only condemn him acknowledge the greatness of his achievements at NASA, achievements that could still set us toward a better future on the Earth and in space. Such forgiveness will be more palatable as well if those who only venerate him can accept, finally and frankly, his criminal complicity with the Third Reich, complicity that is erased from those places where his achievements loom largest.

In a new era when companies like Blue Origin and SpaceX are leading the efforts to return to the Moon and even build a city on Mars—companies that have tolerated, if not encouraged, cultures of overwork, racism, and sexism—the origins of Apollo must be held squarely in view if we are to extract some moral compassing from our first visits to another world and their lessons for the next. In bed, searching the wall's wood grain for more patterns, for sentences, answers, my eyes grew warm to think of our better nature and my stomach clenched. I wanted to believe that we could square ourselves with the violent past that stretched behind this moment by the sea.

Von Braun made the wrong choice to sacrifice the V-2 slave laborers for his dream of species immortality. That some still need to hear this so long after the fact is, to say the least, disappointing. We do not deserve immortality if we continue to put technics ahead of lives. We need an ethical grandeur to guide us. The irony is that the Moon, like any sublime place, provides just that. Places precede us. We need to treat that fact with the humility it deserves wherever we live, on this planet or another. If our species is to last a bit longer by becoming "multiplanetary," as von Braun, Jeff Bezos, Elon Musk, and the poet Nikki Giovanni, among others, have yearned for, we would do well to acknowledge that it's not worth the effort if it brings out our worst selves.

In her poem "Quilting the Black-eyed Pea (we're going to Mars)," Giovanni writes that we go "because whatever is wrong with us will not / get right with us so we journey forth / carrying the same baggage / but every now and then leaving / one little bitty thing behind . . ." Her list of things to leave behind includes torture, lynching, murder, and global war. "We're going to Mars," she writes, "because it gives us a reason to change." So too the Moon.

Seen in this light, what does Apollo mean? At its best, the program offered a sense of wonder, of cooperation, of science and, like it said on the plaque, world peace. Fifty years after the last mission left the lunar wilds, and shorter years till again we go, do we know why we go and what we seek? What does Apollo mean? And, next, Artemis? At least this much: that we fail the promise of the future if we exterminate the past it contains.

CHAPTER ELEVEN

Essentially Gray

Mission Control cap-comm Jerry Carr: *Apollo 8, Houston. What does the ole Moon look like from 60 miles? Over.* [Pause.]

Apollo 8 Astronaut James Lovell: *Okay, Houston. The Moon is essentially grey, no color; looks like plaster of Paris or sort of a grayish beach sand. We can see quite a bit of detail. The Sea of Fertility doesn't stand out as well here as it does back on Earth. There's not as much contrast between that and the surrounding craters.* [Pause.] *The craters are all rounded off. There's quite a few of them, some of them are newer. Many of them look like—especially the round ones—look like hit by meteorites or projectiles of some sort.* [Pause.] *Langrenus is quite a huge crater; it's got a central cone* [peak] *to it.* [Long pause.] *The walls of the crater are terraced, about six or seven different terraces on the way down.* [Long pause.]

My telescope behind a chair, I sit and review the Apollo 8 flight journal transcripts, reading descriptions of the lunar surface by humans flying over it for the first time. First impressions matter, we say, and in this case, it turns out, the Moon did not fare well. The fact of getting there was more important than the there itself. After all, the December 1968 Apollo 8 mission was a huge gamble. It was then the riskiest space mission in history.

With reports that a monstrously large N1 rocket on a Soviet launchpad was being readied for a human shot to the Moon, American officials worried that we'd be beaten again. *First human to reach the Moon* would be added to the Soviet list of *first satellite, first human in space, first spacewalk, first woman in space.* Apollo 8 was to have tested the lunar lander module in Earth orbit. With the N1 report, NASA quickly and uncharacteristically changed its mind.

Instead, the Lunar Module—which was, in any case, behind schedule—would have nothing to do with Apollo 8. (The next flight, Apollo 9, would go on to successfully test the LM in Earth orbit.) Apollo 8 became a sudden, dangerous push to reach and orbit the Moon, involving a whole new set of problems in celestial mechanics, accelerated crew training, and taking the improved Command Module to deep space for the first time. The CM had been overhauled after the Apollo 1 test-pad fire, and Apollo 7 had proved out the new version in Earth orbit. But Apollo 8 would have to rely on the CM's single engine firing while in blackout behind the Moon, both to slow the craft to get into orbit and, more importantly, to fire once more to get back to the Earth. Apollo 8 also would be the first mission to send humans atop the entire Saturn V. The rocket had successfully flown uncrewed tests but had displayed an unsettling effect called "pogo stick," a harrowing vertical oscillation, playful for a child but mortally dangerous for astronauts. Apollo 8 was hurry-up and all-in and a stunning reversal from NASA's developing risk-averse culture.

So it's not surprising how little the Moon—as a place—mattered to the astronauts.

The crew comprised commander Frank Borman, along with James Lovell Jr. and rookie Bill Anders. Borman and Lovell had flown together on the demanding 1965 Gemini 7 mission, which had orbited Earth for some two weeks in a stinky spacecraft with less room than the front seat of the Plymouth Valiant my parents drove.

Of the three, Lovell would go on to express more of the explorer's sensibility. He was easygoing, belying his career as a Navy pilot night-landing jets on aircraft carriers. Famously, Lovell would command the ill-fated Apollo 13 mission. Borman was the archetypal, no-nonsense cold warrior. If Kennedy's challenge had been to build the world's largest and fastest pinewood derby car, Frank Borman would have picked up his Scout knife and whittled a redwood.

Borman was so focused on beating the Russians to the Moon that he didn't even want a TV camera on board. Waste of time. He was overruled, which he later admitted was the right decision. (He was not above taking credit for the famous color *Earthrise* photo that Anders snapped on Apollo 8 which became a widely circulated and iconic image.) Anders himself would describe the Moon as "boring" and "dirty beach sand." Decades later, he remained defensive about his lunar descriptions. Yet Anders had been excited by exploration as a child and had worked with astronaut-geologist Harrison Schmidt to prepare for his duty of photographing as much of the lunar terrain as possible. Anders has vividly described his first sighting of the Moon—as a giant black void blocking the stars. When he saw it, hairs went up on his skin, and the first time he saw light on lunar mountaintops, he scared Borman by blurting out "Oh my god!"

The novelty wore off quickly. Anders grew weary of the craterscape, and both he and Borman were put off by the alien Moon.

In a broadcast, Borman told Earthlings that his "own impression is that [the Moon is] a vast, lonely, forbidding-type existence, or expanse of nothing, that looks rather like clouds and clouds of pumice stone." Metaphors and similes contain tenors and vehicles, the thing or feeling or idea meant to be illustrated and that which does the illustrating, respectively. Borman's vehicle—clouds and clouds of pumice—is surprisingly beautiful but it can't overcome the force of five abstract pejoratives—vast, lonely, forbidding, expanse, nothing—with which he modifies the tenor. The impression is over before the sentence ends. The metaphor is defeated before it can work any magic.

Lovell agreed but with a difference, saying, "The vast loneliness up here of the Moon is awe-inspiring, and it makes you realize just what you have back there on Earth. The Earth from here is a grand oasis in the big vastness of space." Lovell, the transcripts would show, reacted more strongly to the Moon-as-the-Moon— "This is utterly fantastic!"—but his excitement, the awe he expressed over lunar shambles and relic din, would not register with the public, who had absorbed instead those dispatches emphasizing sameness and desiccation: a forbidding Moon of dirty beach sand that Borman said "certainly would not appear to be a very inviting place to live or work."

So people had to wonder, "Then why go?" Public opinion of Apollo and its costs were decidedly mixed in the tumultuous 1960s. When I asked Frank Borman about this, he said, "No, I don't believe that [my description] had anything to do with it. I think . . . it is a very forbidding, foreboding place . . . and I don't think describing it as it really is...is going to make much of a difference."

Yet to the telescopists of the nineteenth century, the Moon's austerity—as it really is—was less a matter of foreboding and more a matter of majesty. Of grandeur. Their scientific training—not unlike that of later Apollo crews—helped them see the Moon as a place on its terms and even as a home.

Borman would not have had the time nor the inclination to read James Nasmyth, who once wrote:

Although it is impossible for a habitant of this earth fully to realise existence upon the moon, it is yet possible, indeed almost inevitable, for a thoughtful telescopist—watching the moon night after night, observing the sun rise upon a lunar scene, and noting the course of effects that follow till it sets—it is almost inevitable, we say, for such an observer to identify . . . so far with the object of . . . scrutiny, as sometimes to become in a thought a lunar being. Seated in silence and in solitude at a powerful telescope, abstracted from terrestrial influences, and gazing upon the revealed details of some strikingly characteristic region of the moon, it requires but a small effort of the imagination to suppose one's self actually upon the lunar globe, viewing some distant landscape thereupon; and under these circumstances there is an irresistible tendency of mind to pass beyond the actually *visible*, and to fill in with what it knows must exist those accessory features and phenomena that are only hidden from us by distance and by our peculiar point of view. Where the material eye is baffled, the clairvoyance of reason and analogy comes to its aid.

This solitary communion, which lauds, as I do, the yin-and-yang of science and metaphor, is a far cry from the necessarily relentless toggle-switch and cap-comm busyness of the Apollo crews. One wonders if future lunar astronauts might yet become in thought, in feeling and in fact, lunar beings, and how that could happen.

Early in the online flight journal transcript, someone has inserted a commentary regarding Lovell's first description of the Moon:

Jim's impressions of what he sees have been replayed many times on documentary films of the Apollo program, especially his first sentence. Though his words are honest, they predate NASA's emphasis on science and geology that led to later commanders, Jim included, becoming accomplished field geologists themselves and enthusiasts of what the Moon's landscape has to offer. Perhaps as a consequence, his portrayal of the lunar surface as grey and colorless set a tone for the public's subsequent perception of the Moon as uninteresting.

Indeed. For the majority of the Apollo astronauts and those in NASA, it was getting there that mattered most, that and not screwing up. Then the view toward home. The Moon wasn't a world that deserved its own attention. The Moon wasn't even much of a symbol for the fliers, for cultural history and poetry were certainly not on the training agenda. (In his address to Congress, Borman called the Apollo 8 crew "unlikely poets, or not being poets at all.") Apollo was first and foremost righteous technic, Uncle Sam's badassery rigged with steady nerves and engineering acronyms. The public heard more about TLI and DSKY than the vistas of the Moon. None of the astronauts were versed in the rhapsodic descriptions of the Moon from selenographers past. None of them had read of William Herschel's preference to live on the Moon instead of the Earth, none of them had spent years in telescopic traverses (though telescope time was crammed in for the crews, including near Tucson, at Kitt Peak) and none of them, surely, had read Edmund Neison. (Some of them did know science fiction, however. Apollo 11's Command Module, *Columbia*, was named after Jules Verne's moonship, and Buzz Aldrin knew the work of Arthur C. Clarke well enough to explain the ending of *2001: A Space Odyssey* to his baffled fellow astronauts.)

In any case, by the time lunar vistas were celebrated—most emphatically by Dave Scott and Jim Irwin of Apollo 15—the public had largely tuned Apollo out.

It's not that the astronauts, some of them at least, were ignorant of their humanistic lack. Take the case of Michael Collins. Collins was cap-comm—the lead radio communicator in Houston—when Apollo 8 was cleared to fire its rockets to leave Earth orbit for the first time and head to the Moon: translunar injection or TLI. Collins said, simply, "Apollo 8, you are go for TLI, over." He wished *so much* for his language to have risen to the occasion, lamenting from then on, privately and publicly, that "I just really wish I had that moment to live over again because I would have said to them, 'Apollo 8, you can now slip the surly bonds of Earth and dance the sky, Apollo 8, dance the sky, you go!'" Embedded within Collins's wish is a line from the poem "High Flight," the one poem all pilots know, in which a flier slips the "surly bonds" of gravity. Of course, had Collins waxed poetic, Borman would have rolled his eyes or worse. Chief of the Astronaut Office, Deke Slayton, might have boxed him about the ears. Collins's revealing confession isn't surprising given that he returned to write a beautiful memoir of space called *Carrying the Fire*. He was one of a small handful of the Apollo-era astronauts with, if not training in poetry, then a poetic sensibility, one revealed most fully after their missions, not during.

More than their uninspiring lunar descriptions, the Apollo 8 crew is remembered for their Christmas Eve broadcast in which all three astronauts read from portions of the book of Genesis. Typical for the time, NASA officials had only told the crew to say something "appropriate" for their historic broad-cast. With no grounding in the Moon's deep role in global cultures, with no sonnet workshops at the Houston simulators, with no awareness of the colorful sweep of telescopic lunar exploration, and with, instead, a public diction and syntax that encouraged speech akin to hard-working drills, Borman, Lovell and Anders were stymied and a bit freaked out. They even considered rewriting the words to "A Night Before Christmas," in what surely would have been most wretched doggerel to hit the universe. Fortunately, a friend of Borman's suggested the opening to the book of Genesis. Still moving, the broadcast is nonetheless ironic: the passage is about God's power, not human curiosity, and it's the Earth not the heavens that get most of the attention. When the crew signed off, they said, "God bless all of you, all of you on the good Earth."

The suggestion was clear enough. Borman, Anders and Lovell agreed with lines from a poet they probably had heard of but whose work they probably didn't know well, Robert Frost: "Earth's the right place for love: / I don't know where it's likely to go better." Certainly not on the Moon, not the Moon as it is. How we see—how we feel—the places we come to goes a long way to determining how we treat them and ourselves. Can we overcome the cargo of our predispositions? Can facts inspire? Can accuracy lead us to lives well-lived? What metaphors do we deploy and can experts in them help us see the Moon? Can awe, wonder, grandeur provide the bedrock for our lunar return?

I believe the story of a famous dead painter, a forgotten French painter and a Moonwalker who became a painter might help us understand the future of the Moon. I've begun to see that we can humble ourselves before the Moon even as we do exalted things and I've begun to see how we might welcome the places we go to—essentially gray or not—so that the place, not our prejudice, predominately shapes us. If so, then place can shape how we shape the place itself. We must jettison the visions of a fictional Moon—I remember mine—and embrace a wonder based on facts. We must fall in love with the real.

☾

For my first six years or so, my family lived in the bland white split-level on Suburban Drive in Indianapolis, where, past bedtime, my sister, Vicki, and

I watched the historic first Moon landing, the ghostly video snowstorm and white plain and black sky of Neil Armstrong and Buzz Aldrin, who, somehow, moved both in front of the Lunar Module and behind it at the same time, as though they were able to pass through solid objects at will, an enviable superpower!

The evening of July 20, 1969, I flew a cardboard lunar lander, purchased by my parents at a gas station, around the dark kitchen where my mother would make spindly casseroles for which she'd stopped caring beside windows to some other impossible life and green mammatus skies. Baron the dachshund circled his bed like a Sputnik, and my misplaced metal planes sliced my sister's foot on the stairs. That night, however, all was forgotten as Walter Cronkite shed a tear, and the astronauts slow-jumped on Motorola's Moon. I flew some more, my mouth a radio by the stove, by cupboards that rose like crater walls of a greater durance. I controlled tiny scouts brave and afraid, soaring over Formica and the cavernous sink whose dishes bore fork-scrape and the surface tension of soapy margarine. I could fly inside my house and disappoint no one.

After moving to the house on Sawyer Street, I was old enough to stay up late and watch weekend sci-fi reruns on Channel 4. They were hosted by a fake vampire named Sammy Terry, who rose theatrically from his casket to introduce films and who talked to a spider named George on a string. I was mesmerized by Technicolor, reds as rich as nuclear fire, greens as thick as our tornado skies, and by the West, where Hollywood rockets stood and above which space became real, more real than the joke name for our town, "Indiano-place." Enviously, I watched the convoys at the gates, the guard stations, and guards themselves with their white helmets, dry mountains in the distance, the discernment of badges, the passing through. Panning, there was the base, its test blocks and tarmac, cranes and gantries, flying wings, reactor cores wired to stanchions that strode like tripods across the tumbleweeds, yesterday's West that I thought would be tomorrow, epaulets and lab coats burning, beakers steaming broken at the climax, the trim girl done screaming who becomes the denouement.

Because the Edmund Space Conqueror proved to be such a tetchy telescope for a young boy's hands, I spent more time dreaming about space than looking at it—and imagining America's golden West—and in this reverie for elsewhere I was encouraged by shortwave broadcasts on a portable receiver from Radio Shack, whose quavering night voices from Radio Moscow and timekeeping beeps from the Royal Observatory, Greenwich, in England I listened to in the dark. It seemed as though they signaled from other worlds. The radio glowed benthic green. Zones were different, explained a chart, and in the garage my father told me that half the globe was already living in tomorrow. When he left us—it was 1971, the summer of Apollo 15—had he gone to that other half of the world?

That was the universe of my childhood Moon. When I imagined standing there, it wasn't the lunar surface of Apollo 11 or the view through a telescope or what Apollo 15 would show us in color. It was a movie Moon, serrated, spired with towers of rock, sharp-peaked like the Tetons I'd seen in library books. A child's Moon, desire's Moon, almost ridiculously rugged. It was Chesley Bonestell's Moon, and it was one that set up so many, including Apollo 8, for disappointment with the real thing.

My lunar surface was the surface of *Destination Moon*, which told the story of the first landing there and for which the artist Chesley Bonestell created its special effects. A silver needle rocket is launched in 1950 from the New Mexico desert on its way to our companion satellite. Weirdly prescient in its premise that a group of private entrepreneurs funded the effort—surely Jeff Bezos and Elon Musk have watched it—*Destination Moon* follows the frantic design and construction of the spaceship *Luna*, its launch that defies government orders and the voyage to, landing on and return home of a crew, a return that film maker George Pal attempted to make dramatic when one astronaut volunteers to stay behind to conserve fuel for the weight-strapped capsule.

The plot didn't matter. It was the Moon that mattered, a retro-futurist, sound-staged sublime of pointed peaks, vertical cliffs, and blasted plains. The flat streets of Indianapolis, the ceaseless corn fields of Indiana that we passed on the way to relatives' houses—this absence of topography I felt more and more keenly as I devoured *Destination Moon* and the other reruns: *Earth vs. the Flying Saucers*, *War of the Worlds*, *When Worlds Collide*. There were, elsewhere, mountains, deserts, spaceships, planets, flying saucers and aliens, who, needing a landscape equal to their alloys and their interstellar drives, would not have landed in boring Indiana's humid plain of soybeans and strip malls, still fly-over for UFOs. At the bus stop by the trailer once, I saw a golden sphere move slowly over trees—a weather balloon—that I tried and failed to make into an extraterrestrial visitation. I clutched my UFO newsletters with their more credible reports to give to Mr. Cunningham, my sixth-grade science teacher, who never gave them back. No one ever showed me the magic of the place I lived, though that owl in the woods beside the trailer park flashed its dark power, and I loved our thunderstorms, the sole sublime of the glacier-crushed Midwest. I dreamt. I watched TV and waited for lightning.

I looked for some sign that there were places both safe and dramatic, where calm and competence reigned over insecurity and rancor, where there were pilots and geologists, benevolent aliens and reliable robots, not a bratty sister, not an absent father, not an exhausted mother so scared that each night after taking more Valium she warned me against the toaster staying plugged in and how each lock was to be tested one more time against the menacing Hoosier dark. When Sammy Terry was back in his casket, I dutifully rechecked those locks and crawled into bed where I traveled as far as I could.

I didn't know it then but the spectacular landscapes of *Destination Moon* were created by an artist who had a profound influence not just on popular perspectives of space but on the people who wanted to get us there. Chesley Bonestell's lunar mountains and fire-scarred craters for *Destination Moon*, along with what was then cutting-edge set-and-camera work, won a 1951 Academy Award for special effects. He gave us the first truly popular views of the lunar surface—often as though *from* its surface—long before we arrived there. It's almost certain the Apollo 8 crew, indeed most of the astronauts, had seen the movie and knew of his space art.

The first attempts to illustrate the lunar surface as though one were standing there, so far as I know, were the newspaper illustrations of bat-lunarians and idyllic Moon ponds that accompanied the 1835 hoax articles about life having been discovered there. A more serious attempt, at least in the English-speaking world, was that of James Nasmyth and James Carpenter in their 1874 book, which included a lunar view of craggy steeples behind which the Earth is eclipsing the Sun. The book's photos of plaster-of-Paris models of the surface seen from above emphasized a crisp sharpness that also influenced nearly every depiction of the Moon that followed until and after the Apollo era. Although nineteenth-century selenographers admitted that rounded objects can cast sharp shadows and that their oblique views of the limb featured rounded mountains, no matter: pointed peaks and vertical faces held sway despite many observers knowing this was not correct.

So inspiring was this fictive Moon of zigzag mountains that even Stanley Kubrick, infamously obsessed with accuracy, gave his own Moon in *2001: A Space Odyssey* the same exaggerated heights and angles of jut and fierceness. But by the 1960s, when Kubrick was filming *2001* in Shepperton, England, we knew from the photographs taken from Ranger, Surveyor, and Lunar Orbiter that, while the mountains on the Moon could be tall and massive, they were rounded by eons of meteorite impacts, edges everywhere softened by strikes into slopes. Kubrick may have disliked every science-fiction film he saw, but Bonestell's lunar sublime stayed with him.

For *Destination Moon*, Chesley Bonestell insisted the ship land at Harpalus crater in the northern reaches of Mare Frigoris and by the Montes Jura that ring the huge crescent "bay" of Sinus Iridum. That way he could dramatically place the Earth in the sky lower on the horizon; surface shots of the actor-astronauts would keep the Earth in the field of view. The visual logic is impressive: from there to here.

Bonestell was, however, selective in his realism and inconsistent in how he spoke of it. For example, he claims to have been angry with filmmaker George

Pal for placing giant cracks on the faux lunar surface. Pal's excuse was that it would give a better depth of field. In another telling, Bonestell says he was responsible for the cracks, which he learned later (he says) could not occur on the Moon. Yet he would have known in the late 1940s that there was no drying surface water on the Moon to leave such features.

The film's long pan shot of Bonestell's made-up Moon remains stirring. The silver, pointed ship lands on its fins, and two astronauts take in the awesome scene from an airlock door on high. Earthlit crater rims are rough, the scattered hillocks are chunky and angular. There's even an arch, a geologic touch that Bonestell would have known was impossible on a Moon free of wind, rain, ice, snow, and streams, any one of which are required to form an arch on the Earth. The sunlit mountains ringing Harpalus are sharp-edged and cut by canyons and brim with precipices. The crater floor is itself cratered—some with white ejecta rays—and long, smooth, almost undulant, sections stretch out in a dark blue-brown. A few rilles curve about. Closer to the ship, there are curious black-green patches—uranium? lichen? algae?—and a dusting of deep yellow—gold? Those huge cracks appear beneath the men like a parched Texas cotton field. The whole scene is a mishmash of visual hyperbole. And the artist knew it.

Born in 1888 into wealth and expectations, Bonestell enjoyed the former and flaunted the latter, becoming not the businessman the family expected but a ne'er-do-well who would drop out of architecture school in order to avoid the math. He became a magazine illustrator, architectural artist, and, eventually, a sought-after Hollywood matte-painter. He worked on *Citizen Kane*. His renderings of the Golden Gate Bridge helped sway public opinion in favor of the project. The gargoyles on the Chrysler Building are a Bonestell touch.

While working these various careers, he recovered an interest in astronomy and taught himself the rigors astronomical perspective and solar lighting as they would exist out there in space, especially on the moons of the gas giants, a self-education that resulted in a stunning 1944 color portfolio of views of the outer solar system in *Life* magazine. No one had seen pictures quite like these. It was as though readers had boarded a rocket far larger than the V-2 missiles about to drop on London and Antwerp. His famous painting, *Saturn as Seen from Titan*, remains a jaw-dropping view. It shows the planet partly lit, the rest in shadow merging with Titan's dark blue-black sky (we'd soon learn that it had a thick orange atmosphere). The planet hangs over Titan's ice-bound surface, the view foregrounding an expanse of sastrugi that leads the eye to snow-dotted brown cliffs, rising like some extraterrestrial vision of an arctic Monument Valley. The cliffs frame the view of huge floating Saturn. A glimmering path of reflected Saturnian light leads us between the high precipice on the left and the smaller massif on the right. In the distance, the Sun brightens a snowfield and a line of cliffs on the horizon. We're meant to step forward. Bonestell would go on to create paintings and illustrations of space stations, other planets and their moons, the Sun, the beginning of the solar system, and more, appearing in major magazines like *Collier's* and *Look* and in best-selling books such as Willy Ley's 1949 *The Conquest of Space.* Bonestell kept painting up to his death in 1986.

"No artist," says Howard McCurdy in his study *Space and the American Imagination*,

had more impact on the emerging popular culture of space in America than Chesley Bonestell. Bonestell did for space art what Albert Bierstadt and Thomas Moran accomplished for the western frontier. Like Bierstadt and Moran, Bonestell's paintings took viewers to places they had never been before. Although the paintings were based on real sites, Bonestell used his imagination to exaggerate features in such a way as to create a sense of awe and splendor. He used light and shadow, as artists had done with the American West a century earlier, to portray space as a place of great spiritual beauty. Through his visual images, he stimulated the interest of a generation of Americans and showed how space travel would be accomplished.

Bonestell turned his gaze to our companion world a year after the stunning *Life* magazine pictorial and five years before *Destination Moon*. In September 1945, only a month after Japan had surrendered to the United States, Americans flew above the lunar surface. We flew in a space craft five hundred miles above Mare Humorum, our eyes spanning 260 miles of sunstruck gray lava, flanked on the south by Doppelmayer—a crater whose mare-side rim has been flooded out of existence—and on the north by the bristly, channeled Gassendi. We flew above the Moon on cheap paper stock printed in full color in an issue of *Modern Mechanix*, which cost fifteen cents and included a feature on the history of radar, ads for correspondence schools, and a how-to column for a homemade drawing compass. We collapsed the distance in a way no telescope on Earth could, next traversing a lunar pole just ten miles above crater ranges of jagged mountains like some otherworldly Wyoming. These are the Moon's mountains, commentators would go on to say, not as they are but as they *ought* to be, the mountains of the Romantic sublime, vertiginous and creased, sharp and tall.

But after years of looking at the real thing through my telescope, I believe Bonestell's Moon pales next to the real one.

Above my desk I have three framed prints from Lunar Orbiter 3, each one a closer view of the huge far-side crater Tsiolkovsky, which features the usual terraced walls of a big, complex crater with an unusual wide floor of jet-black mare lava. The crater looks like a deep, dark lake, and the W-shaped island peak, off-center, rises in several layers and angles, more Appalachian than Himalayan. Still, they are rugged beyond compare. They are a cratered chaos, far out and close in. They do not bore nor disappoint.

As early as the late 1940s, when Bonestell began painting the solar system, the legendary British amateur astronomer Harold Hill was making lunar drawings that showed rounded peaks and gentle rolls quite unlike Bonestell's Alpine fictions. Hill's 1950 drawing of the mountain in the crater Weigel shows a gently sloping feature. A drawing one year later of the mountains beyond the giant crater Bailly, then called the Dörfels, shows rounded summits.

Bonestell may not have kept up with British amateurs like Hill, who published their work in the *Journal of the British Astronomical Association*. When he was a boy, however, Bonestell was schooled on the reality of lunar mountains by a professional astronomer in California. He recalls, "I knew an astronomer, I can't remember his name. He did a model of Copernicus about four feet across, and he had the mountains softly rounded. I was just a kid, and I said, 'Why did you make them so round?' I expected them to be broken and sharp. And he just laughed and walked off. It made me angry . . ." As a ten-year-old, Chesley Bonestell asked for prints of the Moon made with the hundred-inch Hooker reflector. I wonder how consciously he turned to those prints—or against them—as he imagined his views from the Moon.

A 1961 version of the lunar Apennines is a great example of a Bonestell Moonscape caprice: imagined razorbacked ridges and invented sharp peaks. And two depictions of the central peak in Theophilus fill too much of the crater floor, another nod to Bonestell's love of the monumental. A view of Copernicus shows the distant rim more as mountain-serrated not slumped and terraced. A painting of the view from Mons Pico to the crater Plato falsely razorbacks the mountain but faithfully shows the locations and shapes of the nearby Tenerife Mountains. And those jagged, almost prickly, arches, as though southern Utah's national park of that name had been relocated to the Moon, show up again and again, as do weird, biomorphic columns that look like stromatolites. The paintings are all gorgeous as are the tints of this lunar world, from bronzes to grays to the blues and greens of Earthshine on the Moon. They have little to do with the Moon as it is.

"I thought," Bonestell mused of the close-up revelations of the 1960s, "how wrong I was! My mountains were all sharp, and they aren't on the moon." The plea strikes a false note, especially since Bonestell also said of those exaggerations: "I tried to make it just as dramatic as I could." In the late 1960s, perhaps as a mea culpa, Bonestell painted a view of Copernicus showing cosmonauts exploring its foreground and basing his perspective on the famous Lunar Orbiter 2 oblique taken in 1966. Here, Bonestell has softened the edges, smoothed the surfaces, added squat wrinkle ridges, and rounded the central peaks.

His apologists argue that had the painter given us the real Moon we might not have wanted to go. The real Moon was dull. So what if Bonestell preferred a Moon rendered through the lens of dramatic terrestrial ranges: the Alps, the Andes, the Himalayas? Yet the propensity to overlay our landscape experiences and pre-dispositions has had more than aesthetic consequences. Carole Fabricant has pointed out that in the eighteenth century, mountaintops and other high places were "central features of aristocratic landscapes . . . [where] from such heights the . . . spectator, like a lord overseeing his creation, was able to 'command' a view of the country stretching out beneath him and thereby exert control over it in much the same way that the aristocratic class . . . ruled over those on the lower rungs of the social hierarchy." It's not a stretch to argue that Bonestell saw the Moon, consciously or not, as an extension of this heritage. Bonestell's fictive Moon was a gesture toward American dominance of space exploration at just the time when figures like Walt Disney and Wernher von Braun were making that very case. Bonestell's Moon is an ideological Moon, a visualized rhetoric of how we—we, meaning White Americans—could oversee, command, and control this New World, which looks so much like the American West. Even today, Bonestell acolytes defend his choices. Ronald Miller says his "paintings possessed an almost intense believability that was far more important than any mere scientific accuracy they contained." That believability, I've come to think, sacrifices more than accuracy. It makes us prone to imposition.

So Bonestell helped motivate legions of Americans to the Space Race, and when we got to the Moon as it was, we were as disappointed as Frank Borman. The Moon was, Carl Sagan would later say, "A static, airless, waterless, black-sky dead world." In the *New Yorker*, Robert Cooper's piece on Apollo 11 included the headline JUST ONE BIG ROCKPILE.

We see where we are because of where we've been.

The day I came across Frank Borman's high-school letter jacket—hanging in a display at the Kennedy Space Center—the red *T* reminded me with a jolt: Borman was born and raised in Tucson. And I had been doing to Tucson what Borman had done to the Moon.

I had failed to see the rigorous beauty of the desert. At least Borman had better excuses. He was a no-nonsense pilot intensely focused on beating the Russians. (That intense focus took a toll on his wife, Susan, until he recommitted to his beloved.) My failure to see or even just accept the Sonoran Desert was because I felt betrothed to the mountains and valleys of northern Utah, where Kathe and I began our life together. The river we lived next to, with its hammock-perfect willows and its beaver we named Homer and its three mallards we named Huey, Louie, and Dewey, the river with its bony-fingered cottonwood a perch for wintering bald eagles, the meadows beside the banks, the rocky outcrop at the mouth of Millville Canyon I called "Titan," in honor of Bonestell, and how hard Kathe had worked to plant sage and yarrow by the house to replace a lawn and how hard I'd worked to pull invasive dyer's woad over our four acres—all that was bound up in our own intimacy, our love, and our sense of trajectory and possibility. We began there.

In Tucson, we first lived in a too-quiet neighborhood, and, biking to work, I felt as though I had traded the Blacksmith Fork River for the pothole prairie of that desert city's truly shitty streets. I half-consciously feared a career move I myself had made where I felt surrounded by betters instead of peers or friends. The desert's sharp edges—which are everywhere—lacerated me and my persona. It was hot. I was scared. Half-conscious. I even managed to get a concussion within the first year of arriving, just a couple years after my first one. I was miserable and I made Kathe miserable, and the desert was forbidding, except early in the mornings or when gentle winter rains fell. The light was clear but the Sun too bright. Mountains surrounded the city but took too long to reach. The body sought shade like water. What could I grasp that didn't hurt?

The desert's votives were wild animals. At that first house, a bright, mid-century modern, coyotes and their pups frolicked on an empty grass lawn across the street; the little ones nipped and jumped like any puppies. At this same house, whose pool never substituted for the river, Kathe woke me one morning, yelling. We had just returned from our first summer at the cabin. Half-naked, bleary, I staggered to the dining room where, out the window, she pointed to a bobcat on the front stoop, licking his giant paw as nonchalantly as one of our own three cats. The next morning I opened the door and didn't notice him—he was back—till I bent up from getting the paper, and he and I stared a few beats, my heart beating harder, and he was a handsome creature, just staring at me, and I slowly backed up and closed the door. Once I even chased a coyote off who had attacked a neighbor's cat. I bundled it up and walked down the block to a vet.

At the "mud house," the nineteenth-century adobe Kathe lovingly restored and our second Tucson home, where I fell for the Moon, Cooper's hawks perched in the old giant tamarisk. One early hot spring evening, I watched a Cooper's hawk perched on a low, thick branch as it scraped a bloody beak against the bark. The late-day golden Sun lit the red drops like glistening stars made of carbon.

And, though I never found much use for palm trees—even alive they seem like kitsch—when the wind would blow yellow palo verde blossoms into the air, the clouds they made were like a delicate storm, something out of a van Gogh starscape. The indelicate storms of the summer monsoons, their thunder and lightning, gave me back my childhood love of tornado weather. Too jittery, however, I could never let those moments grow into a place.

Meanwhile, Kathe and I made a few friends, the kind that last a life-time. Creatures and scientists and friends—and some pretty good breakfast places—helped me make peace with Tucson during the last few years we were there, and the urban sky grew Moon-luscious.

Whatever comes of our being with—and, soon, again, on—the Moon, we'll need to do a better job than Frank Borman or Chesley Bonestell or myself at accepting a place for what it is. If we don't, we will side with shortcomings and treat it, and ourselves and others, more poorly than we should. Places are their own truths. It's up to us to make peace with that as best we can. They ask that we neither falsify nor bemoan, even when unhappy. Accuracy makes demands, but it can also make caretaking joyful if you are in the right frame of mind and if you are in relation to the real. It was only after researching Chesley Bonestell that I found an artist who preceded him in illustrating space and the Moon both realistically and in relative obscurity. When I found him, I felt relieved.

❋

The French artist Lucien Rudaux, who died not long after Bonestell's first space illustrations were published, took the Moon as it is. He wrote, "An image which is artistically charming, if it is wrong in its proportions, will possess a lesser degree of interest than a sketch which is coarse in execution, but accurate in depiction." Rudaux here sounds like an astronomer not an illustrator trying to make sales. I liked him as soon as I read this.

Rudaux worked in Europe largely a decade or two ahead of Bonestell, who would have the luck of a media explosion in the United States after World War II. Of the French artist, Ronald Miller says, "His portrayals of the lunar surface are especially remarkable. At a time when every other artist was still producing the stereotyped craggy, saw-toothed mountain ranges that dated back to Nasmyth and Carpenter's work, Rudaux was showing us the rounded mountains and rolling landscapes [to be] found by the Apollo astronauts." But, Miller adds, Rudaux's landscapes were "as boring-looking as the Moon itself turned out to be."

Exasperated, I turn to the Moon itself.

How can anyone look at the steep, hulking slopes of Promontorium Agarum on the southeastern edge of Mare Crisium and call it boring? Or the pummeled hubbub of the southern highlands? Or Triesnecker rilles flicking like earnest ideograms? Who could possibly look at the glacis and rays of some craters and not see them as blossoms made of rock? The mountains of northern Utah's Bear River Range swell like the muscles of predators. Their profiles are lunar. Boredom is in the eye of the beholder.

It's also worth remembering that very little of the Moon's surface has been photographed *from* its surface. No one or no camera has perched upon the dramatic mountain rising out of Tycho. No one or no camera has been stationed deep inside a rille. No one or no camera has yet landed at the south pole though that will happen soon enough. There's a moment in every serious lunar observer's sessions when, without trying, you realize you've been seeing the Moon from that eyepiece perspective and, like James Nasmyth, you are imagining standing on the surface. There is a thousand-foot-high wall in front of you: Rupes Recta. There is an even brighter glare: the soil of Reiner Gamma. There are mountains upon mountains that shun terrestrial edges to just shoulder themselves into a sky as dark as a cave but lit by stars. Observers such as Harold Hill, Nigel Longshaw, Sally Russell, and Erika Rix, who have faithfully drawn the Moon in vivid, detailed, and accurate drawings, would likely disagree with Miller.

In a little booklet called *Sur Les Autres Planètes*, published by the University of Paris as part of a conference held on April 10, 1943, I found two Lucian Ruduax black-and-white illustrations of the Moon. One shows the "characteristic aspects" of craters or "petits cirques," with a style that does indeed round and soften edges of the crater rims and a nearby hillock. Foregrounded is a steep interior crater slope with some subdued terracing. Mountains near and far are tall and impressive but not edged like shattered glass. Rilles that run between crater and mountain look like the rilles you'd see in a telescopic view of the Moon, not the cracked bed of a dry sci-fi river. His view of the Apennines is, however, *almost* Bonestellian. Sun lights only the top portion of an overly steep rendering of the western face; yet here too the peaks are not nearly as punchy as others have made them. In the shadows at the base, a rounded hill lurks next to a wide rille, presumably Hadley though possibly it is part of Fresnel.

"We only see a strangely diverse terrain . . . immense plains . . . [that] show numerous differences in level, folds and crevices . . . an uneven and chaotic appearance: there are mountains in isolation or in groups, and more primarily, innumerable formations that resemble volcanic craters of all sizes," he writes.

Ruduax noted that "the mountains project immense shadows: at first glance this gives the impression of a hill with exaggerated height with steep summits standing like sugarloaves. This isn't the case." The slopes are not Alpine-sharp, he explains, and "many seem to have flat or domed summits." The steep mountain slopes on the Moon are not nearly as Bonestell imagined.

Had Rudaux been more popular, asks Miller, "Would we have been so anxious to visit the moon?" Strange question. Many who worked on the race to get us there had been inspired by Chesley Bonestell, but President Kennedy was not reading *Conquest of Space*. He was looking at reports from the Central Intelligence Agency and poll numbers (though he did fall under some spell when he witnessed test-firings of the Saturn's massive F-1 engines). By the time dramatic photographic surface views were available from Apollo 15 and 17, the public had become indifferent. NASA didn't place highway billboards showing the tiny Apollo 17 lander *Challenger* in the Taurus-Littrow Valley—photographed from two miles distance with the massive South Massif in the background five miles farther; the lander was set among the rolling, steep sweep of it all like a cottage in the wilderness. Such a billboard would have been a good idea. Look! That's what the Moon is!

Even from 250,000 miles away, through my ten-inch telescope, the Apollo 17 landing site is wildly evocative. Separated by badlands from the featureless crater Littrow, the region looks like Space Age mesa country, covered by a white tide of light so strong it makes the gray luminous. Here is the giant mountain the Apollo 17 astronauts called North Massif. It looms eerily over the valley floor across from the next dolmen, South Massif, while, at the entrance to the valley, the mountain Ching Te rises like an insistent elder from a chair.

Given Bonestell's fictional views, which were mimicked by so many, and given NASA's failure to understand the importance of *how the astronauts would describe what they saw*, and given the relatively poor quality of early live TV images from the lunar surface, it's clear that enthusiasts for spaceflight—irony of ironies—primed the public for disappointment. It didn't help that both scientists and broadcasters were largely reliant on obtuseness and cliches as they spoke of our companion world. Somehow, in our first journeys to the Moon, we missed it.

❀

Among many astronauts, something did change. They found the Moon and its valences upon returning to Earth. Al Worden came back and wrote poetry. Ed Mitchell came back to create an institute devoted to consciousness and the paranormal. Charlie Duke became a Christian missionary, as did James Irwin. Although tough-guy Alan Shepherd cried when he was walking on the Moon, it didn't change his demeanor. How much more impressive it would have been to have heard him confess to his tears instead of watching his frat-boy stunt of whacking a golf ball on the Moon. They weren't prepared, that alpha-male generation, not for expressing a place and their place in it. This is less criticism than a plea that the astronauts going back to the Moon be trained not only in tasks but in looking—and even in poetry.

Years ago, when I landed at McMurdo Station, Antarctica, to take part in a scientific expedition, I had studied the continent, knew what snow was, and had my boots on. No one expected a speech but I was speechless on arrival and I needed profoundly to pee. Articulation came later, though not without its costs. It took a long time to notice that the blue ice we camped on looked like an eye riddled with cataracts. And I went there as a writer.

Beside bodacious flying, Apollo, what need for eloquence? It came, in moments. After Neil Armstrong's first memorable words, he compared the flat lunar plain to high desert country in the American West and called it desolate. The aptness of this description was weirdly fussed up when he called it "pretty," a word so grossly inadequate it's almost laughable. Buzz Aldrin, after a pause, said, "Magnificent desolation." This was as apt and spontaneous as Armstrong's diction was labored.

The many diversities of modern astronaut crews may enable more eloquence. Shuttle astronaut Story Musgrave, like Worden, returned to write poetry and even obtained a literature degree. Space station astronaut Nicole

Stott has become an artist. Memoirs have been written. But all this happened after the return to one-G. Perhaps the crews selected to return to the Moon should get field training in iambic pentameter, excursions in the history of exploration, exercises in egressing abstraction and landing on imagery instead. They could do worse than to study Apollo 12's Alan Bean and the entire mission of Apollo 15.

Alan Bean, who—and this is really, seriously, deeply uncharacteristic of the 1960s military—took art classes while he was a test pilot. He became obsessed with Claude Monet. Bean returned to paint the most accurate and spectacular and personal images of the Moon's surface yet made. Bean had an artist's eye while there—seeing how the Moon's subtle grayscale changed depending on the angle of sunlight and angle from which it was viewed. He captured the tight horizon and ceaseless pockmarked varied undulations all "jumbled about," he reported. In later paintings, he added colors—colors that aren't there. This was no Bonestellian fiction, however, for, unlike Bonestell, Bean wasn't trying to pass off the colors as real nor were they embellishments. For Bean the colors were ways to express subjectivity and feeling. They were metaphors.

In his painting *Fast Times on the Ocean of Storms*, a Moonwalker isn't walking but stots a few inches above a brown-gray and rough surface, right by a deep crater whose black interior echoes a black sky. There are faint black lines, an almost crackling sense of texture that belies the absolute precision of lunar surface photographs from Apollo. But the color and terrain are faithful, and, from afar, the impasto gives us a sense of the Moon almost bursting with the physical energy the astronaut brings to his stride.

When Bean tapped into color, it was to express, as one painting put it, *That's How I Felt on the Moon*. In the work, a space-suited figure, legs apart, arms held out slightly from the side, stares through the landscape-reflective gold visor. A blue-green palate dominates, as though Bean is emerging from ocean shallows, from the submarine terrains of Yves Tanguy. Overhead and behind a dark blue and purple mottled sky holds thick. It is a beautiful painting, one in which literal accuracy rests nicely with an emotional accuracy foregrounded by the title and the colors. The painting depicts, the colors compare.

"We are used to Earthly landscapes," Bean once wrote, "where the sky is usually brighter than the surface. But on the moon, all the light in the scene reflects off the ground, not the sky. In the way that a dark asphalt parking lot looks bright if you stand under a streetlight at midnight, this surface looks bright." It's not all one pitch, however, because the Moonwalkers all experienced variations in brightness and hues—grays, browns, tans—depending on whether they were looking upsun, downsun, or across. "When I paint the moon," Bean said, "I adjust the values to make the moon look like it appeared to my eye." Some scenes featured, for example, "warmer yellows or oranges or reds, to give the feeling it's hot down there, which it really is . . . You have to understand exactly what *is* true for a situation, or what you see as true, and then decide in your art how close to adhere to that truth." The Moon of Bean was more the Moon of Rudaux (and poets) than of Bonestell.

Bean, who commanded a Skylab mission for fifty-nine days, spoke often of happiness and love. The latter helped the former. The work that Bean left, at its best, combines technical skill, physical accuracy, and emotional truth. That dedication to the real would reach its Apollo apex in the summer of 1971 on a mission when three humans felt grandeur again and again because they had trained to feel it, not through Monet's impressionism (though that might have helped too) but through what many of the astronauts truly loathed: studying geology, a field that at first they could not have guessed would become both adventure and homecoming.

CHAPTER TWELVE

Grandeur

Could it be a Lunar flight/Is one small step towards home?
—**Apollo 15 Command Module pilot Al Worden**

They were students, apprentices to evidence of flowing liquids—from New Mexico's Rio Grande Gorge to the ancient lava and gasses at Hadley Rille along the base of the lunar Apennines. In New Mexico, before they reached the Moon, they drove a jury-rigged version of the Lunar Rover nicknamed "Grover" and they stopped along the edge of the gorge, sampling rocks with scoops, taking pictures of cliffs and vistas, using walkie-talkies to radio geologists in a tent, as though they were on the Moon and the scientists were at back-room consoles in Mission Control. They were Apollo 15 commander Dave Scott, Lunar Module pilot Jim Irwin, and Command Module pilot Al Worden, all accomplished fliers. Irwin and Worden were space rookies while Scott had flown on the Apollo 9 mission to prove out the Lunar Module and, earlier, had been Neil Armstrong's copilot when their Gemini 8 capsule made the first successful orbital docking. The mission was aborted, however, after the capsule spun out of control due to a malfunctioning thruster; Armstrong got the Gemini under control in time to prevent the astronauts from losing consciousness and dying.

As the crew of Apollo 15 prepared for their July 1971 flight to the Moon, which would be the first of three heavily scientific missions capping the program, the training, as always, focused on the spacecraft, the flight, engine burns, switches off and on, dials, fuel pumps, waste dumps, oxygen pumps, CO_2 scrubbers and sextants. And they, with other astronauts, had been sitting in classrooms, enduring many boring lectures about rocks. Rocks! (Here's a memorization sample, as NASA's William Phinney recalls: "Hypidiomorphic granular, porphyritic with medium-grained gray phenocrysts.") The drone of stuffy teachers felt like a hangover and about as useful. At least a hangover meant you'd had a good time.

The science they were to attempt on Apollos 15, 16, and 17 demanded changes to what NASA had thought sufficient for, well, for picking up rocks. From boredom to wonder, Scott, Irwin, and Worden fell under the sway of the Moon so when they arrived in orbit and when Irwin and Scott landed on its plains, the men felt the frisson of recognition and mystery, the embrace of home, and the inspiration, even fear, of the wild. Better teachers changed attitudes about looking at rocks, picking up rocks, and looking at rocky landscapes: The crews began to see places and rocks as more than objects. They saw them as stories.

The Rio Grande Gorge—and other sites, from Iceland to northern Arizona—were a kind of simulation, places to fly over to discern impact and volcanic topography, places to drive across, scanning for typical and unusual geology, places to practice rigorous Moonwalk schedules. These field sites had no beige walls, no chalkboards, no shuttered blinds while someone showed a slide of an olivine thin section for the umpteenth time. Places have elbow room, sky, and ground. They have brio. Getting the astronauts out of the classroom meant that instead of memorization there was visualization, places that could tell someone of catastrophic asteroids smashing into the ground, of lava erupting in stratospheric columns or oozing in roiling red plumes, of sediments deposited in gentle shallows to compress and dry over many epochs to be cliff-raised, as if in a slow dream of change. Deposition, uplift, erosion, eruption, impact: the stories of the land and rocks—rocks as big as a mountain, rocks as small as a grain of sand . . . the stories are epics, invisible to the untrained eye. They cannot be told, let alone appreciated, unless you have a vocabulary for seeing. And the best way to learn that vocabulary is where you need to see. Places, if you let them and if you look, fit into the moments of our lives. They can become challenges. They can be welcomed as gifts or, in retrospect, understood as such. They become years, history. They can be quiet human homes and more-than-human adventures. At their best—at our best—that can happen all at once.

I am standing beneath a midmorning, waning summer Moon. I am scuffing about a scruffy, gravely, rolling desert that gives way to chunks of pale orange rock and the darker cliff—lineated in both the vertical and the horizontal—of the Rio Grande Gorge. The river is sun-glint, a thin crystal hundreds of feet below that somehow measures time by ignoring it. The gorge opens into a crumbly black chaos of slope, of sheer, of outcrop, of dead-grass terrace. The bulk of its rock is volcanic, the river having later cut its course. Beneath a cloudy scrim, a golden eagle, and beneath the eagle, piñon and juniper. To the east, the outline of Guadalupe Mountain there-afters into gray. Other mountains look like low, distant banks of cloud.

They stood here, Scott and Irwin, pretending to be at Rima Hadley, that mile-wide, thousand-foot-deep former lava channel, dimensions I keep in mind when it wavers in my telescope like a strand of dark cirrus. The basaltic Rio Grande Gorge—between Dead Cholla and La Vista Verde—is a good lunar analog at 650 feet deep and between a thousand and four thousand feet across. They came to the gorge on Thursday, March 11, and Friday, March 12, 1971, a dry, cold late-winter overnight.

Before the astronauts arrived, there had been aerial photos taken to plan traverses, hauling of science gear and luggage, reservations made for lodging. They had to check out the big white backpacks they'd wear on their simulated Moon walks. The night prior to the field days included a discussion of what to expect from a geologist who had sussed out the terrain and what the crew should do—usually quickly—at each stop. It wasn't just for Scott and Irwin. The backup Moonwalkers for Apollo 15 were Dick Gordon and geologist Harrison "Jack" Schmitt. Everything was in duplicate: two Moonwalking crews, two support crews, two geology teams, two communications systems, two separate astronaut vehicles, and two different capsule communicators (cap-comms).

The crews drove their Grovers to planned checkpoints and took photos with a color gnomon to show detail and coloration, they took wider shots for context, announced the bag number, scooped up rocks or soil samples, dropped them into the bag and sealed it. They kept logs of what they had collected. They tinkered with scoops and tongs—sometimes the hand tools needed tweaks before they were lunar-ready—and they described what they saw and did, which was recorded, transcribed, and became part of the post-trip debriefs to improve everything from the to-do checklists they'd wear on the lunar suit cuffs to just being more precise with their questions and descriptions. "Crew," says geologist William Muehlberger as an archetypal example, apparently re-quoting geologist Dale Jackson, "you told them this and this and led them off on this wild goose chase, when you should have phrased it this way . . ."

This was a far cry from the early 1960s, at the beginning of Apollo, when the mere suggestion that astronauts pick up rocks to bring to Earth was met with incredulity. Back then, for most at NASA, it was all about getting there. Now the destination itself was of interest.

NASA scientist Uel Clanton recalls,

There were only two or three people at MSC [Manned Spaceflight Center] . . . who were really thinking or planning any sort of science-type activity on the Moon. I rather vividly recall one of the earlier conversations with people really high in the structure where the suggestion was made that one might want to pick up some rocks from the Moon and return them. And the question was, "Why?" And after an explanation . . . the people involved seemed partially satisfied. When again the comment was that on the next mission we would probably also wish to pick up rock samples . . . the question again, in astonishment, was "Why? You have already picked up one set, doesn't that tell you all you need to know about the Moon?"

A rejoinder came from MSC engineering director Max Faget: "It wouldn't look very good if we went to the Moon and didn't have something to do when we got there."

Meanwhile, the scientific community was not thrilled with sending humans to the Moon. Robots could do the job—or, at least, some of the job, as the Soviet Lunokhod program would show. Figures such as Clanton, Gene Shoemaker, Jack Eggleston, and others were fighting to have astronaut-driven science taken seriously: at one point, there wasn't even room in the Command Module to properly store samples. After the thrill of landing, science would need to be the justification. But how to prepare for that?

Starting in 1967, newer astronauts included scientists and they began to help shape the patchwork and tedious geological training that had proceeded them. Some efforts seem a bit humorous in retrospect, like the sandpit at Cape Canaveral with craters poked in it—that would promptly fill with water at high tide. Political tensions also saturated the training development because the US Geological Survey, tasked to help Apollo, faced opposition from personnel at NASA. Eventually, there were more lively classroom sessions in Building 31 at the Manned Spacecraft Center, show-and-tell rock specimens, a simulated lunar surface, a simulated "Moon Room," and replicas of the LM in order for astronauts to practice getting out and getting back in. Astronauts used sampling tools and performed other tasks in parabolic flights with brief episodes of lunar-equivalent gravity and underwater in a pool, also approximating low gravity. These physical challenges helped stoke an interest in just what all this was for, and some astronauts began to embrace Apollo's science goals.

Mostly, the wilds called. Led by Apollo 8 veteran James Lovell, the Apollo 13 crew privately paid to have Caltech geologist Lee Silver lead them on a field trip—during the astronauts' vacation time—because previous instructors and training had been less than helpful. (An exception may have been USGS's Eugene Shoemaker, who led early trips to which some astronauts responded well. The abrasive geologist soon became exhausted, however, with NASA's "funny kind of feudal kingdom.") Silver even cooked the meals for the Apollo 13 crew.

Silver was a former professor for the only trained geologist in the astronaut corps—Schmitt, who would be picked to replace another astronaut on Apollo 17 in order to get an academically credentialed geologist up there. Schmitt was pivotal in getting the Astronaut Office to take geology and science more seriously and getting better teachers involved. He also lobbied to get Mission Control personnel on field trips so they'd have a better understanding of what the astronauts were experiencing as temporary lunar geologists. Confronting the later miasma of budget cuts and public apathy, Schmitt even argued for spectacular final missions, such as to the far side and lunar poles. He was vetoed. But spectacular would describe the Hadley-Apennine site of Apollo 15.

With Silver and other geologists, things got more interesting in the field with "the Moon game," an exercise that placed astronauts in a geologically interesting locale with little background on the site, forcing crews to react as though they were on the lunar surface. What would they find? What was typical? Unusual? What was the story of the landscape? Crews competed to best each other in answering those questions. So geology became a contest among the naturally competitive fliers. Throughout the corps there were astronauts who loved the fieldwork or, at least, the chance to get away. Buzz Aldrin, himself a scientist, said he gave "rapt attention as the instructor talked about things that took place eons before man existed on this Earth." Anders liked geology, but we've seen how quickly he grew bored with the Moon. On the other hand, Collins did not believe in the extent of the geology coursework, and Al Shepard simply didn't give a shit and didn't care if everyone knew. Nor did Deke Slayton, chief of the Astronaut Office.

Dave Scott did. So did James Irwin and Al Worden. Lee, who already had agreed to train some crews, didn't want to take on Apollo 15 due to his busy schedule at Caltech. Eventually he relented and took the crew to the Orocopia Mountains in California in the spring of 1970. From May 1970 to June 1971, Apollo 15 trained on seventeen—seventeen!—field trips, from the Nevada Test Site to northern Minnesota, from Medicine Hat, Alberta (where they witnessed the explosion of five hundred tons of TNT to make a crater they studied), to several sites in northern Arizona and to Moon-like and isolated Iceland, which was Dave Scott's favorite field site. It was in the San Gabriels of Southern California where Silver showed Scott and Irwin the mineral they could find on the Moon: a piece of snow-white anorthosite, that remnant of the magma ocean whose primordial extent was hardly understood.

"I loved them," Scott said of the field trips. "I loved to be outdoors. It was a chance to get away from the simulators and other hardware for three days at a time, a chance to have a few beers in the evening after a hard day in the field."

In William Phinney's history of Apollo science training, Scott admits,

> In the beginning, we had little idea of what was expected of us. When asked to describe what I saw on one of the first trips to Oro-copia . . . I got not much further than saying, "Boy, there's a lot of stuff on the other side of the hill" . . . soon we were describing . . . granite, basalt, sandstone, or conglomerate, and its shape as angular, sub-angular or rounded . . . Lee Silver was an inspiring teacher and a really nice guy. He made learning entertaining.

It helped that Scott had an interest in history and archeology.

It wasn't all smiles. Silver remembers Scott fondly too, but was keenly aware that the commander had an unbelievably full portfolio for Apollo 15. "The mission commanders were training me. More than once, David Scott barked . . . 'Do we really have to know this? Do we really have to do this?'"

One thing Scott did have to do—he wanted to—was put a new hatch on the top of the Lunar Module so he could open it upon landing and just look around. Which is exactly what a geologist would do: get to a high point and survey the surrounds. Silver said this SEVA—or Stand-up EVA—was "splendid." Slayton disagreed. A design change this late in the process was, admittedly, dangerous, and Slayton just couldn't see the value of Scott's pro-posal. Nor did he like Scott's suggestion that the mission trade weight for extra camera equipment and a new rake to get smaller samples. Eventually, Scott got it all. He and his crew were all-in on doing the best science possible, but it would not be easy, especially for the Moonwalkers.

"We had to make instant analyses, and decisions on the scientific value, of the objects we found, without the time to relish their meaning," Scott recalled in his memoir *Two Sides of the Moon*. "In effect, we would have to rely on instinct and training in picking a sample and would have only about five seconds to look at it, and maybe ten seconds to describe it, before bagging it and moving on."

The speed of the mission would be helped by the addition of the rover. The Lunar Rover would vastly extend the ability of the astronauts to make traverses, carry equipment and gather more samples. An actual rover would have collapsed under the weight of the Earth's gravity, so the USGS made its training "Grover" instead. Each Grover took three months to build and was so cheap—just $1,900—because technicians scavenged parts from old cars and even airplanes. The steering mechanism is from a Renault, the wheels from an Olds Tornado, and the Grover used landing gear parts from a World War II–era B-26 bomber. Driving the Grover literally set the pace for geology training. Geologist Don Wihelms writes that Lee Silver would run "along behind Grover to see what the astronauts were seeing so he could judge their observations."

Geologist Farouk El-Baz, who mentored Command Module pilot Al Worden, said the astronaut took his training to heart as well, soon learning the difference between the appearance of lunar craters and Earthly volcanos. "The training particularly of Al Worden was a great pleasure, because he soaked [it] all in. He would give up much of his free time to get more of my instruction. By the time he was ready to fly . . . He showed his ability during the first orbit by conveying the names of the craters he was going over—stunning the guys at Mission Control," El-Baz recalls. Worden said El-Baz became "like a brother."

In his memoir, *Falling to Earth*, Al Worden uses words like "awed," "magical," and "majesty" to describe his training areas and similar language to describe the view of the Saturn V rocket from his T-38 jet and the prospect of space it embodied: "Enormous," "transcendent," "something wonderful." The Apollo 15 astronauts seemed to carry a substrate of poetic feeling for the sublime despite an overburden of necessary work. The geology training honed that even more.

Worden writes,

> As Farouk animatedly took me through the trajectories I would be flying . . . my appreciation of the lunar surface grew. He allowed me to get know geology by really feeling it, not just by memorizing. We studied every tiny detail of the craters and other features . . . Not only would I come to learn their names, but I would also understand what was special about each crater, what I needed to look for in detail, and how to describe it in ways that would help the scientists listening back on Earth.

To make photography, like geology, a form of muscle memory, Worden took to carrying a camera with him everywhere he went. The lens became an extension of his eyes and hands.

It wasn't all work. There were campfire steaks marinated in gut-rot whiskey. There were deceptive rocks from other locales that weren't supposed to be in the current training area (Schmitt had no trouble picking those out and calmly telling the trainers just where the joke samples came from.) There were plaster rattlesnakes and real ones and walks to the edge of active volcanos and encounters with bears and shots of cheap Mexican vodka. Even in Houston, to lighten the mood, folks would put Godzilla toys on the surface screens of the Moon that the simulated LM flew over. Change direction, head in for the landing, and there's Godzilla staring at you.

They didn't find a monster on the Moon. What they found was better than science-fiction creatures and wished-for mountains. They found alien wilds that were intensely, undeniably real. They also found a sense of home. "The lunar maps began to feel as familiar as my home street from childhood," Worden wrote. "When you drive down a familiar street, you know what is coming up soon and remember details such as who lives in which house. The moon began to feel the same way to me, even before I traveled there. The moon became a friendly place."

☽

To that friendly place, Apollo 15 launched at 9:34 A.M. on July 26, 1971, entered lunar orbit three days later, and stayed in orbit till August 4, safely splashing down three days later (with a heart-stopping moment as one of the main parachutes failed). It was the first truly scientific crewed expedition to another world. It carried almost three times more weight for experiments on the surface—some twelve hundred pounds—than Apollo 14. For the first time, Apollo 15's lunar-orbiting Command Module had an entire instrument bay stuffed with experiments, four times more than the orbital science equipment of Apollo 11.

There were other matters throughout 1971. Gas cost forty cents a gallon, and Charles Manson was convicted of murder. The war in Asia featured more intensive bombing by the United States of North Vietnam, and upward of five hundred thousand antiwar protestors marched that spring in the capital. The second Earth Day took place. Conflicts in Pakistan and India and Bangladesh continued, and, in a little-noticed development, the first Starbucks opened in Washington State. The *New York Times* began to publish the Pentagon Papers. The Soyuz 11 crew died upon reentry after occupying the first space station in history, and the first e-book was "published" on a computer at the University of Illinois. President Richard Nixon announced his historic trip to China. In August, race riots rocked Camden, New Jersey.

Against such a backdrop, the fourth Moon landing began to pale. NASA put together a public guide to Apollo 15 before the mission but even its author suggested parts of it were "tedious." He was correct. The fine art of communicating the detective work and excitement of science was still largely undeveloped, though guest TV appearances by former astronauts and sci-fi celebrities like Arthur C. Clarke helped spice up network news coverage that fewer and fewer people watched.

Still, the questions Apollo 15 would perhaps begin to answer were compelling. Would they find ancient ejecta from the massive Imbrium impact that Grove Karl Gilbert wrote about in the nineteenth century? Would they find samples of the ancient lunar crust? What would they discover at Hadley Rille? Was it formed by flowing water—unlikely—or lava or gasses? One researcher, knowing the crew could not descend into the rille to reach the other side, wondered if some kind of retrieval crossbow could be used to sample materials from across the canyon. The idea was nixed, but what great television that would have made! And, finally, what would they learn of the Moon's origins?

Before the answers would come the view.

> *Worden: Okay, Joe. And you'll be interested to know that there's a very thin, crescent Moon in front of us.*
> *Cap-comm Joe Allen: Roger. We've been suspecting that all along.*
> *[Pause.]*
> *Worden: And it may be thin, but it's big . . .*
> *Scott (onboard): . . . Look at that, pal!*
> *Irwin (onboard): Did you . . .*
> *Worden (onboard): Oh, I didn't . . .*
> *Scott (onboard): (Laughter)*
> *Irwin (onboard): Ohh!*
> *Worden (onboard): . . . dream any . . .*
> *Scott (onboard): Look at that! . . .*
> *Scott: Hello, Houston, the Endeavour's on station with cargo, and what a fantastic sight.*

To which Apollo 14's Alan Shepard complained, "To hell with that shit, give us details of the burn."

To which, in 2004, Dave Scott said he didn't care:

> So, yeah, people didn't really like hearing all this esoteric stuff. But it's so beautiful, you just can't avoid it. You can't pass it up. You

want to tell everybody . . . But when you get that appreciation and understanding of what you're looking for on the surface, then you can really enjoy it. And without that, it's just like Apollo 8—Plaster of Paris boring. Yeah, they missed the point . . . That's because they didn't have an opportunity to learn it and understand it, and we did. So we come around the corner and looked down at the Moon and here's all these things we've been told about in reality. And they're much better than the photos. And you say "Wow, look at that stuff. Wow, look at Aristarchus. Look at Humboldt, Copernicus, all that stuff. Wow!" It's even better than these guys who had lectured to us and got us stimulated. So it's all back to the guys who teach you.

Apollo 15's first precise flight descriptions would presage many such observations, made mainly by Scott and Worden before the landing then Worden while on station over the Moon. This was Apollo 15's mode: frequent exclamations of abstractions like "beautiful" and "spectacular," with geologically precise recitations and the occasional metaphor. Apollo 15's enthusiasm was palpable even if much of the description was too scientific for the lay audience to follow.

> Scott: And Houston. We're over Mare Crisium at the present time, and the sights are really striking. I guess some of the interesting things we've noted is the variation in albedo from white to dark gray with many variations of gray in between. And many times, this albedo change appears without any significant change in topography, other than perhaps a—a mountain ridge or a chain or a wrinkle ridge or something, but there are many . . . variations in the albedo all over the surface. I guess our general consensus is that it's gray. We haven't noticed any brown yet.
>
> Cap-comm Karl Henize: Excellent. If I'm not mistaken, this is probably the first time men have been over Crisium. [Pause.]

Scott: *I guess that's probably right. [Pause.]*

Scott: *We have everything from the very old subdued craters that are almost completely washed out to the very bright fresh ones which have interior walls of almost pure white. [Pause.]*

Henize: *We're lapping it up down here. Keep talking if you feel like it. [Pause.]*

Scott: *Another interesting fact that—that we've all noticed is that it—it looks like a great desert across which we've had a number of dust storms. And, in many places, you can see the—the tracks or the swirls across the surface which looks like the—a great dust storm has been blowing across the surface—primarily indicated by the albedo change. But all over Crisium, you can see the streaks, which obviously are from impact at some point or another, but the impression we get is that the [swirls are the] result of a dust storm . . . And, you know, as we look at all this after the many months we've been studying the Moon, and learning all the technical features and names and everything, why—when you get it all at once, it's just absolutely overwhelming. There are so many different things down there, and such a great variety of land forms and stratigraphy and albedo, that's it's hard for the mental computer to sort it all out and give it back to you. I hope over the next few days we can sort of get our minds organized and get a little more precise on what we're seeing. But I'll tell you; this is absolutely mind-boggling up here.*

Henize: *Gentlemen, I can well imagine that a foreign planet must be a weird thing to see . . .*

It was. When Worden described different shades of gray over Mare Sereni-tatis, Henize was impressed. "Al, sounds like you're seeing a marvelous amount of detail up there." Worden replied, "Well, after—after the King's training, it's almost like I've been here before." The King was Farouk El-Baz's nickname. When the Command Module came up over the Apennines, the crew was both moved and bereft of words. Scott tells Henize, "Houston, as we cross out of Serenitatis into the Apennines, why, it's just—unreal. You know, those are very poor descriptive terms, but the—mountains jut up out of the 'ocean' here in great relief. I'm sure the guys who've been here before can probably sit down over a cup of coffee and tell you. But the relief is really pervasive." A half hour later, Mission Control would say: "Farouk and company are ecstatic about what you've told them so far and that all of the SIM [Science Instrument Module] bay experiments are looking to be in excellent shape."

As they came to know their new home with their own eyes, both Scott and Irwin were reminded of snow-covered landscapes. Worden remembers Scott being "almost misty-eyed" when describing the Moon, its "smooth and rounded" mountains and how there was "enough light being reflected off the sides of the mountains . . . to supply some light down on the landing site. And the rille is quite distinctive as we pass right over it."

"Beautiful," Karl Heinze replied. "That must be an eerie sight in a half-light."

Much of the recorded transmissions and internal conversations were, of course, technical—the requisite monitoring of position, status of consumables, buttons put to rights. Also, the mundane. Irwin to Scott, privately, "I'd like to take a crap if I can work it in, Dave."

After undocking, Irwin and Scott headed toward the surface. From orbit Worden photographed everything he could, described everything he could, and greeted Houston's Mission Control after each far-side radio blackout with: "Hello, Earth. Greetings from Endeavor." He did so in multiple languages, as in "Allo, Terre. Salute de l'Endeavour." This was another task the King had helped his protégé with.

Apollo 15's orbital swath described a kind of bow-tie shape, including on the nearside east most of Mare Crisium at the northern edge and Langrenus at the southern edge, narrowing over Apennines and the Hadley landing site, widening again to the nearside west to include Aristarchus and south of Struve to dissecting Riccioli farther south. A similar bow tie described the far side orbital coverage, though farther to the south and included most of the two-ring impact basin Hertzsprung in the east and the crater Tsiolkovsky in the west.

That latter would impress Worden over and over again. After the mission, he'd say, "That central peak is a monster . . . You look over the horizon, and you don't see the [near] rim until you see the central peak. Then, once you see the central peak, you realize that there's a rim there. In other words, you realize that the central peak is sitting off by itself . . . It looks like a big slab that's been stuck up on edge."

Honed to look for signs of volcanism, Worden frequently mistook impact craters with dark deposits as volcanic cinder cones, but his observations were still incredibly useful. He noted changes in colors, steepness of walls, the edges of lava flow fronts, rockslides, freshness of cratering, including places I know well like the shallow crater Picard in Crisium, which Worden could see has six rings on its interior and is browner with a darker halo than the lighter brown of Crisium. They were finally seeing browns. None of this can I see from my orbital station on Earth.

These seemingly arcane details are part of the ongoing puzzle of working out the Moon's long history, place by place by place, age by age by age. Worden had some moments of accidental metaphor too, as in his mission debrief when he described how the fan-shaped ejecta from the crater Proclus makes it appear that one is "flying above a haze layer and looking down through the haze layer at the surface . . . like it's suspended over it."

In his memoir, Worden called the first views of the Moon "weird," "eerie and dramatic," "immense," and saying that "it felt scary to be grazing over mountains and valleys which now filled our windows . . . The moon looked ancient, battered, pockmarked—and dead. I didn't feel a sense of foreboding, but of lifelessness." The surface was "disturbingly close and sharp" and its "variety was fascinating: faults, swirls, wrinkles, powdery dustings, and features that looked weathered by Earth-like oceans and dust storms. Rivers of ancient lava rippled across the barren plains. I reported with excitement on subtle surface flows, patterns, and variations in colors and shades."

He was in "awe" looking at Aristarchus, over which the Command Module's mass spectrometer measured leaking gasses, though the crew would see no lights or red events marking visible transient lunar phenomena. (The mass spec also picked up Worden's urine and fumes from the CM.) Sun angles and changes in the spacecraft's position offered endless views and changes of subtle color. Meanwhile, the mapping camera combined a laser and photos to get distances, angles, and images. "I knew photos could never capture what I observed. Neither can these words," Worden wrote. And "thanks to Farouk," Worden felt "strangely comfortable—I knew this place."

While doing sit-ups with a device called the Exer-gym or running in place, Worden would listen to a small tape recorder—Frank Sinatra's rendition of "Fly Me to the Moon" (of course), Judy Collins, and even audio excerpts from the journal of James Cook, captain of the original HMS *Endeavour*. He talked to Scott and Irwin on the surface, joshing the latter about using his soap. Worden also listened to the Moody Blues, a favorite band of my father's, perhaps taking in the final portentous poem on their groundbreaking album *Days of Future Passed*: "Cold hearted orb that rules the night, / Removes the colours from our sight. / Red is grey and yellow white, / But we decide which is right, / And which is an illusion."

In orbit, Worden knew the answer. He let Houston know in one brief, overlooked exchange. When Henize read Worden the news, the pilot responded.

Worden: That's your world right now.
Henize: That's right; that's our world.
Worden: Our world's up here right now, Karl.

☾

And when Dave Scott and Jim Irwin flew the Lunar Module *Falcon* toward the Moon, their world was neither "up here" nor "down there." It was *right there*. They flew in steep and low, not just passing over the Montes Apenninus but flying *between* peaks. The LM pitched at twenty-five degrees, much steeper than previous missions, Scott took *Falcon* over a twelve-thousand-foot ridge then *through* a pass—like a jet threading a canyon—with the fourteen-thousand-foot Mons Hadley and the nearly eleven-thousand-foot

Mons Hadley Delta looming above them on either side. It was such a tingly approach that Scott later joked that they felt "almost as if we should pull our feet up to prevent scraping them along the top of the range."

Scott came in long, seeing craters he'd not learned in simulation—as well as a view of the gaping Hadley Rille—and made eighteen precise corrections largely in tense silence. As *Falcon* descended, it kicked up dust—"like looking through a thick fog"—just before touchdown. "Okay, Houston," Scott reported. "The Falcon is on the Plain at Hadley." Blinded by the dust, Scott had not seen a small crater on which two of the four landing pads had touched, putting the spindly craft partly over the crater's edge. Thus tilted, two Moonwalkers in their temporary homestead landed on the Moon on July 30, 1971.

Two hours later, Irwin and Scott made sure their suits were functional and safe. They depressurized the LM, and Dave Scott opened the rooftop hatch that he had argued so vigorously for. It would be the first and only SEVA—the "stand-up extra-vehicular activity." Standing on the engine cover beneath him, Scott spent thirty-three minutes photographing the surroundings and describing what he saw to scientists in Houston.

"It was like an exhibition of exquisite images by the great photographer Ansel Adams," Scott wrote later. In the moment, it was a geologically precise exercise, with Scott noting a small rise that blocked his view of a cluster of craters he knew was just over there as well as shallow lineations in the worn St. George crater. "All of the features around here are smooth," he reported. "The tops of the mountains are rounded off. There are no sharp, jagged peaks or large boulders apparent anywhere. The whole surface . . . appears to be smooth." The terrain was nonetheless rolling, cratered and appearing as though covered with a dust, again, like a heavy mountain snowfall. Perhaps the suggestion of a winterscape made sense to Scott, who would later say that the closest he'd gotten to the Moon on Earth was a week in Antarctica.

After the SEVA, Scott and Irwin had a more mundane but welcome gift: They would be the first lunar crew to doff their spacesuits. They'd need that respite because they faced nearly three days of precisely scheduled, arduous exploration. Here's one sample of their EVA schedule: On their planned third Moon walk, at a stop labeled station 13, some three hours and twenty-nine minutes into the EVA, Irwin and Scott would have fifty-three minutes for a "multiple objectives stop at end of North Complex between Chain Crater and 700-m crater." They were to observe a "160-m crater on western rim of the 700-m crater; the 700-m crater; Eaglecrest Crater; Scarps; Based on the characteristics and accessibility of each of these features, the following tasks should be completed at the discretion of the crew: Documented sampling; Panoramic or stereo panoramic photography; Possible drive core tube; Exploratory trench; Soil sample; Targets for 500mm photography; Penetrometer."

In all, there would be three EVAs of between five and seven hours each, totaling 18.5 hours. They'd cover some seventeen miles with the rover. They'd collect 170 pounds of samples and deploy science instruments. They'd be at their lunar outpost for sixty-seven hours. What they'd find in those nearly three days would thrill the Moonwalkers, their colleague orbiting above them, the King and Lee Silver and the other geologists and even the capcomms. If during the mission, much of the public may have been put off by the geological descriptions that excited everyone else—then decades later we can learn from the science and the language of Apollo 15, gleaning lessons not only about how we talk about exploration but, maybe, just maybe, how exploration can become, on that full-circle globe in the sky, a kind of homecoming.

On July 31, James Irwin compared the Moon—"it's beautiful out here"—to Sun Valley. Dave Scott deployed rhetoric, perhaps not entirely successfully: "As I stand out here in the wonders of the unknown at Hadley, I sort of realize there's a fundamental truth to our nature. Man must explore. And this is exploration at its greatest." It was also exploration with the sharpest color TV from the Moon and exploration without front-wheel drive. (The TV from Apollo 15 was in crisp color, a vast improvement, but by then networks and audiences had been losing interest.) Scott and James Irwin began to motor around the Moon for the first time, sampling soil and rocks, taking photos and describing in detail exactly what they saw. But their balky front wheels were a challenge. Despite the rover's driving snafu—it would mysteriously correct itself for the next two EVAs—Scott exclaimed, "Man, this is really a rocking-rolling ride, isn't it?"

Like the LM, the battery-powered rover itself was a tricked-out beast, as rugged looking as the LM appeared delicate. With a nearly sixty-mile range but constrained to a walk-back distance of six miles, the rover featured a low-gain antenna, a 16mm camera, a buddy umbilical system bag, a rear payload pallet, under-seat bag storage, a lunar brush bag, the control console, the lunar surface TV camera, a high-gain antenna, steering and speed controls, and, of course, brakes.

The suits alone were like little spaceships, with an oxygen supply, carbon-dioxide scrubbing, heating and cooling systems, and they included a checklist pocket, tongs, a cuff checklist, a chronograph, a 500mm camera, a marker pen, penlight, a sample container, core tubes and hammer, collection bags, a core tube cap dispenser, a 70mm camera, a twenty-bag dispenser, and a scoop. There was a water tube and fruit sticks by the mouth as well. Exploration at its greatest requires a lot of stuff.

From the open seats—imagine lawn chairs with seatbelts—the astronauts kept up a running commentary.

> Scott: Yeah, man. I wish we could just sit down and play with the rocks for awhile. Look at these things! They're shiny, sparkly! Look at all these babies here—gosh! Man!
> Irwin: Come on, Dave. There'll be a lot of them, let's go back.

Then the astronauts began a series of discoveries that electrified them.

> Scott: There is one boulder! Very angular, very rough surface texture. Looks like it's partially . . . Well, it's got glass on one side of it with lots of bubbles, and they're about a centimeter across. And one corner of it

has got all this glass covering on it. Seems like there's a linear fracture
through one side. It almost looks like that might be a contact—it is!
Within the rock. It looks like we have maybe a breccia on top of a
crystalline rock. It's sort of covered with glass—I can't really tell . . .
    Irwin: It looks fairly recent, doesn't it, Dave?
    Scott: Yes, it sure does! It sure does! And I can see underneath the
upslope side, whereas, on the downslope side, [regolith]'s piled up . . .
Boy, that is really something. Hey, let's get some good pictures of that
before we disturb it too much . . .
    Cap-comm Joe Allen: Roger, Dave and Jim . . . And it probably is
fresh; probably not older than three and a half billion years.
    Scott: Can you imagine that, Joe? Here sits this rock, and it's been
here since before creatures roamed the sea in our little Earth.

Scott hammered and hollered "wowee" and the sound of the hammering
was carried by his suit. Earthlings could hear sounds of the rock's resistance
carried from the Moon. What had they found? Ejecta covered by fiery remains,
a rock of ages with a glassy sheen made from a rain of liquid fire. Their excitement echoed back from Mission Control, everyone was in the spirit, so much
so that the astronauts soon pretended to have a seatbelt problem in order to
go off schedule and collect what Scott would call a "beautiful rounded piece
of scoriaceous basalt." It would be dubbed "Seatbelt Basalt."

On the Moon the astronauts were enthused, saying things like: "This is
unreal!" "The most beautiful thing I've ever seen." "Oh, look at the mountains,
Jim, when they're all sunlit. Isn't that beautiful?"

As any astronaut will tell you, however, working in a spacesuit is incredibly demanding, and the Apollo suits were hard to move in and the gloves
were very stiff. For Jim Irwin, the great difficulty on the first excursion was
a failure in his drinking-water system. Nothing came from the water tube
when he tried to suck on the opening. But he told no one in order to save the
EVA from being canceled, which it would have been, instantly. When they
unsuited in the LM, Scott saw how badly dehydrated his crewmate was. It was

a moment that would haunt both men, each wondering then and years later if the incident had contributed to Irwin's irregular heartbeat on the Moon and his three heart attacks on Earth. The last one in 1991 killed him. It was almost exactly twenty years after the mission.

☾

August 1, 1971, on the Moon . . .

"My, oh my!" Scott said. "This is as big a mountain as I ever looked up."

On a fifteen-degree slope, the rover started to slip downhill. Irwin moved to hold it in place while Scott collected a rock. Irwin would long to reach the summit.

They were in thought—and body—lunar beings. A recounting of Moon walks should take into account walking on the Moon. Once, I had myself rigged up to a device at the US Space and Rocket Center that mimicked the Moon's one-sixth gravity. It felt like swimming and crab-walking and trying to do ballet for the first time. I kept my arms out front, waving, and my legs bent and walked on my tiptoes. It was—awkward, then, after a couple of minutes, weirdly peaceful. I simply could not take my steps for granted. I was almost literally grounded in the present. I bounded, mindfully.

Walking heel-to-toe, our natural terrestrial gait, is harder on the Moon than the lunar loping the astronauts engaged in. Dave Scott compared it to trying to walk on a trampoline. Moonwalkers would lean forward to start loping along then lean back, heels kicked down, toes up, to stop. As Alan Bean reported, if you fall, it "progresses so slowly . . . that there is plenty of time to almost turn around or to catch your footing . . ."

Things were just slower. Buzz Aldrin explained in an article from the *Lunar Sourcebook* that while the Moon's gravity is one-sixth that of Earth, "Objects seem to weigh approximately one-tenth of their Earth weight [and are] easy to handle . . . Once moving, objects continue moving, although their movements appear to be significantly slower . . ."

And what they saw was truly alien: Visibility and color—grays, tans, browns, whites, black—all depended on the height of the Sun and if a Moonwalker was looking away, across, or toward the Sun. Shadows show only cross-sun, and far objects appear much, much closer than they are. Craters beyond one hundred feet were not readily discernible. They might see lunar horizon glow, the Sun's light scattered by high, thin, invisible concentrations of lunar dust.

Scott writes—and one wishes he'd had the time to speak like this from the Moon—that

> as we climbed the slope at the base of Mt. Hadley Delta, the spec-
> tacular panorama spread before us both took me by surprise and filled
> me with wonder. At sea level on Earth the horizon is roughly twelve
> miles away, but on the Moon it is only about a mile-and-a-half . . .
> From our elevated position several hundred feet up, for the first
> time we could see much further [sic] than that . . . In the distance
> was the gentle valley of Hadley Rille snaking across the landscape,
> surrounded by undulating crater-pocked terrain. In the foreground
> and to one side we could see our temporary home the silvery, spider-
> like *Falcon* squatting like a small insect on the vast Hadley Plain
> several miles away. Looming above us to the east was the majesty
> of the 15,000-foot Mt. Hadley Delta—something else entirely. The
> smooth flanks of the high mountains had taken on a golden hue as
> we moved slightly later into the lunar morning. Without the cycles of
> freezing and thawing to crack rock, the tops of the mountains were
> not rough as they would be on Earth, but were instead smooth and
> undulated, clearly framed by the dark dome of the sky.

The lunar mountains were stranger than anything Chesley Bonestell dreamed up.

Their eldritch smoothness can be attributed to older impact-deposited mate-rial being exposed by more recent impacts that shove the material up. This overburden, as scientist David Harland explains, sloughs down the mountain slopes. "The slopes of the massifs are so thick with this talus that slumping tends to erode away craters . . . There is, therefore, very little 'rough terrain' on a mountain slope, and this helps create the impression of smoothness." The summits do, however, tend to have "knobby terrain" that's clear of the overburden.

But Irwin and Scott had only seconds—here and there—to absorb the weirdness, the sublimity, the grandeur of where they were and what they were doing. Checklists beckoned. Houston called. They were the first geologists, albeit not by profession, on another world. As such, they were keeping sharp eyes on everything, from pebbles to peaks.

Like a rock they saw sitting on a kind of outcrop. Scott relayed his excitement to Joe Allen.

> Scott: And there's another unusual one. Look at that little crater here, and the one that's facing us. There's a little white corner to the thing . . . OK, there's a big boulder over there down-sun of us, that I'm sure you can see, Joe, which is gray. And it has some very outstanding gray clasts and white clasts, and, oh, boy, it's a beaut! We're going to get a hold of that one in a minute . . .
> Irwin: Oh man!
> Scott: Oh boy!
> Irwin: I got—
> Scott: Look at that.
> Irwin: Look at that glint!
> Scott: Ah . . .
> Irwin: Almost see twinning.

Twinning is the pairing of large crystals and something that suddenly suggested—anorthosite.

> Scott: Guess what we just found. [laughing] Guess what we just found! I think we found what we came for.
> Irwin: Crystalline rock, huh?

Scott: Yes, sir. You better believe it.

Cap-comm Allen: Yes, sir.

Scott: Look at the plag [plagioclase] *in there.*

Irwin: Yeah.

Scott: . . . Oh boy! I think we might have ourselves something close to anorthosite, 'cause it's crystalline, and there's a bunch . . . It's almost all plag. What a beaut . . .

Allen: Bag it up!

Scott: Ah! Ah!

Irwin: Beautiful.

It was Bag 196 with sample 15415: the Genesis Rock, as the suddenly interested press called it. The geology team in Houston was as ecstatic as the Moonwalkers. Though it turned out not to be a fragment of the ancient lunar crust, it was an incredible confirmation of the anorthosite-rich magma ocean and one of the oldest samples from the Moon, ever, at more than four billion years old. And it was something else: an objective correlative for the excitement and wonder of purely scientific exploration. Right then and there, Irwin and Scott proceeded to gather a more wide-ranging set of lunar samples than the other Moonwalkers before them had on their entire missions. They did so in ten minutes. No robot could have compared.

If that EVA showed how quick and how smooth lunar field geology could be, their last traverse on the Moon showed the opposite. There was the matter of a stuck drill meant to pull out what would be, if they could budge it, the deepest-core sample of regolith ever. But the drill would not budge. Scott grunted and pulled. The men were overworked and sweating, and their fingers—so hard to move in the giant, inflexible gloves—were turning to pulp, black with bruises. (Irwin writes that he'd cut his fingernails back "just as far as I could" early on, in order to lessen the pain on the EVAs; he encouraged Scott to do the same but he didn't.) A few hours before they had to lift-off, standing beside his com-mander, James Irwin exhorted Dave Scott to not give up. Irwin joined the effort as they squatted beside the drill and propped its handle with their shoulders. It was finally enough—it came free. The core included nearly sixty distinct lunar layers. (Taking the core apart for transport back to Earth then became a longer-than-expected chore because the tool meant to section the core had been incorrectly assembled). Irwin and Scott were relieved—and wiped out and frustrated. By now, Scott had lost nearly twelve pounds and hurt his shoulder.

The core sample had been a trade-off. The Moonwalkers had to forgo an entire area, the North Complex of craters. But before they left the Moon they explored more at Hadley Rille, where the astronauts provoked some nervous worry from Mission Control. "When we dismounted from the rover and began to take samples, we found that the lip of the Rille was a far gentler slope than we had expected and the footing was firm," Scott recalled, "though this was not how it seemed to those following our progress in Houston."

At the rille, they saw—no one else had or would—layers of lava bedrock, an untapped chronology of lunar vulcanism to die for. Irwin and Scott sampled what they could, working fast, and even descending into the rille a little—they could not see the bottom. Cap-comm Joe Allen wondered where they were. Scott told him there was no precipice: "Oh, gosh no, it slopes right on down here . . ." Allen was relieved. The last thing anyone wanted was a Moonwalker or two slow-falling hundreds of feet to their death, unable to climb out.

Irwin said he felt at home on the Moon, and neither man wanted to leave. Instead of falling into Hadley Rille or looping down like low-gravity mountain goats, Scott wanted to drive to the bottom. They were melancholy beneath the busyness of collecting, driving, communicating, and photographing. Fortu-nately, Scott's desired drive into Hadley Rille was never going to happen. He may have remembered a comment from geologist James Head, who had told Scott at one of those lively evening geology dinners: "Oh, *yeah*—Jesus, you could spend the rest of your *life* down there!" Then the humor of the answer set in.

How good had the training been? Andrew Chaikin relates that Lee Silver and his cohort had had difficulty convincing other geologists that the

Apollo 15 Moonwalkers were up to the task. Chaikin writes of one sampling moment when Scott and Irwin demonstrated they understood the importance of what Lee Silver had taught them: Don't just pick up rocks. See where they are. See the context. Sample not only the rock but the context itself. Scott described the boulder with the glassy top, sampled it, collected the soil around the boulder's bottom, rolled the boulder over and obtained unperturbed soil beneath. It could be dated by its lack of accumulation of the solar wind, thus showing the age of the boulder's position. One geologist was so enthused that night at dinner that, as Chaikin relates, he loudly commented, "'Why, they did everything but fuck that rock!'"

Back at the LM, getting ready to climb in and take off and dock and head home, Scott and Irwin had several tasks. Among them, they placed on the surface a small figurine called the *Fallen Astronaut* and a placard with the names of deceased members of their American and Soviet tribe. For sparse TV audiences back on Earth, Scott dropped a falcon feather and hammer. They landed on the surface at the same time, proving that Galileo was right about the acceleration of objects under gravity. "Nothing like a little science on the Moon," Scott chirped. Then, true to form, he began to convey details he saw on another peak but Joe Allen interrupted. The Moon walk was over. The astronauts had to load up and go. Quoting a fictional poet from a story by the science-fiction author Robert Heinlein, Allen said to the Moonwalkers, "We're ready for you to come again to the homes of men on the cool green hills of Earth."

And yet the Moon *had* become their home, they knew it so well and loved its details and vistas, their spacecraft their households. "All I knew in those moments was that I had come to feel a great affection for this distant and strangely beautiful celestial body," Scott recalled in his memoir. "It had provided me with a peaceful, if temporary, home."

Gray, yes, but also tan and white and black and golden. It was deep brown. It was even green, in its subtle, mineral-world way. You had to "open up the mind," Scott said decades later. The philosopher Sandra Shapshay calls what the Apollo 15 crew experienced the "thick sublime": an appreciation of the more-than-human that was deeply informed and moved by facts.

Apollo 15's most important discovery was that the Moon is not essentially gray after all.

☙

There are two well-known legacies to Apollo 15, one short-lived and, frankly, absurd; the other, enduring among planetary scientists.

The first legacy of Apollo 15 wasn't its remarkable achievements in both human exploration and scientific knowledge. Instead, the crew had taken postal covers aboard for a private dealer to sell after the astronauts had retired, the funds to go to their children's college educations (astronauts were paid fairly meager military salaries so deals, like free cars, were quietly part of their livelihoods). The plan wasn't against NASA policies—except the unwritten one: no bad press—so when, contrary to the agreement, the stamps hit the market right away, controversy erupted. NASA was embarrassed and angry, and the astronauts were abashed and investigated. While exonerated of anything illegal, they would never fly in space again. The men refused any funds after the fiasco, which admittedly they had brought on themselves, and, many years later, Worden got NASA to return the stamps to the three crew members.

If the political overreaction to the affair seems ridiculous now, that's to the better, because Apollo 15 showed us that the Moon was scientifically knowable and thoroughly exhilarating. Among other things, the mission discovered magnetic anomalies within craters, recorded subsurface heat for the first time, detected variations of materials inside the Moon, found a wider-than-expected range of aluminum deposits and extracted the first core sample, taller than an astronaut, with its detailed stratigraphic record. And they'd found the Genesis Rock.

Worden ran fourteen experiments and conducted multiple photography runs from orbit. Irwin and Scott deployed sixteen science packages on the plain at Hadley Rille. Before heading home, the crew deployed the first small satellite from a Command Module over the Moon, and Worden performed the first deep-space walk to retrieve packages from the science bay in the CM.

Worden was the first person to see the entire Earth and the entire Moon on each side of him. He was ecstatic.

Farouk El-Baz says of the mission's achievements,

> It was the first mission to have a rover allowing us to send the astro-nauts to collect surface samples [away] from the landing point—a mountain front that was different from anything we had sampled. It was also the first mission to have a Scientific Instrument Module . . . bay . . . that carried a host of sensing instruments and a high-resolution camera that was seven feet long and could see down to one-meter [resolution] on the Moon. [And] the high-latitude location [of the Command Module's orbit] allowed us to examine a large swath of the Moon's surface—with much of the lunar terra [highlands] on the farside and of several maria on the near side—including sensing their chemical composition.

Indeed, altimeter and X-ray spectrometer data for aluminum agreed so nicely that their plots matched: aluminum increased with lunar elevations, another indication that this relatively lightweight element was deposited by the magma ocean early in the Moon's history.

"The diversity of the samples that were collected by the Apollo 15 astronauts was astounding," says Ryan Zeigler, the Apollo sample curator at NASA's Johnson Space Center, who specifically points to the Genesis Rock and "the first sample of pyroclastic glass from fire-fountain eruptions."

Apollo 15 points us to the future as well, says Kate Burgess, a geologist with the US Naval Research Laboratory. She explains that "the training of the astronauts that allowed them, as former test pilots, to recognize the importance of [the] Genesis rock while on the Moon" demonstrated the importance of geological field training, which "is probably among the most lasting legacies, as I know it is influencing the training of the current Artemis class."

As the mission came to a close, Apollo 15 paid tribute to the scientific training that NASA had fully embraced. Lee Silver was let into Mission

Control, given a headset, and stood by a console. "Hey, Dave," he called. "You've done a lovely job. You just don't know how we're jumping up and down, down here."

> *Scott: Well, that's because I happened to have had a very good professor.*
>
> *Silver: A whole bunch of them, Dave.*
>
> *Scott: That's right. As a matter of fact, so many of them, it's just hard to—hard to remember it all. But we sure appreciate all you all did for us in getting us ready for this thing. And I'll tell you, I think Jim and I both felt quite comfortable when we got there, about looking around and—and seeing things. I just wish we had had more time, because, believe me, there is an awful lot to be seen and done up there.*
>
> *Silver: Yes. We think you defined the first site to be revisited on the Moon.*
>
> *Scott: Well, as we go around in lunar orbit here, I can look down—and I could just spend weeks and weeks looking. And I can pick out any number of superb sites down there which would take you several weeks to analyze on the surface. There is just so much here. To coin a phrase, it's mindboggling.*
>
> *Silver: Beautiful, Dave. Thank you so much.*
>
> *Scott: Yes, sir. I hope someday we can get you all up here too. I—I think we really need to have some good professional geologists up here. As a matter of fact, good professional scientists of all disciplines, not only in lunar orbit, but right on the surface, because you all would just really have a field day . . . There's just so much to be gained up here.*

There is so much to be gained on the Moon. The crew of Apollo 15 attested to that, again and again, and how their grounding in science gave them a deep appreciation for the Moon as a world on which we can be householders.

If Dave Scott, James Irwin, and Al Worden did not quite have the capacity to bridge enthusiastic abstractions—spectacular, mind-boggling, beautiful—with their exacting geological details—angles of slope, placement of clasts, numbers of lineations—they show us that with additional grounding in metaphor, astronauts have a chance to be artistic communicators while they are on the Moon and not just after.

That said, there was a hint of more-than-usual creativity even before the crew arrived at the Moon. To make traverses and maps of the surface clearer, they applied their own names to craters and landmarks. There was Dandelion (from Ray Bradbury's *Dandelion Wine*), Earthlight (an Arthur C. Clarke novel), Exupéry (the pilot author), Icarus, Kimbal (the protagonist of E. E. Smith's Lensman series), references to J. R. R. Tolkien, and even to the wine that Jules Verne's lunar crew drank after launching, St. George.

If only a poet had accompanied Lee Silver and the crew at the Rio Grande Gorge, around the desert campfires, to give the briefest of lessons on haiku and fresh imagery, on the use of such allusions as they were already utilizing in those landform names, on the internal wiring of a metaphor. The Moon must have poets among its explorers. In a way, on Apollo 15, it did.

The morning after Worden's return to his home, he was shocked when he stepped out to get the paper: There was the Moon. He was beginning to grapple with the scope, details, and feelings of the Apollo 15 mission—and his return from space. "NASA never trained me in public speaking," he said. Certainly they didn't train anyone in creative writing. Whereas Alan Bean came to his Moon walk with a personal interest in painting—and training in it—Worden was unprepared. But he wrote, sometimes in the middle of the night, alone in his postdivorce Houston apartment.

Mindful of war and poverty, uncertain as to our ability to crack through the crust of ego and politics, Worden looked at the surface of the Moon both as intensely real and as a parable for possible destruction: "The moon must teach us . . . not only of age and geology, planets and solar puzzles / But of life, else we end up like her." (Worden also wrote of the personal costs of Apollo: "But, God forgive, the friends erased.") In "Moonscape," he called the lunar surface "a fantastic procession," with each crater "a lesson." He wrote, "There's an order implied by the jigsaw of features . . . In the order of time the moon is our book."

If a reader can set aside his neophyte's love of exact rhyme, Worden's intuition begins to come through: Metaphor is a quest and convergence. It is a chance at exploratory precision. A metaphor is a test flight, revealing more than known at takeoff. If it succeeds, it lands on ground it also excavates and grows from. It evolves. Its dimensions can become enormous while staying folded, intimately, like an eternal origami crane.

To convey a place we must be open to it, or, really, suffused by it, learn it, and set aside what should be set aside. And more, we must find ways to speak it beyond familiar words used anywhere in any context. One reason we've put the Earth in crisis is because we use cliches.

To be a good human requires living the paradox of the marvels of adventure and the clarities of home. Requires the sublime nature of place inform the humble act of householding. Al Worden makes it clear in his poetry and his memoir that he felt at home in zero gravity, at home flying over the Moon the King had made so accessible and exciting to him, and at home as he wrote his naive and often moving poems.

God, of course, so—that's it! Al Worden, you showed me, I knew it, but it must be felt, like a soaring in the chest: Homecoming is wherever one is. Householding is wherever one is. Keeping home is the home keeping you.

"Sometimes, while I sit and enjoy the good company of my family," Worden reminisced in his later years, "the moon will slowly rise above the trees. I generally don't pay it much thought. But occasionally I am reminded of my brief glimpse into infinity while alone on the moon's far side. I still have lingering questions about what I experienced. The answers won't come in my lifetime. That will be your job."

If, as the writer Brian Aldiss suggested at the time of the first Moon landing, that Apollo 11 was something like "autism brought to perfection," then by the time of Apollo 15, the obsessiveness had given way to an openness that implied a question Aldiss also asked: "How should a human being live . . . ? Does it make other people feel good as well?"

There is now a whole school of psychology and neuroscience devoted to positive emotions, including awe. We're learning that awe—call it the sublime, call it grandeur—fosters happiness, empathy, kinship, and a sense of purpose and meaning. ("You want to feel insignificant? Go behind the Moon sometime," Worden told a reporter in 2019.) In some places, such as China, the catalyst for awe tends to be experiencing kindness in a communal situation, whereas in the West, it tends to be an individual's encounter with the more-than-human. In either instance, researchers have yet to determine how long the positive effects of experiencing awe last and what they portend over a lifetime. For all the attention given in the space community to "the overview effect"—an epiphanic form of the sublime experienced while in Earth orbit—I've met more than one astronaut who was no angel. I've met more than one tree-hugger who was an asshole. Experiencing the sublime is no guarantee of a wholesome personality. Yet Michelle Shiota of Arizona State University and others have found that there are "awe-prone people" who are "particularly comfortable with revising their mental representations of the world." Other work suggests that the awe-prone are more creative. Those that experience awe more frequently also tend toward being, well, better beings. Our homes, our householding, our homekeeping, must not be machines for living but organisms for care. The sublime can provoke care wherever we are.

The corona of the Sun seen like a wavering ghost above a dark lunar mountain, the crater-rimmed broad valley that curves about the southern edge of the elongated crater Schiller, so flat and wide it's like an embrace waiting to be filled. Green minerals that glint in hot Sun on an outcrop on the Moon.

To explore is to gift a place with curiosity. To make a home is to submit ego to place. In both endeavors, we need grandeur and tenderness and the words to say it all.

CHAPTER THIRTEEN

The Empty House

It is as if we were alone in an empty house.
—Norman Hoss, *The How and
Why Wonder Book of Stars*

The day before I drove to Mount Wilson, California, Kathe and I browsed in an antiques store.

Neither one of us is fond of clutter, so it's somewhat paradoxical that we enjoy antiquing as much as we do. Kathe found a candlestick while I looked at old landscape prints. Then I saw a globe tucked in a corner, a world between the angle of two walls, the Moon, which was the color of desert sand, smaller than a kickball, and gorgeous. The Crams Moon globe had a classic serif typeface in all-caps for the names of the maria and a sans-serif, perhaps Helvetica, in small size for every other lunar feature. Both were italicized as if to emphasize: *place, place, place.* In all-caps and bold the sans-serif type announced the landing sites for Apollos 11, 12, and 14. So the globe was made sometime in the months of my parents' dissolution. I picked it up. Someone had a run a string through the south and north poles—there was no stand—and to secure it were emblems of domesticity: two white buttons. I happily bought the Moon.

After a lunch of shepherd's pie, we stopped at a used bookstore, where on a table between two old wing chairs was *The How and Why Wonder Book of*

Stars, its garish cover featuring an orange-and-fuchsia Sun. The book was about everything in space, not just stars, I knew, because I'd had a copy as a child. In between its oversized, glossy covers were stars above caves and temples, the Sun's mysteries through time, the Earth and the Moon, even the possible eventual disappearance of all matter, everywhere.

"I remember every illustration," I told Kathe and the cashier to whom I handed the book. The Moon in spangled space by a film-strip-colored Earth. The Earth a black dot in profile over the tremendous Sun. The dark lines of a crater—Wargentin, I now know, the brimful crater with the impression of a bird's foot.

I'd take it with me to California, where, if the skies were clear, I would observe the First Quarter Moon through what was once the largest telescope in the world, the Mount Wilson sixty-inch reflector in the San Gabriel Mountains above Los Angeles. I would use what was the first truly modern telescope to look at the surface of the Moon, which is ironic for at least two reasons. First, the telescope was intended mostly for nonvisual instrumentation. And, I knew, as Charles Wood once wrote, "Almost nobody, then or now, looks at the Moon with a large telescope with their eyeballs." But I would, I finally would.

Kathe and I drove home, but we weren't in Tucson anymore. Over the past two years, we had lived in Salt Lake City, leaving Casa Luna behind and taking up in a 1964 glass-and-brick box house overlooking the city. A backyard ravine, when the oaks leafed out, was a cozy hammock haven, and I had reasonable views to the south and west for lunar time. We'd moved during the COVID-19 pandemic, and I had largely finished up my work at the university long-distance before retirement. It was an anticlimactic end to more than three decades of teaching. But we had returned to the smell of mud, the calls of chickadees, lakes beneath mountains, mountains beneath snow. Already I, and we, had traveled back to Tucson a few times for work and for friends. It felt easier, more comfortable, to be in the desert. Its lessons abided. We still had the cabin in Logan Canyon, and we spent time there with our friends in Cache Valley when the weather warmed. I felt newfound calm rooted by mountain water and the sound of mountain water.

I turned off our usual route—Kathe wondered why—on to Blaine Avenue where, it turned out, Apollo 15 Moonwalker Jim Irwin had lived in a modest brick home during his high school years. When I learned that he too thought of Utah as his home, I felt a surge of gratitude, for him, for the Wasatch, for his walks on trails I walk. I'd looked up from his memoir, *To Rule the Night*, and saw storm clouds over H Rock, just above the neighborhood, and the benedictions of virga. I felt the surge of time, my first telescope-view of Copernicus that sent me on this journey, the peace of all terrain, here and there, lives to savor that the histories gave me—Fontanelle, Gruithuisen, Pickering, Cameron, more—and the last times Kathe and I looked into the eyes of our dear departed cats and my father's look when the door closed, the telescope of four yards and the Space Conqueror still beside my desk along with oh-so-many Moon books, and at home, on a cushion our new cat, Yoshi, not yet introduced to Moongazing. Irwin's book sat lightly in my lap. Everything sat lightly now, which was remarkable, but also cored deep, as though the air vaulted out of bone and the lungs had made amends. Grosbeaks, like maria, provided examples. I wished that I had met him, Jim Irwin, that I could tell him of a trail we have shared.

The eyepiece of a telescope is one small part of the optical system, the final passage before the eye takes it in. Without this small thing, the large and magnified-larger image would be a smear of light. We need, it seems, something small, even delicate, to make sense of that which is bigger than us. Breath makes sense of air. Home makes sense of place. A yard, of the wild. Childhood, of life. We have to ratchet the focus of each into sharpness we may not know we need.

At home, I wiped the telescope down, the black metal gleaming in soft early spring light. I put the *How and Why Wonder Book of Stars* into my backpack, along with charts, atlases, and observing lists. I was getting ready for my lunar pilgrimage and, as I did, I recalled my trip a few months before in which I had hoped to see the first test flight of our new Moonshot, the one called Artemis.

More than fifty after years after the last Apollo flight and with the start of the new Artemis Moon missions, historians still argue, as I knew, about the meaning of the last time we went to the Moon. Was it a geopolitical gamble that helped stifle the Soviet Union? Was it an exemplar of crowning technology? A waste of money best used to alleviate poverty and the host of problems on Earth? Or a force for educational and industrial inspiration? Probably all the above, in degrees, and it was, as I had learned, the pinnacle of a dark story about modern rocketry. Despite the unsettled legacy, we are preparing to return.

In late summer 2022, I kayaked through the Merritt Island National Wildlife Refuge to a beach bordering the Kennedy Space Center in Florida, hoping for a wild, close-enough view to see the first Artemis launch in person. I was there to cover the event for a magazine—and, well, just to see the world's most powerful rocket, the Space Launch System, fling an uncrewed Orion capsule loaded with test equipment to orbit the Moon. A proof-of-concept for the SLS—overdue, overbudget, and not-yet-launched. A proof-of-concept for Orion to fly to lunar space. What could I glean from Artemis I? What are the questions, provisional answers, the possible meanings as this new program—led by the United States but with many global partners—reaches for the Moon this time to stay? What do we bring with us to our steadfast companion world?

As I paddled, vultures skulked along the shore, manatees surfaced like gentle greetings and the red plumes of standing cypress flowers foreshadowed rocket flames. I looked through my binoculars at the SLS across the water, seeing its factual solidity, its size: as tall as the Statue of Liberty, a white liquid-fuel core-stage flanked by two deep-orange solid-rocket boosters, a triple pillar shedding vapors like clouds off their metal skins. Pelicans wheeled. A flotilla of dozens upon dozens of sailboats, speedboats, and houseboats gathered like white animals of the sea. The sky was clear, in the way that skies in Florida are clear, which is to say, hazy-blue. For a few days, there had been thunderstorms and bouts of torrential rain followed by insipid, insistent scrims of humidity, so thorough and so endemic they became subcutaneous. The air was a room-temperature plasma. In Florida, raindrops fall as fast as cyclists peddle. But it's a great place to launch rockets because being closer to the equator adds momentum to your capsule.

Egrets stood white and still as sculptures while osprey taunted them with hauls of taloned fish. Jays perched on the edges of antennae dishes, and there were updates from the launch center. This or that delay. A hold in the count-down. I sweated, waited, and wished.

A former NASA engineer who had witnessed four Apollo launches once shared a story with me. At the Apollo 11 liftoff, he said, all talking stopped five seconds before ignition. Then the viewing stand went from "pin-drop quiet" to the "second-loudest human sound ever." Only an atom bomb was louder. To demonstrate his point, he pounded his chest and staggered, drawing stares from tourists at the US Space and Rocket Center. You don't *hear* such sound. Rather, he nearly whispered, "It strikes you. The mind is effectively shattered."

I wanted that. I didn't get it. Because of fuel leaks, weather delays, and other problems, the rocket didn't fly while I was there. So, a few weeks later, I watched the late-night launch on television. The high-definition video and speaker-sound was compelling—this was a rocket producing more thrust than the Saturn V—and the success was a relief to me and others who hope for a human return and peaceful settlement on the Moon. But the delays were a reminder that the SLS cobbles together new and old tech (the latest in computers, say, along with modified space shuttle heritage systems) and that the program failed to motivate firms to stay within deadlines and budgets, a system called "cost plus." Simply put, the tab for higher costs was picked up by the government—the taxpayers. This approach is now being replaced by fixed-price contracts, which are exactly what they say. No exploration of a place, let alone domestication, is free of foibles, human and otherwise. If we make it to the Moon under Artemis—and if the Chinese build their planned base—can we ensure that complex systems, including those who wear dress clothes or flight suits, cooperate? What does this lunar return proffer?

Not much, say critics, some of whom argue that we should wait for cheaper and better systems or skip the Moon and get to Mars or just stop spending money on human spaceflight altogether in an era of climate change and systemic inequalities. Critics cite the Artemis price tag, $93 billion, though NASA comprises less than a half percent of the federal budget, and companies with NASA contracts provide jobs in all fifty states.

Still. How do those of us who believe that human lunar exploration and science are not luxuries but necessities answer the critics? The easy response is that, in fact, we have barely explored the Moon, humans having traversed just a few miles of its terrain. It's also possible for our species to do more than one thing at a time, such as exploring space *and* crafting a fair economy.

The more complex responses offer deeper meanings, not only of Artemis, but of the entire existential endeavor of our species. We live on Earth—what writer and anthropologist Loren Eiseley called "the sunflower forest"—and some of us explore and even live in "the star fields" of space. These days we need both the humility engendered in the sunflower forest and the visionary toolmaking inspired by the star fields.

Where do things stand? Over the past few years, NASA, in association with commercial firms, has been privatizing the delivery of science instruments to the lunar surface. NASA is buying science-instrumentation room on landers and rovers owned by companies like Lunar Outpost and Astrobotic. In other words, NASA is paying for a service, not a spacecraft. That (hopefully) budget-friendly approach will include cargo for humans and humans themselves, as NASA won't "own" its future Artemis landers but lease them while the astronaut corps stays squarely within the agency. (Private astronaut landings on the Moon are not a thing—yet.)

The 2022 uncrewed Artemis I mission successfully tested the SLS, Orion, and the European Space Agency service module, sending Orion into lunar orbit and returning it safely to splashdown. The plan now is for four humans to fly on Artemis II in late 2024, putting our species in "cislunar" space for the first time in more than a half-century. (Cislunar space is that between the Earth and the Moon.) Artemis III will land the first woman and first

person of color on the Moon—at the dramatic, long-shadowed, mountainous South Pole region—in 2025 or later. The Artemis III mission will
launch using the SLS, then the Orion capsule will dock in lunar orbit with
SpaceX's Starship to descend to the lunar surface. Blue Origin also has been
contracted to develop a lunar lander for later missions. Artemis IV and V
will take place nearer to 2030 and begin construction of Gateway. The
plan is to have Gateway, a small Moon-orbiting space station, serve as
the docking point for Orion and the lander. Gateway could be a rich scientific outpost for lunar and other research and a jumping-off point for
deep-space missions (though many think it's a project largely to sate certain
congressional districts). Some believe that eventually the SLS and Orion will
be replaced by Starship. At about $4 billion a launch, SLS costs more and is
not reusable as Starship aims to be. With twice the thrust of the Saturn V,
Starship aims to the most powerful rocket ever and the first completely
reusable one. But the first Starship that launched blew up; it also damaged
woefully inadequate and poorly regulated launch facilities. The damage
extended to a wildlife refuge, causing some to wonder why, if Elon Musk
wants to make life multiplanetary, he seems not to care for it here on Earth.
As of this writing, conservation groups are suing the government for more
rigorous oversight of SpaceX launch facilities.

Yet the symbolic importance of Artemis III cannot be overstated. It is more
than a lunar return. The age of White-only astronauts is only barely passing.
Out of 360 NASA and international partner agency astronauts, fifteen are or
have been African American, sixteen have been Hispanic, and seventeen Asian;
one is Native American. The current NASA astronaut corps is twenty-three
men and sixteen women. Some right-wing pundits erupted at the goal to land
the first woman and first person of color on the Moon. They claimed it meant
NASA was putting gender and race before all else, as though the United States,
and the astronaut corps, did not have a long history of doing just that in favor
of Caucasian men. Compensating even partially for that systematic prejudice
by having highly qualified astronauts from marginalized communities work
on the Moon is a fair and inspiring start.

They will have plenty to do. The monthlong mission will have the two Moonwalkers on the surface for a week that includes four traverses. Two other crew will stay aboard Orion. NASA's science goals include exploring the terrain, understanding how the ice and regolith mix, collecting samples, and conducting additional science experiments—like Apollo, but in a far more dangerous area. Later missions will include a rover.

"Artemis III is the first in a series of missions which is expected to culminate in the construction of the Artemis Base Camp, humanity's first permanent field station on another world, by the end of the 2020s," as one agency document puts it. Several smaller camps may be used instead of one. Their residents will study how to "sinter" regolith to prevent miniature dust storms from landing and launching rockets—a significant concern is to preserve the delicate exosphere for study and to protect people and infrastructure from damage. They will study the behavior of materials in low gravity, how to use regolith for construction, how to extract and use water ice to grow plants (including in the regolith itself), as well as distill oxygen and hydrogen for breathing and for fuel. This is an order of magnitude more ambitious than Apollo.

Then there's China, rapidly becoming America's main competitor in space. Famously indifferent to international laws, pragmatically in favor of forced labor, rigidly antidemocratic yet oddly nimble for a state-controlled quasi-capitalist economy, China has made huge strides, sending rovers to Mars and the Moon, including the first to the lunar far side, and it continues launching taikonauts to build a space station. In the 2030s, China seeks to build its International Lunar Research Station. Geopolitical competition is a theme once again.

American public opinion is broadly supportive of a range of space activities and settlement. While Artemis isn't seen as a top space priority in recent polls, a 2019 Pew survey found strong support for NASA across the political spectrum, with a whopping eighty-one percent approval rating. In a deeply polarized US political climate, Congress has embraced Artemis in rare bipartisan fashion though funding may lag, forcing NASA to make budget trade-offs, like possibly canceling Gateway. Still, Artemis, spearheaded by a Republican administration, has been embraced by a Democratic one.

We are returning to the Moon, this time, it seems, to stay. It will be anything but easy.

In the mid-1890s, the *Pall-Mall Gazette* in England ran illustrations by Fred T. Jane in a series called "Guesses at Futurity," which were visual speculations on life in the year 2000. Number 7 was "Interplanetary Communication. Gold Mining on the Moon." A partially buried and shaded glass tunnel ran from the foreground past two lampposts and into a complex of enclosed bridges and towers. Above the usual craggy peaks, there hung a half-Earth and an oblong satellite that bristled with antennae. While the architecture is more late Victorian than twenty-first century, the reason for the base—mining—is closer to the mark. Instead of gold, today it's the water.

We're following the water, not only as an economic but as a scientific gift. Lunar water ice preserves a chemical genealogy dating back to the formation of the solar system. Such knowledge will help guide our search for life elsewhere. Ancient cometary ice can further our ability to understand the nature of the outer solar system, and the composition of other lunar materials will help refine our models of the Moon's formation and structure. The far side beckons as a site for radio telescopes. Shielded from the noisy Earth, the far side could be an ideal place to study everything from the magnetic fields of exoplanets to signatures of the early universe. Others dream of infrared telescopes located in dark polar craters, instruments so sensitive that they could detect seasons and *weather* on worlds around other stars.

The place astronauts will call home will have to protect and comfort them. The Moon's surface is beset by sweeping sheets of radiation, from the Sun and other stars, even other galaxies, near-light-speed particles that slice through unprotected viscera, cutting DNA molecules like rice paper and leading, eventually, to cancerous mutations. Long-term lunar dwellers also could suffer from the lingering effects of extended exposure to one-sixth gravity.

Moondust is a huge problem. Apollo astronauts were vexed by the sharp-edged powder, which got under their fingernails and into their noses, lungs, mouths, and eyes. Apollo 12's Alan Bean said residual dust in the LM cabin "made breathing without the helmet difficult, and enough particles were present . . . to affect our vision." The stuff is like "silty sand . . . [but] sharp and glassy," according to the *Lunar Sourcebook*. Coughing and itching are nuisances, but simulated long-term exposure, in one study's words, revealed "significant cell toxicity in neuronal and lung cell lines in culture, as well as DNA damage." Mitigating the dust is a significant challenge, but astronauts and their equipment could be protected with invisible electrodes that activate what researcher Carlos Calle calls an "Electrodynamic Dust Shield"—shifting electric fields that keep the dust from sticking to a surface.

Located at the south polar region, a lunar community would have access to nearly perpetual sunlight for power but might have to bury buildings in regolith for radiation and micro-meteorite protection. Companies like ICON are experimenting with 3D printing of heated lunar regolith. Japanese researchers are considering a trial-run lunar "base" populated first by robots. It might not be a bad idea. Questions abound about the ease of working with lunar compounds, including how to extract oxygen directly from the regolith. None of this has yet been demonstrated on the Moon. It probably will be harder than we imagine.

A more permanent settlement might take us back to our first homes: caves. Altamira 2.0. Although many lunar rilles are collapsed lava tubes, there are others whose roofs remain. Housekeeping in a lunar lava tube removes much of the risk of surface habitation: You are protected from micrometeorites, radiation, temperature swings, and surface dust. Some lava tubes have partly collapsed roofs—skylights—that can function as entry points. Researchers have provisionally identified nearly a dozen lava tubes, including one in the Marius Hills that goes on for miles and miles. There seems to be a system of connected lava tubes near the crater Mairan. These are, unfortunately, quite far from the lunar south pole region where precious water ice waits. Yet some tubes could have ice in them too. No one knows for sure.

Wherever they are on the Moon, astronauts will find that the only elbow room will be in a spacesuit. Science-fiction dreams of capacious domed crater cities must wait. The living space will be tight. And small groups in confined spaces inevitably experience poor sleep, along with stress that can lead to moodiness, tempers, anxiety, mild to severe depression, even decreased dexterity. Which suggests that lunar denizens would benefit from something that both monks and researchers have demonstrated is helpful: mindfulness. Defined as consciously focused, nonjudgmental awareness of the present, mindfulness can be cultivated through everything from three-minute breathing exercises to years of meditation. Studies on mindfulness-based practices among astronauts are relatively scarce, but psychologist Francesco Pagnini is changing that. A researcher at Milan's Catholic University of the Sacred Heart and an associate researcher at Harvard University, Pagnini tells me that while NASA isn't formally incorporating mindfulness or meditation into astronaut training—not yet, anyway—there is of course the consensus that promoting well-being is foundational. Some astronauts do in fact already practice mindfulness. Though the astronaut corps is a far cry from the test-pilot machismo of the early Space Age, Pagnini explains that that space travelers are still loath to admit to boredom or stress. Doing so could endanger their flight status. Still, he thinks that mindfulness, meditation, and relaxation exercises will be part of training within a few years. It's no longer about just toughing out a mission. "Now we talk about thriving [in] space," he says.

To that end, Candice Alfano, a University of Houston psychologist, has found that overwintering residents of Antarctica benefit from "savoring." This can include things that dwellers in isolated communities have relied on for eons: song, dance, rituals, and festivities. To sustain that in space environments, mission planners need to allow astronauts something they are typically reluctant to offer: spare time. Famously, Apollo 7 commander Wally Schirra defied Mission Control when that flight's schedule became impossible to manage. Skylab astronauts did the same. It's better now, but there will be enormous pressure on the Artemis astronauts, even on longer missions, to fill every moment with tasks. Mission planners need to respect the emotional need for savoring. It can also have a practical benefit: it could make astronauts better able to communicate their experiences creatively and authentically while in the moment for those of us on Earth.

Then there's the green world. In 2015 and 2016, Scott Kelly grew the first flowers in space—zinnias—and the photographs he took of them rival those of the James Webb Space Telescope for their beauty. Like other astronauts conducting experiments with plants, Kelly became absorbed by the sprouts, the leaves, the blossoms. Before astronauts dine on a few lettuce leaves grown in hydroponic racks, they eat up the views, smells, and textures of these reminders of the Earth.

Recently, researchers at the University of Florida did something no one has ever done before: They grew plants in alien "soil." Using actual lunar regolith collected during Apollo, scientists Anna-Lisa Paul and Robert Ferl successfully raised *Arabidopsis thaliana*, a delicate cress with tiny green leaves. Although the lunar-soil cress plants did not grow as vibrantly as the control group, the fact that they grew at all left Paul, in her words, "astonished." There, in the photos, are *green leaves* birthed from dark, Moony stuff. Someday, we'll be watering cress planted in Moon dirt with melted Moon water.

By needs we take the green world with us. Not just for salads but for psyches. The green world, the sunflower forest, savannahs, jungles, riverbanks—this is where we came of age as intelligent primates. Green feeds our being, not just with calories but with meaning. It softens the metal edges. Cover a lunar lava tube, seal it, pressurize it, melt some local ice: we could have gardens running riot in tunnels on the Moon, vines gripping basalt, Pickering's revenge.

Can we grow enough on the Moon to lessen or even eliminate terrestrial supplies? A team of Chinese researchers survived on their own for two hundred days in a sealed habitat, raising crops, recycling water and waste all while producing breathable air. At the University of Arizona, the Controlled Environment Agriculture Center features the Lunar-Mars Greenhouses, corridors in which hydroponic sweet potatoes grow wildly—much of the plant is edible, a good choice for space. It's also viny and green—welcome sight for nature-starved humans. Such space farms will provide calories, recycle water, and breathe out oxygen. Researchers are looking at how fish and algae can be raised and harvested in space; both are sources of protein and provide recycling capacities. There's even talk of genetically engineering space food (including protein-rich insects) with the radiation-and-desiccation fighting DNA of hardy microscopic tardigrades. Insects may be more than protein: a Korean study found that caring for crickets improves mood and affect.

We need green on the gray Moon—violas, prickly pear, millet—because if we are to live on other worlds, we must borrow from and care for this one. What the poet Dylan Thomas called "the force that through the green fuse drives the flower" can charge our ethics as we learn to live more carefully.

In 1961, Philadelphia's Fels Planetarium director I. M. Levitt had a "Moon Room" constructed to show visitors how we'd live there in the future. The circular Moon Room—made of molded plastic—featured all the comforts of home: TV, recliner, built-in desk, an ultraviolet lamp for simulated sunlight and more. Comfortable enough! Giving the astronauts as much agency as possible in designing everything from their sleeping pods to utensils will be important in the Moon Rooms to come. Skylab astronauts, for example, loathed how small their mealtime utensils were. (Industrial designer Raymond Loewy was a pioneer in helping NASA see that creature comforts were necessary for Skylab's astronauts, our first long-duration space residents. Loewy convinced NASA that Skylab needed color schemes, muffled circulation fans, and, against engineers' wishes, *windows*.) The International Space Station has come a long way since.

"In all of the excitement surrounding the prospect of human settlement of the Moon," writes philosopher JS Johnson-Schwartz, "it is easy to forget that it will be incredibly difficult for humans to eke out an existence there . . . lunar settlers will have to devote a considerable portion of their energies to the provision and distribution of the basic necessities of life." In the long run, if the Moon proves to be a place we can live and work, some of our worst tendencies may reappear and this needs to be acknowledged and minimized. Johnson-Schwartz cautions,

> Whoever is tasked with overseeing life support production and distribution systems might succumb to the temptation to use their position to extort others by controlling the flow of breathable air or consumable water. Privacy may be virtually nonexistent . . . And the need to maintain a population that is both large enough to sustain the settlement but not so large it strains life support systems will place possibly unwelcome bounds on the reproductive autonomy of settlers.

Lunar humans—lumans?—will demand, design, and deploy community standards and facilities that will make habitats into homes and homes into communities. Johnson-Schwartz also sees the opportunities in talking about these issues now: "We should set the bar high instead of low when it comes to offering lunar settlers lives that are worth living." It is a long trek from an Artemis Base Camp to "the first lunar state," but we have the opportunity to map such a trek now. The ultimate lunar spin-off technology might not be low-impact mining of rare Earth elements. It might not be hydrogen rocket plants for ships bound to Mars and the Asteroid Belt. It might be the yet-perfected art of solving problems without creating new ones. It's a task as tall as the mountains of the Moon.

The San Gabriels—rising between the Antelope Valley and Los Angeles—are steeply faceted, dissected, rugged mountains, like a more widespread version of the Catalinas outside of Tucson. I'm driving to Mount Wilson from the Antelope side, inhaling the thick, sweet perfume of white bark California lilac and catching glimpses of distant, rolling ridges and peaks, many still with snow.

I am also smiling. This April day is bright and fine, the forecast is for clear nights, and I'm going more slowly than the caravan of sporty BMWs would prefer. I ease into a dusty pull-out and am not bothered. They take the swerves, I wait. Eventually, I make my way up the road to the Monte Cristo Campground where a road-cut across from the entrance makes me giddy.

"White! So white!" I exclaim at the stretch of exposed rock.

For there, rising a few feet above the pavement, is dazzling anorthosite. Having checked with the California Geological Survey, I knew this section of the mountains contained a swath of anorthosite but seeing it is another thing entirely. It was here that Caltech's Lee Silver brought the Apollo 15 prime and backup Moonwalkers, along with three other geologists, to see a terrestrial sample of what, on the Moon, would be the clinching evidence for the primordial magma ocean. I pull into the campground, park, stretch my legs beneath budding sycamores, then walk through Mill Creek as it sheets water over the access drive. I cross the mountain road.

The anorthosite is already strewn about so I consider, discard, and make my final decisions—one sample is the size of a date, two the size of fists. The exposure gleams under the high sun so that even with my sunglasses on, I squint, and here beside roots slowly cracking rock to blossom, golden and blue, and to green leaf, I hear Mill Creek rushing by the road and remember Lacus Temporis, the Moon's Lake of Time, how in moments like this, time steadies and flows. I am the child aiming his telescope. I am a man who loves the Moon. By the truck, I wash the rocks in running water and they darken to gray-white, planes and flecks of crystal glinting like ice and stars.

The astronauts spent almost five hours along Mill Creek on November 19, 1970, and more time the next day in Soledad Canyon and at Mount Parker, refining their abilities to recognize anorthosite and discern kinships among various types of rocks. Where they camped, I do not know, perhaps right here. That night, the Moon was a few days past full with the Apennines fully lit. The overcast over Los Angeles broke up for a time in the predawn hours. Perhaps from their sleeping bags they stirred to see Moonglow lighting anorthosite on Earth.

It's not long after that I arrive at the Mount Wilson Antennae Farm—a portion of the six-thousand-foot mountain that bristles with SoCal's TV and radio infrastructure—then turn on a curvy drive where white domes peak out among pines. At the shed-like office of the Mount Wilson Institute, I'm given my key and shown to the Monastery down the way. It's always been called that, since it was built in the early twentieth century, this long, two-story, metal-roofed building with narrow rooms where astronomers stay. My home for the next three nights is a welcome cell, my corner of the capsule: dresser, shelves, tiny desk, single bed. Down the long, window-fronting hallway are a couple of bathrooms, a storage closet with a microwave, the laundry room. Only one other person is here, an employee, so as I look out my window, past burnt pines, down slope, toward the city, I feel embraced by time and quiet.

After unpacking and setting up my telescope in the lot beside the kitchen building, I walk the woodsy grounds—warm sun, chilly shade—past fir and cedar, canyon live oak and sugar pine. Stellar's jays and mountain chickadees—part of our backyard in Salt Lake—squawk and buzz. The ocean I look for is hidden by smog and clouds of the marine layer, but mountains loom to the east. Ravens glide.

It was here in 1889 that W. H. Pickering came to test telescopes, he of the gardens and insects of his preferred living Moon. Pickering left, but George Ellery Hale came, and his work helped to change humanity's view of the universe. The wealthy astrophysicist had built what was then often called the world's largest telescope, the forty-inch refractor at the elegant Yerkes Observatory in Wisconsin. In California, he would outdo himself. First were the solar telescopes, larger and larger, then Hale built the sixty-inch reflector in the early twentieth century.

Refractors can be, as we've seen, plagued by chromatic aberration—false colors—and their optical tubes are extremely long. They'd reached the terminus at Yerkes. Yet reflectors had suffered a bad reputation, largely due to their mirrors being made of polished metal not glass. This speculum metal had many drawbacks, including rapid tarnishing and excess flexing due to temperature changes. The last such telescope, the Great Melbourne, was a forty-eight-inch instrument in Australia and preceded the Yerkes refractor.

Hale recognized that the technology for casting and polishing glass mirror blanks proved ready for the sixty-inch. It took nearly two years of specialized grinding to achieve the mirror's shape and polish, then, horribly, it was scratched. Because the technicians had gained so much experience, however, work to correct the flaw took just four months. The mount was another technological leap; the triangular base is massive, the size of a small truck and weighing seven tons. A fifteen-foot-long polar axis, about which the scope turns, weighs 4.5 tons. And the metal-truss telescope tube—eighteen feet long and 6.5 feet wide—sits in a five-ton, cast-iron fork. To precisely move all this required a four-ton, ten-foot diameter steel float that included a space for 650 pounds of liquid mercury, the pool of which supports most of the weight. The only hitch? Sweeping up leaky mercury from the observatory floor. (The problem was corrected.) There were giant gears, motors and struts—all of which were tested (including after the April 19, 1906, earthquake) before being hauled by mule and truck up the tortuous road from Los Angeles. The dome for this technological marvel was itself cutting-edge. Nearly sixty feet wide, it has a twenty-four-inch air space between interior and exterior shells to keep the facility cooler in the day.

On December 13, 1908, the sixty-inch saw first light—its first observing run. It had twice the light-gathering capacity of any other scope and was able to divert light to a separate spectrograph. The telescope featured different configurations for visual use and the ability to be kitted with instrumentation like cameras, thus moving astronomy squarely into the twentieth century. From 1908 to 1917, the telescope was the largest in the world. Hale would next build the hundred-inch, just down the way from the sixty, then the two-hundred-inch reflector on Palomar Mountain.

With the sixty-inch, Harlow Shapley showed that the Sun was not in the center of the Milky Way as once thought. Shapley also thought that the entire universe was contained in the Milky Way, but, with the hundred-inch, Edwin Hubble found that Andromeda was a separate galaxy. The universe grew. Then, with the hundred-inch Hooker telescope (named for a funder), Hubble famously discovered the expansion of the universe, a fact that brought Albert Einstein to the mountain. Later, I'll use a porch corner at the end of the Monastery for sipping coffee and reading, exactly where Einstein sat. Though less glamorous in the annals of astronomy, the sixty-inch also helped map the pre-Apollo Moon, of particular interest to me of course. It's a bit overshadowed by its bigger sibling.

Among the domes I walk, pausing in front of the sixty-inch, where I will seek things I have not seen before—such as the elusive rille in the graben-bottom of Vallis Alpes, a rille that Pickering himself discovered—and things that I have seen many times, only magnified, bigger—*bigger*—like sunrise over Alphonsus, where, on the sixty-inch, Alter thought he'd detected outgassing from its central peak, part of the weird history of transient lunar phenomenon. It's unreal, the whiteness of the dome, as bright as sunstruck anorthosite, and the doors into the domes—simple human entry to a portal. *All the white domes of Mount Wilson:* the sixty- and the hundred-inch, which is much more massive as you approach it on a walkway, and the smaller domes of Georgia State's Center for High Angular Resolution Astronomy, devoted to combining visible and near-infrared data from six telescopes as though it were one giant thirteen-hundred-inch mirror. I walk past two concrete piers on which were mounted a 1920s experiment that recorded the speed of light with unbelievable accuracy. Mount Wilson has rightly been called a "temple of science." I bring no scientific expertise, no about-to-happen historical breakthrough, but I do bring the love at the heart of science and art: that of seeing itself.

After dinner, I look through my telescope—but miss a goal, my second sunrise on Hypatia, something I have seen just once, when, years ago at Casa Luna, I was thrilled by a ray of light shining through a notch in the crater wall. On the interior of Hypatia's floor grew a flame-shape as though a candle were being lit in slow motion, from nothing to a white-silver something. It is one of the most beautiful things I have ever seen. I will see that sunrise again someday. I move on from Hypatia. And I'm *here*, where the seeing is famously steady.

My view of Valentine Dome confirms it. The terminator is just to its west, with a low Sun lighting the hills on top of this volcanic blister in the north-western Sea of Serenity. Valentine is illuminated but looks very dark, lending a spangled sense of texture to the three sun-tipped large hilltops and half-dozen smaller ones, none which I'd seen before. While I cannot detect a narrow fault that runs across the top—I'd fail to see it tomorrow night too—the small lit peaks give the shield volcano a profound dimensionality. I stare and stare then scan into brighter sun, where the familiar, grand trio of Theophilus, Cyrillus, and Catharina impress with fine steadiness, including the revelation of a small hill in the gully north of Cyrillus: a little bright patch new to me. In Theophilius, the always-impressive central peak reveals on its southeast base etched ridges and fanning shadows. Then, farther south, and not expecting it, I see facing vertical lines on a long peak in Fabricius, gullies never noticed.

This old world keeps offering anew. I gather air, Moonlit and dark, into my lungs, and it smells of pine trees and clean rock.

To whom does the Moon belong? Who gleans its gifts?

Let's use the word deliberately: one need not believe in the divine to see what sustains us is a gift. To those on the lunar surface, the Moon's gifts will be many—from scientific insights to vistas atop the Malapert Massif, a 16,400-foot mountain from which astronauts will see an 11,500-foot cliff and the rolling cratered country of the south pole. That's where they'll descend into pools of deepest dark to scrape and lift to sudden glint the water ice as primeval as creation itself. The work will be delicate, metallic, industrial, scheduled, routine, thrilling and, one hopes, loving. It will be precise. Precision is a form of reverence if it includes gratitude.

The Moon belongs to no one. This, despite a faux educational film, *The Moon of Earth*, featured in *The Simpsons*, that intones, "The Moon belongs to America and anxiously awaits the arrival of our Astro-Men. Will you be among them?" In truth, the Moon belongs to all of us. Like the sky, it is a common good, but, also like the sky, it is governed. Well, somewhat governed. This is where the future gets a bit cloudy.

The Outer Space Treaty, signed by more than one hundred nations, including the United States and China, declares that outer space cannot be appropriated or owned by a country and that it is open for exploration and use by all. But the OST doesn't cover corporations, which, during the Cold War, were not seen as viable agents in space. The treaty also isn't clear about ownership of materials like lunar water ice or conflicts over an area that two entities are trying to use. There's the matter of science versus industry. Scientists may wish to sample and preserve certain materials while companies and space agencies may prioritize using them as "resources." (Attentive readers may have noticed I've been trying to avoid that word, preferring the neutral "materials" or the more humbling "gifts.")

Meanwhile, the decades-old Moon Treaty, which attempted to set out firmer rules, has not been signed by the United States or other space-faring countries. Like it or not, it is a nonfactor. A nascent effort called A Declaration of the Rights of the Moon says that our companion world has "the right to exist, persist and continue its vital cycles unaltered, unharmed and unpolluted by human beings." This too will almost certainly have no policy influence, certainly not on China, though the declaration could have the salutary effect of raising awareness.

The best comparison to terrestrial analogues, I learn, is not the sky but the high seas—owned by no one, traversed by all who can, and whose gifts can be gathered for profit and use within reasonable though sometimes hard-to-enforce frameworks. In fact, President Obama signed a law in 2015 that gave American firms rights to any materials gained from outer-space mining for either their own use or to sell. To set the precedent, the small Colorado company Lunar Outpost agreed to sell to NASA—for one dollar—any lunar regolith scooped up by one of its future rovers. Other companies have agreed to sell as well, but not for one dollar.

Within the space community, there is a large streak of libertarianism that chaffs under and scoffs at much, if any, regulation, even though governments remain the main drivers of space activity and those governments (at least most of them) are answerable to the public. So regulations, standards, and agreements for lunar exploration and use are crucial. Some already exist. Others may be needed. So if the Moon belongs to us, it means the first thing we do is talk with the lawyers.

Michelle Lea Desyin Hanlon, who directs the space-law program at the University of Mississippi, has said that critics of the mix of public and private partnerships in lunar exploration have missed at least one foundational fact: that the United Nations Universal Declaration of Human Rights includes the right to own property. Ownership does not mean rewarding recklessness, however. Guardrails from corporate fines to international sanctions must be at the ready if activities get out of hand, especially because, as she says, "The Moon is something special. The Moon has been with us our entire existence."

We can start with protecting cultural heritage sites on the Moon. To that end, Hanlon formed and leads For All Moonkind, the only nonprofit devoted to protecting lunar heritage from Soviet robotic landing sites to Apollo exploration zones. That organization successfully led the charge to pass a US law guiding NASA and others how to protect historic lunar spacecraft. It was stripped of enforcement teeth, but it's a start.

Hanlon thinks that eventually we will need to "step . . . back" from the traditional concept of ownership and to "think of something new." That shift may be a long time coming. So, for now, the Outer Space Treaty reigns supreme, and, despite something of a dearth of international space law, it's not, she says, "the Wild Wild West out there. The outer Space Treaty is our Magna Carta . . . and . . . the fundamental principle of all of space, no matter where you go, no matter what you're doing, is that it is the province of all humankind. It is free for exploration and use by all." This includes Article 3 of the OST, which says, in her words, "You can't just go into space and violate all the norms of behaviors that we have created here on Earth." And Article 4 notes that celestial bodies, like the Moon, must only be used for peaceful ends.

"There's no law that says you can't mine the moon," Hanlon adds, but any activity on the Moon must proceed with "due regard" for others' activities.

That means information has to be shared ahead of time with appropriate parties. The process for doing so and for one party to object to another is murky. Unresolved, and this is important, is the question of whether due regard includes only those present on the Moon or embraces people on Earth for whom the Moon is sacred.

To help astronauts, agencies, organizations, and companies in steering due regard, another space-law expert, Mark Sundahl at the Cleveland State University Global Space Law Center, is spearheading the Article XI Project with the University of Luxembourg's Antonino Salmeri. Their focus is on "how best to share information about lunar activity" in order to "ensure transparency regarding the peaceful nature of space activities; to avoid harmful interference; and to enable operators to act with 'due regard' with respect to the interests of others." They propose lunar activities be shared with the UN Office for Outer Space Affairs.

Sundahl and colleague Jeffrey A. Murphy, of the Cleveland Marshall College of Law, are optimistic. They write that the OST and a liability convention for space activities is enough, at present, as a guiding agreed-upon vision for lunar activities. Part of the reason why is something called the Hague International Space Resources Governance, which has established twenty "building blocks" to help the international community in crafting agreements from activity databases to protecting scientific areas of interest. Sundahl points to the efforts of the nongovernmental organization the Moon Village Association—which has gathered actors from around the world, from multinational aerospace firms to policy experts based in the developing world—to continue to discuss these issues. Other groups include Open Lunar, the JustSpace Alliance, and the Outer Space Institute, though the extent of their policy influence is a matter of opinion.

Ultimately, it's in no one's interest to sow conflict on the Moon. But what about billionaires like Elon Musk or a country like China, who have said regarding Mars and the Earth's oceans, respectively, "I'll just do what I want"?

"The Moon and other celestial bodies are not subject to national appropriation by claim of sovereignty by means of use or occupation, or by any other means," Hanlon says. "A state can't claim, but maybe a private individual can. And note the language. It says national appropriation. Well, you know, Elon Musk isn't going to go appropriate the Mars for a nation. He might appropriate it for himself and that's not a claim of sovereignty. So we really don't know exactly what this means." She quickly adds that the OST holds nations responsible for "for all activities of their nationals in outer space whether it's by non-governmental agencies or not."

Sundahl notes other concerns. If you announce you intend to explore or work in a certain lunar area, will other parties respect that before you arrive? Can domestic launch licenses be withheld from those seen as violating norms? While public/public and private/private disagreements likely can be resolved through existing means, what about a disagreement between, say, a Western company and the Chinese National Space Agency?

Then there are the Artemis Accords. As of this writing, nearly thirty countries have signed on to this US-led set of protocols, from France to Nigeria to India. The Artemis Accords are the bedrock of the US/global partnership for the program, emphasizing cooperation in the belief that science helps foster peace. Partners will deploy interoperable systems, pledge emergency assistance, register activities, release scientific data, protect historic sites, use the OST to guide "resource extraction," address orbital space junk around the Moon, and provide information about activities in order to avoid danger and conflict.

Nonetheless, Hanlon says, "The technology is going to move a lot faster than the lawyers." After all, people are still arguing about "what constitutes a celestial body." Will "safety zones" function as de facto national appropriations? Will any violations of the OST lead to serious consequences? Will enforcement come from a coalition of the willing, from the UN, or from the various actors who might be on the Moon? Can we assign work zones on the Moon using an international approach, much as we assign global radio frequencies?

In the long run, it's possible that a lunar economy dominated by private corporations or a mixture of public and private activities may never fully mature. The lunar future might be one closer to that of Antarctica, in which a continent—the Moon as the eighth continent—is, by treaty, devoted almost entirely to science. It would be foolish, however, to count on that to keep the Moon—and ourselves—from being unduly scarred by thoughtlessness.

☙

At my little wooden desk and outside by rocks and trees, I have reviewed my targets several times—there are far too many for what will be about four hours of observing—and I have watched a yellow warbler flit in the pines by Einstein's porch. I've seen the Moon rise like solid cirrus from the snow-peaks of the mountains, stalked lizards on the hot duff at the bases of soughing pines, witnessed white domes almost fusing with white light, imagined an entire city beneath the clouds.

The geologist Don Wilhelms wrote that he "relished . . . telescopic observing [of the Moon] more than anything else I did during my career." At the Lick Observatory, he listened to the sounds of ocean wind, violins on the radio, the creaking of gears as the dome moved, the whir of motors as the telescope slewed toward moons fat and slender. I will have such data soon.

There's a knock on my Monastery door. It's one of my guides for the night, retired Jet Propulsion Laboratory astrophysicist Tim Thompson. Active both with the Mount Wilson Institute and the Los Angeles Astronomical Society, Tim is smart, sweet, gregarious, and wry. With a long white beard, he has the welcome mien of a wise teacher. I immediately like him. We fall into a walk about the grounds. Tim shows me more than I've seen or understood, including the Monastery library in the kitchen building, where he has me sit in another chair—Einstein sat here too, of course—and lets me browse the books, from Alfred Noyes's *The Torch-Bearers* (its epic poetry about science was inspired by his trip to Mount Wilson) to technical titles on the Sun. I find just one book on the Moon, one of Patrick Moore's old guides, and glance at some paintings on the wall, pretty fair landscapes by Albert Michelson, he of the light-speed experiment.

We walk on to enter the hallowed space of the hundred-inch Hooker reflector. It *is* huge. Standing beneath the dome reminds me of having stood

in the Pantheon. I'm awed by the sheer size of the Hooker reflector and its
history, knowing that here the evidence of the Big Bang was first accreted,
photon by photon. I'm charmed by the wooden podium at which the telescope
operator worked, with its old-timey phone—a separate earpiece—and a beautiful wooden clock. As I gape, we head out, Tim turns out the lights, and we
walk to the sixty-inch under darkening skies, here, on the biggest backyard
I've ever observed from.

Through the door, then up metal steps illuminated by red lights to the floor
of the sixty-inch. Waiting for me is Tom Meneghini, the executive director of
Mount Wilson Institute and a retired tuna-fleet captain, and David Hasenauer,
a retired optical engineer. They've come to supervise the night's operations,
keep things safe and sure, and to enjoy a night with the telescope. I pass a table
set with coffee, hot water, snacks, and try to take it in. We small talk, I set
down my laptop, they bring a chair for when I'll be sitting to note-take. I look
up at a telescope four times longer than my own—nearly twenty feet—and
six times the aperture. This is a dark temple whose roof is slit open to the sky.
Moonlight pours in. *I thought I was ready.*

Charts, atlases, my computer open to the target list, David is telling me
he's printed it out for me too, there, on the yellow ladder I'll be climbing to
reach the focus—the eyepieces are as thick as a forearm—and Tom and Tim
remark on how steady the air is tonight.

"I have some grails I want to see," I announce, and wonder about playing
the soundtrack I'd compiled, everything from that adagio in *2001* to Bach
Cello Suites to Steve Reich. I don't and I hesitate. "I think I'd like to review
my targets before I look," I say, to David's dismay. I want some long prelude,
some introductory movement before I bend to the Moon. An overture perhaps?

"Oh, no, get on up there!" he encourages. David has gotten the Moon in
the eyepiece, put his laptop with a Moon app on the ladder too, and gestures
me on. Time is short, though I won't be to my room till one A.M. So, feeling
uncertain, I go.

As my companions chat, I inhale deeply, climb—the ladder clatters on
the floor—while David talks me through where and how to place my arms

and legs. He says the focus knob is slow, says the controller for the scope is on my left—relief! I can slew the big reflector on my own. On either side of the eyepiece plate are mounted several outlets where astronomers used to plug in their surplus World War II electrically heated flying suits for cold nights. I won't need that. The air is mild. I'm in a T-shirt, jeans, a jacket to come.

"Go on," someone says.

I lean. The Moon looms—features swarming in ways I recognize but confusedly—and I know, suddenly, that unless you've looked at the Moon through a ten-inch telescope for years, you can't know the hugeness of it all, the sheer detail that will make my search for several tiny rilles both easier—they're all there!—and that much harder because the Mount Wilson Moon swells with unapologetic sharpness, with oodles of craters that pop like declarations then trail off like half-finished sentences only to shout again. My eye moves from one sunstruck peak to another histrionic crater wall, another, another, and just where, where am I looking? I'm more overwhelmed than moved. Why is—wait, is that where Purbach is? Slew. What is above Ptolemaeus—brief excitement at the floor of that gray world—is that, no, Alphonsus's rim is in the wrong . . . Slew. Left, right, up, down, it's all a jumble. Slew long, ah, there's Cassini, but counter to my muscle memory. Simple. The orientation is different from my scope. I throw my reading glasses on and grab my trusty field map, consult, okay, glasses off and look again, no less flustered.

"It's huge," I say, something like that, resorting to conversational cliche because of my sudden worry that I'll waste a lot of time trying to grasp the altered directions of this telescope. *Ah, just breathe. Ride the terminator. That will calm you down, that will take you home.* And I do. And I am.

Dear Stöfler, hello! Ginormous craterland of the south, hello, and there's Faraday with the mass that fans from its imposed crater wall. There's oblong Heraclitus with its dividing ridge. Slew, look, map. There's the trio I expected! Walther, Regiomontanus, Purbach—pause, look closely: *you were confused because the mountaintop crater in Regiomontanus is still shadowed.* God, the debris in Walther is so crisply factual, the broken rim dividing Purbach from Regiomontanus so deftly real. David has the hundred-millimeter eyepiece in, getting 240-mag with this mirror the size of a walk-in closet. Higher magnifications will come. *Yay, yay, hell yay,* I mutter while other chatter sprinkles the dome's arching air. My Moons align: mine, the map's, the telescope's. *Look. Look—*

—remember the sharpness of Copernicus that night years ago in Tucson? Remember the thrill of Hadley Rille seen entirely for the first time, there in the second desert backyard? Remember Rimae Goclenius witnessed from the cabin meadow, the cliffs of Logan Canyon on either side of you? You are standing on a yellow ladder in a hallowed dome with new friends talking and a quiet old friend that has kept you company, kept you steady even when the air wasn't, but this mountain air holds fast, a Moonscape anchored to my sight.

The crater rims of Arzachel and Alphonsus are lit like rough, ancient crowns, like Greek gold-leaf crowns, and Ptolemaeus looks soft as mud and pocked, count, one, two, up to twelve, give up, little saucers, little shallows, ghost craters only a few of which I've seen before and that now seem so many, so fluid, that Ptolemaeus becomes a pond and each ghost crater rim is the dying mark of a ripple, water striders of the Moon! Fish nipping at Pickering's bugs! The floor is awash with static motion.

"Dave!" I say, hurriedly. He had photographed sunrise on the crater floor while Tim and I had walked the grounds.

"Yes, Chris?"

"Did you see the gray bars at sunrise?"

He knows exactly what I'm talking about, answers yes, so I scoot down. He shows me on his phone—Dave is expert at holding his cell phone to the eyepiece—and there are the soft gray pillars of light that cross the crater's floor at sunrise. I remembered being so impressed by this at the cabin the first time I saw them that I drew them, poorly, and counted: I saw, what? Three or four light bars where in this photograph, depending on how one divides them, there are five.

There's a William Blake illustration of a child at the base of a thin ladder leaning on the Moon, captioned "I want! I want!" There's a couple about to be left behind. The parents? Is this, as I believe, an innocent desire, the desire to discover? Can I climb desire so that its frustration is no loss, its culmination less victory and more a kind of rest?

I want. I want. I head north knowing firmly where to go. The Alpine Valley, that gorge in the Alps that divides the southern portion of the range endcapped by Cassini's goofy eyeball from the northern portion that curves like claw gripped around Plato, which remains hidden in the night.

"Jesus," I whisper as I pass the Montes Apenninus, looking like mountains out the airplane window, my Wasatch on the Moon, past frozen wave-crests of crater-radial ridges, like those of Aristillus, and lava flows topped with crescent hills never before seen and the black-top crater atop Mons Piton and I skip my eye beyond to see in lunar day Aristoteles and Eudoxus flanking the eastern defiles of the Alps. Their rims rise up. Crater chains make two dark broken streams.

What if I don't see it? I've been wondering for weeks, after having been told I would catch that narrow rille that skedaddles down Vallis Alpes. What if, after booking time on this giant telescope, making this journey, planning with maps, e-mailing with seasoned observers who have had no trouble seeing the rille with much smaller instruments, what if it eludes me? I'd have to let it be, accept, *Here is all the rest of the Moon.* You know. Because the Moon is one way to talk with death and a way to take up life. *Relax. Either I see it or I don't,* I whisper, and I defocus then refocus on the Alpine Valley, which cuts its sedate divulsion in the crust. I stare. Really, really hard. My eyes tear up and I grip two fists and rub them and step away from the eyepiece, holding on to the yellow railing, and I lean back in.

Later, I'd realize it was like the first time I'd found a meteorite in the wild—I had to look in a pausing sort of way, almost deliberately willing the realization of sight to keep disbelief and potential disappointment at bay. Looking paused and poised into one steady *nooooo* . . . but then *yes*: a white line, just that, just so, a white line—slightly silvered?—catching in the eye like silk-strand glint in the wind's periphery. The rille that runs down Alpine Valley, it's like that—I look, I lean back, back in, it's still there, not the black gash I'd expected, no, it's spottily intermittent at its narrow end, which begins in Mare Frigoris then, more than halfway down, toward where it debouches into Mare Imbrium (still in night), the line steadies, like a waver becoming a tone, the rille curving in one strong section past the shadow of its nearer cliff and plunging into the not-yet-sunrise dark of the valley's mouth.

"Gentlemen! Tell me if you see this!" I command, and, one by one, Tim, Tom, and David climb and look and see the same. Relief and elation cohere like the rille itself coming cleaner in its course.

For weeks, I'll look at David's photos from my night and see the same, white as more than intimation. Not long after getting home, I'll show Kathe, knowing she'd be—as almost anyone would be—underwhelmed by this nuance. I will say, excitedly, "How can I explain this? In my scope, at high power, the Valley is as wide as the end of an open set of tweezers, it's that small! But with the sixty-inch it was like a fat pencil and the rille was like one of the ridges of the pencil, only white." I will admit she's right when, after her congratulations, she offers: "They both sound rather small."

For a moment, the entire rille pops into view, from beginning till its white line crosses the terminator. Dave and I talk earnestly, and he calculates that I'm seeing objects about a third of a mile wide, about eighteen hundred feet wide. Five American football fields. That's something I can take in. That's how many seconds of cycling? Humanizing scale. *The rille is a third of a mile wide.* At its final curve into the darkness, the white line shows a shiver's worth of shadow. I see its side in shadow, I perceive its depth.

Rima Alpes once glowed like a line of prairie fire.

At some point I had turned music on—"Is that from *2001?*" someone asked—and now Tavener's "The Protecting Veil" soars with the encircling red light that glides up the enclosing dome. I pause, step down, munch a snack, drink more water, use the restroom, pass a locker labeled HUBBLE, put on a jacket and just stare at the telescope staring, the blue-trussed tube and its base of circles and curves and plates, the engineering of movement made moving, like some futurist architecture only this time dedicated to silence and peace, not noise and rancor. I stand beside the most beautiful telescope in the world, one the best had used. If I could put my arms around its girth, I would.

☾

What if, with your own eyes or through a friend's binoculars or through a telescope borrowed from the public library, you could see the lights of a lunar settlement or, perhaps, the scars of strip mining? What if we knew that such scars were for the putative fusion power source helium 3 in order to provide electricity for everyone on Earth or what if, instead, the scars were for cracking hydrogen out of rocks to fuel a billionaire's flight to Mars?

Would the lights enchant? Inspire? Would they disgust? Would the scrapes of lunar dragline excavators remind us of human potency—I, for one, find large dams and power plants weirdly attractive—or would these spark indignation?

During a November 2021 online meeting about the ethics of lunar exploration, one researcher said, "The atmosphere of the Moon and the look and feel of Shackleton Crater and the [South Pole] peaks are not of interest to most people. But put a strip mine that shows on the 'Man in the Moon,' [and you] will get 7 billion people opposed to the activity."

Massachusetts Institute of Technology aeronautics and astronautics doctoral candidate Alvin Harvey, who is Diné, has said that "for a lot of Native Americans, certain things like the Moon have a spirit that should be respected. If you're outside looking at the trees or the mountains, you've got to treat them with a certain respect. That's something that indigenous people, based on their own experiences, can contribute to the conversation."

When I spoke to him, asking him to amplify these comments, he emphasized the need for "kinship and relationality . . . understanding one another, how interconnected we are across knowledge systems, asteroids . . . and stars." In his own work, he applies this to a project that is starting to explore, among other things, Indigenous protocols for space travel and how to use Indigenous research methods to show how we can "travel to space in a good way"—and talk about it to children. More immediately, he and his fellow MIT students are, with "kindness and firmness," working to increase the presence of Indigenous faculty at that institution. Harvey is also interested in how the social norms of Indigenous peacemaking can be used in the context of long missions, isolation, and simulated missions on Earth. He hopes to organize the first all-Indigenous analog space mission ever.

While taking care not to claim he speaks for all his culture or other Indigenous peoples, Harvey does point to the need for spacefaring entities to bring "different [Native] knowledges" and

> that connection with the cosmos into your mission profile, your methodology, your grander thinking . . . What does it mean ethically to do this and procedurally to do that? It requires you to have those relationships to Indigenous people . . . It really means to be in community, in conversation, in an authentic and grounded relationship with all of its challenges and all of its nuances. I think people, especially in the commercial space sector, and I'll even say at NASA—and there are great people at NASA who know how to do this—but more broadly, at the upper levels, it's the same problem we have at MIT, the lack of personnel, the lack of subject-matter experts, that have the lived experience to help guide these institutions, these missions and to guide a process of building these relationships and sustaining them.

He acknowledges that "there's a fear that if we do that, we'll say no to everything—and that might be the case and we'll have to have hard conversations about what it really means to respect the Moon." At the same time, Harvey says,

> I feel there's also an understanding that our people, especially our young people, Indigenous people I've had the honor to serve or be around—they love NASA. They love space. They love the idea of being able to dream and travel there. I think that's such

an inherent thing to us because, well, the cosmos, the stars, the Moon, is something that connects all of us, especially Indigenous people, and those are our ancestors . . . our grandmothers and grandfathers, and I think we want to visit them too. You want to see your grandma; you want to see grandpa.

There's a caveat, however. "When you go to your grandparents' house, you don't to tear-ass through there," he says. He laughs as he says that, but the point is deadly serious:

> We do want to participate in this and how can we be empowered to begin to understand how we go to space and do this in a way that isn't reinforcing these kind of harmful procedures and tactics that we've seen again and again on Earth that harms everybody. How do we go to the Moon? How do we visit our grandmother? How do we do this in a way that is not being exploitative? Being aware of all of that and having Indigenous people involved in these conversations and empowered is really important.

Those conversations "will take a lot of bravery on the part of people working at SpaceX or Blue Origin, at NASA, to make that stand." But it's necessary, Harvey emphasizes, because "every space mission that's ever been, that will be done . . . will in some way have an impact on an Indigenous person, their land, their community, their lives, in either harmful ways or maybe allow them to dream but at what cost too?"

So this isn't just about policy or datasets or software or environmental impact statements, important as those are. Harvey echoes my own sense of the role of emotion in this, saying that

> being able to understand the importance of one's feelings, your heart, your spirit, that those aspects of the sublime, that go beyond how we are trained so hard to be scientists and engineers, to use

our brains, logic, structure, linearity and all of that, to be able to consider the fluidity of emotions that drives a lot of our actions . . . I think being able to have that be part of these conversations about science and engineering and journeying to space is really important.

Right now, we need to agree on some basic tenets for treating the Moon respectfully, whether that respect implies divinity, intrinsic secular value, or the instrumental value of wisdom in restraint. These might be ways to not tear through the lunar house if we dwell there.

Not littering will be easiest. Agreements can be struck to avoid cluttering lunar orbits the way we have the Earth with "space junk." Policymakers are already looking at this. The Artemis program also can set an example by not disposing of bags of human waste and other trash on the surface, which, because of fuel and space constraints, Apollo did. (Ironic: part of the cultural heritage of the Apollo sites are Moonwalkers' shit.)

Environmental contamination from dust, gasses, and chemicals is another threat. Such materials will last for eons on or under the surface of the Moon. What are best practices to ensure such spills do not occur? If they do, how will they be remediated?

For the time being, lunar exploration and harvesting of water ice will take place in or near permanently shadowed craters near the south pole. (The very drama of that landscape may help awaken public interest in seeing that it not be scarred, despite what that one researcher suggested.) There are no legally binding set-asides, however, whereby, apart from minimally invasive scientific exploration, areas would be free of commercial extraction. That should change. Sooner rather than later, space-faring entities and governments can—and should—agree to set aside portions of the Moon as protected, develop criteria for minimally invasive operations on other parts of the Moon and designate a scale for what activities can take place in what we might call water-ice farms. We have a chance to guide our lunar presence with multiple values in balance. Scientific infrastructure will likely be more acceptable than large, ugly industrial zones. This is perhaps somewhat akin to viewshed protection on Earth.

In those areas where industrial activity is taking place, we should consider what the facilities look like. Beauty is in the eye of the beholder, and, while we may criticize Elon Musk for many things, SpaceX's Starship is pretty. We need to hone an aesthetic sense of lunar infrastructure. Companies will have a chance to develop lunar factories, landing and launch sites, and homes that blend in with and complement the Moon's stark beauty: more Petra than Quonset.

The first step in addressing these concerns is to talk. And "to talk" is easier said than done, pun intended. As Judge Learned Hand suggested, reasonably, "Justice is the tolerable accommodation of the conflicting interests of society, and I don't believe there is any royal road to attain such accommodation concretely." Of the Moon, much remains unsaid.

Elizabeth Frank, chief planetary scientist at the company Quantum Space, has worked both in pure scientific research and with business interested in investing in space activities. She notes that terrestrial "mining companies across the globe are increasingly incentivized to work with impacted communities above and beyond any regulatory boxes they must check. Their method of engagement is called social license to operate, defined as the 'ongoing acceptance and approval of a mining development by local community members and other stakeholders that can affect its profitability.'" There are no communities on the Moon, Indigenous or otherwise. Yet "if a company does not provide pathways to input perceived as fair by external stakeholders, the mining activities may be interrupted by protests, for example. The Moon is more complicated because in some ways, every human is a stakeholder of lunar activities." She tells me that the fiasco of SpaceX launching all those dark-sky-ruining Starlink satellite constellations is a caution. "If SpaceX had reached out to astronomy community, then maybe they could have avoided problems."

While she says the "space is cool" mantra is one she shares, it doesn't excuse ignoring social impacts. Many engineers and entrepreneurs suffer from lack of exposure to ethics education and do not understand the social norms and abuses that have affected so many. "There's almost a naïveté among the companies. I don't think the aerospace industry has fully wrapped its head around it, and there aren't any incentives to do that yet," she tells me. Despite the Artemis Accords, she notes, purely private enterprises are not beholden to them. This, along with the OST, are "too high level for engineers and tech people . . . They're not trained to think about these impacts." Frank started to speak out because of this lacuna. "We need oil and gas and mining," she says, drawing on a terrestrial example. "But the mining industry learned too late on Earth. Mining has learned the hard way. Aerospace has not learned about this on a pretty big scale."

As of now, NASA has an Office of Diversity and Equal Opportunity. And the White House's document on a "National Cislunar Strategy" includes this passage: "Many unresolved issues regarding space exploration cannot be solved through engineering alone, such as the guiding ethics of human expansion into space, long-term cooperative models for space development, and equitable governance structures for space communities. The United States will enable and support research in the social sciences to advance our understanding of these issues. This research will also encourage the scientific community and broader public to think more deeply about humanity's long-term future in space." This is laudable but needs to be put into practice at the agency level and below.

Casey Dreier, chief advocate for the nonpartisan, pro-science organization the Planetary Society, emphasizes that the sheer difficulty of "in-situ resource utilization . . . may actually defuse conflict." Whether it's utilizing regolith for construction, sintering lunar dust to a landing-pad sheen or extracting water from ice, "ISRU is unproven and difficult to do." Dreier thinks we need to respect "the pitfalls of extrapolation and particularly in an alien environment. We don't respect the alienness of the Moon." That strangeness is part of the "space is cool" mantra Frank mentions. And Dreier says that wonder functions as the real reason for practical policy, that the larger, more visionary goals of exploration inhere even as "we can't acknowledge the real reasons because policy focuses on acceptable reasons," like making rocket fuel out of lunar ice. Policy is about the practical even if those who write it are motivated by dreams. Perhaps those dreams need to be articulated more often in order to steer policy away from mistaken assumptions or, worse, reckless disregard.

Alex Gilbert, who has a background in nuclear power regulation, environmental protection, and space resource governance, is completing a PhD at the Colorado School of Mines. There, he is a fellow of the Payne Public Policy Institute, while also working for Zeno Power, a start-up interested in commercializing radioisotope power-generating systems. He's been a fellow with Open Lunar and has taught part-time at Johns Hopkins. And when we talked, he emphasized from the start that the Moon "is a dead place. It's toxic, which is exactly why it needs environmental governance. It's very fragile. There's no natural way to mitigate pollution . . . [or] severe regolith destruction." Right now, however, discussions along these and other lines are largely at the diplomatic level. Technical consultations are necessary to actually advance best practices, just as Frank suggested.

But the two major spacefaring nations interested in the Moon can't talk to each other at the agency level. NASA is banned by law from consulting, let alone cooperating with, the Chinese National Space Agency, which is, as Gilbert points out, an opaque, quasi-military agency. Security concerns have understandably dictated this approach, but something will have to give. So, if in the next few years, launches and landings cause regolith dust to settle on someone's solar panels or the blast effects damage another operator's equipment, Gilbert says, "The main tool is shame." No one wants to be seen as the first bad actor on the Moon.

While China spurns the Artemis Accords, and while some think, perhaps overdramatically (but perhaps not), that the Accords will transfer the worst of capitalism to the Moon, the fact remains, says Gilbert, that the Accords have made much bigger table for global voices. Signatories have a say in setting standards and enforcing them. The international coalition will be vital in establishing basic governance within existing treaties and in revisiting when that governance needs to be altered.

In the longer run, Gilbert is one of those voices in the middle of the space community that says "space can't just be for billionaires." Space tourism near the Earth is one thing, but human dwelling on the Moon must be more than joyride or plunder. It will require public voices from across the political spectrum and around the world, including the voices of those who worship the Moon, in order to guide human activities in a fair, transparent way.

For US activities, there may be a path toward avoiding the worst. Gilbert and others have written about the National Environmental Policy Act, which President Nixon signed into law and which requires federal agencies to submit Environmental Impact Statements regarding their activities and oversight. Activities in outer space have been exempted from NEPA, but astronomers are suing the Federal Communications Commission for more regulation of satellites like Starlink. Should they win, Gilbert notes, the door may be open for scientists and others to have legal standing and could invoke NEPA for lunar activities sponsored by or financially supported by the US government. Libertarian-minded space advocates despise this possibility. Gilbert counters that an advantage of NEPA—or a similar approach—is that it makes operations more transparent. As long as taxpayer dollars involved, that seems critical.

Thoughts of NEPA lead to bigger things. In a reflective mood, Michelle Hanlon articulates it for me: "We're all human beings. We're one species . . . that is where we should end up in in fifty or a hundred years." She pauses. "*Star Trek*. Yay, I mean it."

☻

Seen from a distance, the domes of Mount Wilson are "no larger than the small white dome of shell / Left by the fledgling wren when wings are born," wrote the English poet Alfred Noyes while he was in California. At nightfall, he thought, the distant white tops became like the very stars the telescopes observed. In 1917, Noyes was invited to Mount Wilson for the first light of the hundred-inch telescope. With astronomers and engineers, technicians and mechanics, the popular poet saw Jupiter and a Galilean moon. So inspired was Noyes, that he stepped outside after his turn at the eyepiece and decided there and then to write his celebration of scientific progress, *The Torch-Bearers*, the first volume of which, *Watchers of the Sky*, begins with the prologue poem "The Observatory." Drawing on martial imagery—understandable given the Great War—Noyes compared telescopes to weapons battling ignorance. Science was a struggle for good, for knowledge. If the metaphor seems less appealing now, "The Observatory" has an ethos worth pondering. He quotes a poeticized letter from an astronomer, presumably George Ellery Hale, saying,

> *Even to-night*
> *Our own old sixty has its work to do;*
> *And now our hundred-inch . . . I hardly dare*
> *To think what this new muzzle of ours may find.*
> *Come up, and spend that night among the stars*
> *Here, on our mountain-top. If all goes well,*
> *Then, at the least, my friend, you'll see a moon*
> *Stranger, but nearer, many a thousand mile*
> *Than earth has ever seen her, even in dreams.*

Like Noyes, I had come to the sixty-inch as though to "some great cathedral dome," here on "the dark mountain with its headlong gulfs," but unlike Noyes, I had not "lost all memory of the world below." Indeed, such memory is needed more than ever on the mountaintops of science.

It was the world below that slowly alienated Apollo astronauts from their own views of the Moon. Over the years, I've been able to ask them—Fred Haise, Walt Cunningham, Michael Collins, Al Worden, Dave Scott, Charlie Duke—if they still look at the Moon. They answered no. Scott and Worden had both kept looking through small telescopes for a time after Apollo 15, Scott from London and Worden from Texas. Scott didn't remember what kind he had, but Worden told me he'd loaned out his Edmund Astroscan—a small telescope with a ball-shaped base you can cradle in your lap—and never got it back. Only Duke still had a telescope, but demurred, "I'm just too busy in the evenings." I felt pained after I'd learned these explorers no longer looked at the Moon. Indeed, Haise almost scoffed at my question. I had imagined not only that they still looked but they had communed with its places and its histories too, just as I had. I had imagined bonding with Dave Scott over geology, then looking at our favorite places through my telescope.

It is love that keeps me awake at the eyepiece tonight on Mount Wilson. All these years to this moment, my legs sore, my neck stiff, my hand gripping the controller and pushing red-lit buttons to wheel over the Moon that stranger but nearer fills me as I fly a few hundred miles above, an astronaut bound to Earth, smelling cool air and seeing, as we near midnight, the slight scrim of wildfire smoke. The dome quiets more and more. Now and then motors grind as Tom moves the observatory slit. The gears I control whir as I navigate—

Shocked by Delauney, the central mountain ridge dividing the dark heart like an aorta of light.

More rilles at the northern end of the Apennines than I've ever perceived, a wicked tangle.

Rimae Atlas, crackling outward from dark swathes under bright sun, a static inversion of lightning.

At Rimae Triesnecker, the Moon's most raveled rilles, I lose count: nine, more, no, twelve, like the cracked glaze of a Ralph Blakelock painting.

Here, the mountains in Walther looking like a tooth. Here, Rima Hyginus outspread like a vulture's dihedral, craterlets in a row, pitting the rille in little stanzas of collapse, twenty or more, shadowed edge, bright edge, depth again, and excess. Here, Abulfeda's bull's-eye white-halo crater, and all night I return to the craterlets of Ptolemaeus. I look for another volcanic feature, the Ina depression, that irregular mare patch in the Lake of Felicity—of Happiness. There—a slightly lighter shape? Maybe?

"I'm starting to see a shimmy," Dave says.

Scan! Just go! But gears slow and stop. Have I broken a telescope? Keep pressing the button, the gears will catch up. I scan the entire surface. Petavius in high sun, Mare Crisium, Proclus, Yerkes. High sun, the lunar limb notched with rim and peak and the black thereafter.

My last views are sunrise over Alphonsus. I'm glad to see this, a bit of homage to Dinsmore Alter. The central peak bright and flanked by a truncated soft bar of gray light (itself cut in two by shadow) and a bright bar across from which faint floor craters begin to catch the dawn.

It's been, as David would later tell me, "a spectacular and frankly uncommon night." It's nearly midnight, and I will dream of rilles and craterlets a third of a mile wide, blown about the Moon like pollen.

☍

The day after my lunar orbit I stayed at the Monastery and rested, alternating among naps, packing, and remembering the night before. My last evening there, I set up my ten-inch scope in the parking lot again and looked as night came on. I was curious—would I feel disappointed to step down to my more domestic aperture? No, it felt like coming home. I hugged my telescope—since it's small enough to hug—as I thought of where it had taken me, from disillusion to mountaintop, from bewildered beginner to happy apprentice. I took my last views from Mount Wilson. The Sun plunging into Eratosthenes. Baroque scree. The sickle-shaped vent whose eruptions made Hadley Rille. And with the Sun low enough on the dark badlands between Sinus Aestuum and the crater Schröter, I thought that perhaps I might find one more place that had eluded me.

I have come to learn the Moon as a place—as a world of places—as I have come to learn the Moon's place in human consciousness, from scratches on bone and rock to our prayers, from philosophical and religious beliefs to the developing scientific view, which, after Galileo's telescopic observations, tore through so many assumptions like a meteorite. Places on the Moon have become temporal cairns. Albategnius reminds me of Galileo. Mountain gouges remind me of Grove Karl Gilbert. Hadley Rille, of Apollo 15. The Aristarchus Plateau, of the checkered history of transient lunar phenomena. I have traversed the worst of humanity as I cataloged the erasures of modern rocketry's criminal origins. And I have been moved by the beauty of a rocket launch. I have spoken with astronauts who walked on the Moon and I have paddled in sight of our new Moon rocket to see the beginning of new lunar explorations. Scientists, astronomers both professional and amateur, historians, policymakers, even artists, have given me their time, perspectives, and expertise. Once strapped into a rig to mimic one-sixth gravity, I have traveled far, but I was always coming home to the newly welcome quiet wherever we are, so much made to grace by the light of the steadfast Moon.

Looking at the low light on those dark hills and ridges north of Schröter, the air crystalline, I focused on what topography I could make out. I had tried before. I defocused the view, almost imperceptibly. Yes! The fanning ridges that Franz von Paula Gruithuisen had proclaimed as "colossal buildings and gigantic ramparts," the Wallwerk, the lunar city he was certain he'd discovered, there it was, a city on the Moon! It lifted gargantuan embankments like something spinal, something veined, a fossil body, some kind of lunar trilobite. I focused and now could see the shapes, most of them, matching this V in the eyepiece to the drawings I consulted. How much we yearn for company, I thought, how much we dismiss what we already have. We might be less lonely if we simply looked around. Yet our lives are not designed for pause. Yet without it, one cannot come to lifesaving wonder or the other clarities. And without those, we risk everything. Perhaps that's one reason I am so smitten with science. At its best, it uses methodology as secular liturgy to describe what we can know. This is, literally, wonderful.

A few years ago, I visited Tempe, Arizona, because of something called LROC, the Lunar Reconnaissance Orbiter Camera. Lunar Reconnaissance Orbiter may be, for the wider public, the most underappreciated space science mission ever. It launched in 2009 with several instruments, perhaps none more famous than its camera, which has been obtaining breathtaking views of the entire lunar surface for years now. Based at Arizona State University, LROC's Science Operations Center can be visited at the midcentury modern Interdisciplinary A Building. I had just wanted to see. There are lunar photos, informational posters you can take, videos, and a glassed-in area where a visitor can watch the science team sitting at their computers. Behind the glass are models of rockets, a Moon globe, and screens. Quiet work, from that perspective, but exploration nonetheless.

When those explorers leave their low-lit office, they surely blink, as I did, in the palm-tree sunlight of the built desert. LROC provides resolution down to one meter—about a yard—and has been studying polar illumination to help us understand what areas are permanently lit or shadowed or mostly so. The work allows scientists to develop 3D maps, understand landforms, and look for previously unseen hazards at possible landing sites. It also produces views that LROC personnel rightly call "exquisite," "beautiful," "mighty," and "spectacular," all adjectives that would be at home in the Moon books of the nineteenth century.

Quite apart from the ethical and policy conundrums that face space exploration—really, for example, we need to insist on launch facilities that don't harm wildlife—we must redesign our lives for occasions of wonder. The dopamine bursts of commodified triviality that mark our days do not count. Hypercapitalist work expectations may have eased during the pandemic but still threaten well-being. Everywhere, income inequality and systemic oppressions shorten lives and doom them to poverty and strife. Designs for and experiments in living on the Moon may seem incalculably remote from such pressing concerns. They are not.

Recently, astrophysicist Erika Nesvold was talking about her important book, *Off-Earth: Ethical Questions and Quandaries for Living in Outer Space*. Does settling space enhance or diminish individual liberty, enhance or diminish collective good? Will private settlements be run at the behest of shareholders, be regulated by terrestrial governments, evolve their own civic structures?

During her interview on the *Planetary Radio* podcast, she laughed, "I think it's always important to recognize emotional motivations within ourselves, right? There's still the ten-year-old inside me who just thinks it's neat." Those motivations are a form of intelligence. Nicole Mann, the first Native American astronaut (Wailacki of the Round Valley Indian Tribes), says, "Whatever your background is, whatever your race, your religion or anything you can be, [children] can share in that joy."

Childhood wonder connotes preservation of its causes.

My former student and colleague Dr. América N. Lutz Ley, a professor at the College of Sonora in Mexico, puts it this way: "A potential approach for the future is to benefit from this natural feature in children and make everything in our power to keep it through life; that could be easier than trying to recover it later. Maybe the most important thing to know here is that most of us *already had it*." This is more than rhetoric: neuroscientists have found that the brain regions responsible for memory are the same for imagining the future.

Childhood, memory, and wonder are part of who we are, part of being a family. Vinay Kumar Srivastava says that community is family, and that the Indian ethos of *vasudhā kutumbakam* can help guide us: "The entire Earth is my family." Kumar Srivastava does not see modernity's tools as necessarily evil, not if they are guided by renewed respect, expressed through folk beliefs, religious practices, or, one could add, facts and loveliness brought forward by science.

Now I am exploring language again, the terrain of connotation and metaphor. Behind systems are the words that make them. What we say is who we are. Behind policies are cliches like the foundations of old abattoirs sunk in mud. *Colony. Frontier.* The brilliant Russian rocket theorist Konstantin Tsiolkovsky famously said, "Earth is the cradle of humanity but one cannot remain in the cradle forever," a metaphor so pithy that its advocates miss the flaw. *A planet is not a piece of furniture shaped by the creatures the cradle holds.* Its connotation of permanent exile from Earth is appalling. Equally so, other metaphors comparing space exploration to conquest simply reinforce the spirit of men haunted with power. Even Loren Eiseley's space-critique metaphor—that we are mindless spore-bearers—is soaked in power, in this case that of mindless nature.

There are other metaphors for our relationship to space and to the Moon, which, to be honest metaphors, must link all realms of human habitation. The poet Nikki Giovanni speaks brilliantly of quilting. I think of housekeeping, of homemaking. I still smile at Buzz Aldrin's joke/not-joke when, crawling ass-backward from the Apollo 11 Lunar Module, he says he won't lock the door. "That's our home for the next couple of hours and we want to take good care of it."

Here's the full passage from *The How and Why Wonder Book of Stars* that took my breath away when I was reading it by an observatory dome: "Let's say that we are standing in a place where there are no buildings or trees or mountains to block our view of the sky, for instance on the deck of a ship in the ocean. The sky will look to us the same as it looked to the earliest [humans] on earth . . . Beyond our ship all we can see is water and sky. It is as if we were alone in an empty house."

Where Tsiolkovsky's cradle implies leaving, this layered metaphor from a children's book implies voyaging and householding at the same time. The danger of the journey and the rigors of home and, in both, surprises, disappointments, lessons, joys. When we leave home, we take it with us—the best of it—and we look to make another. There are no frontiers. There are only places. The danger is how routine blunts the sharp edge of wonder.

Writes Brian Aldiss in *The Shape of Further Change*, his forgotten book on the Apollo Moon landing, technology, and dwelling: "I own [my house] in the way that the Moon and tree shadows own me: by falling across me and influencing me intensely . . . The intense love one has of one's house is really a love of life itself."

There is water on the Moon! When we drink it, we'll toast the Moon. When we breathe its component oxygen, we'll breathe deeply and give thanks. When we light its component hydrogen to fuel a rocket launch, we'll honor our long history with fire. We will share. We will feel the love one has for home.

Metaphors in hand, wonder before us, we know that just as the Gilded Age gave way to the Progressive Era, the Age of Billionaires will fade as we tend the common good while fostering innovation. These are not mutually exclusive, but to do both will require fewer opinions and more thinking. It will require the work of listening, the art of argument, and the gift of compromise. We must reteach ourselves the lessons we try to pass on to children. If Jeff Bezos builds his road to space, then a lot of people will be able to travel it, people

who may be less inclined to practice hypercapitalism, people who see cooperation, not competition, as the key to equity and progress. In other words, the
billionaire race to widen access to space may be the undoing of the system
that creates billionaires in the first place. We might create a commonwealth
in which we finally face another great task of our civilization: how to sustain
a place between abundance and fairness where it is possible to live without
regret. Nostalgia for an Edenic past and anticipation of a limitless future are
equally false. To be alive—to dwell—is to compromise. Rather than bemoan
this fact, we should beautify it. There is a middle ground to be found here,
perhaps on the rim of Shackleton Crater.

Behind and beneath ideology and behind and beneath economics is the
child, who, ideally, was loved, who expressed needs that were met, who was
nurtured and challenged, who felt wonder that grows into the nuances of adult
curiosity. Ernst Bloch once wrote that "if human beings have grasped themselves and what is theirs, without depersonalization and alienation, founded
in real democracy, then something comes into being in the world that shines
into everyone's childhood and where no one has yet been—home." You may
disagree with Bloch's politics, but his insight remains. Home is where our
child-heart dwells, safe, curious, and in love.

I think about the hardships of the selenographers, driven to map, to discover, to name, with their tiny refractors and cold nights, with their obsessiveness and failing eyesight, their struggles. Cartographers of what the gothic
shadows, the black stalactite shadows, told them, observers of lunar mountains
almost sedate in their awful imposition and the endless library of detonations
that goaded them. They ravened the surface for pips and dusty ravines. Night
after night after night.

William Wordsworth writes (using the gendered language of his day)
that "the Child is father of the Man." Lord Byron's "So We'll Go No More
a Roving" might seem a rebuke to this sentiment but it's not. Lord Byron's
poem recognizes that, ideally, we grow out of youth toward wisdom and restful
creativity. Youth is not childhood—it is childishness—and wisdom without
wonder is merely antiseptic. It lacks kinship and joy, two traits that are part

of science and the best kind of exploration and domesticity. Lord Byron does not rebuke any of this, only selfish "roving." He sees that the Moon is not for lighting adventure but for lighting love, and I have seen that its light is adventure.

When the Moon rises between buildings or over trees, it's not just a beautiful sight: it is an archive of human feelings and material truths. The Moon is more than a rock. It's a story. It's many stories. The Moon signals what came before and what comes after, and I can't wait to look again when the next lunation begins. There is much to see. Known by reason and held by love, the Moon invites our gaze and our gasps, the sights and breath of awe, the glow and rhythm of dwelling itself.

ACKNOWLEDGMENTS

F irst and last, I thank my wife, Kathe Lison, for bearing with all the ups and downs, for still being in love with me after more than two decades, for everything from her humor and clarity to her teaching me that, with the exception of the Moon, the best light at night comes from candles. I love you, plum.

I am delighted to have landed a patient, enthusiastic, and superb agent in Elise Capron, and a patient, enthusiastic, and superb editor in Jessica Case at Pegasus Books. Thank you to everyone at Pegasus Books, especially my publicist Julia Romero and copy-editor Jocelyn Bailey. Maria Fernandez did a wonderful job with the book's design. Thanks also to proofreader Jessica Bax. And Charles Brock's jacket design blew me away.

Gratitude to Meera Subramanian for introducing me to Elise, and to Mike Branch, longtime pal who has published with Pegasus and who inexplicably vouched for me.

The authors, historians, artists, astronauts, scientists, and others whose published efforts and interviews I relied on are listed in the Works Consulted and/or referenced informally in the text. Such work has been a source of deep pleasure and learning. I bow to them all.

Several people gave of their time to read my proposal, chapters and/or the entire manuscript. I thank Kathe Lison, John Price, and Jennifer Sinor for reading the proposal. I thank Earl Swift and Michael Paterniti for the same and for the gift of their proposal endorsements. Different chapters were reviewed by Barb Cohen, Anthony Cook, Bob O'Connell, Tom Dobbins, David Hasenauer, and Elizabeth Frank; I appreciate your time and feedback. Robert Garfinkle, author of the extraordinary *Luna Cognita* reference work, and Tim Swindle, former director of the University of Arizona's Lunar and Planetary Laboratory (and fellow Antarctic meteorite hunter), each read the entire book and called me out on errors. Thanks to Kate Northrup for much-needed affirmations after an early reading and the same to my Logan peeps after another presentation. Kathe also read the entire manuscript with her careful eye. Thank you all.

Several people have helped me on this lunar journey. In particular, I thank Chuck Wood and Bill Hartmann for fielding a lot of scattershot questions. I thank Eric Douglass of the American Lunar Society and the good folks with the Royal Astronomical Society of Canada for granting my observing diplomas.

I also am grateful to the following for help along the way, from recommendations to advice and more: Dalila Ayoun, Steve Simms, Mike Dinsmore, Andrew Smith, Jon Willis, Paul Bogard, Craig James Wood, Timothy Thompson, David Hasenauer, Tom Meneghini, Richard Bell, Renee Marchiano, Robert Cohn, Art Cosgrove, Pagnini Francesco, Dirk Schulze-Makuch, Bonnie Buratti, Sabra Minkus, Selene Cameron Green, Sheri Carina Cameron Katz, Mike Dickerson, Leonard Srnka, Peter Schultz, Cameron Hummels, Robert Morehead, Hakan Kayal, Maria Kelson, Patrick Douglass Cline, Story Musgrave, Farouk El-Baz, Friedrich Horz, Kate Burgess, Michelle Lea Desyin Hanlon, Mark Sundahl, Alvin Harvey, Elizabeth Frank, Alex Gilbert, Andrew Smith, Charlie Goldberg, Amy Clark, Christian G. Appy, the Utah Historical Society, Dave Scott, Al Worden, all the other Apollo astronauts who answered my question about whether they looked through telescopes at the Moon, Elizabeth Hora, Troy Scotter, Jake Bleacher, Ryan Zeigler, Jessica Heim, Corey Gaught, Amy Clark, David Reioux, and Nick

Snow. Laura Mortensen and Sarah Gordon at Utah State graciously translated the Rudaux booklet. Alexander Huber provided a copy of the Thomas Gray poem in translation.

If I have forgotten anyone, I beg forgiveness.

Any errors that remain are my own.

For my family, I give thanks: my stepmother Judie Cokinos, who always asked how the book was going; Jeff, Phil, and Karen and their children and grandchildren for the love they gave Dad; my niece Jessica, her guy Sean, and nephew Connor; for the Wisconsin crew of Tiff and Mike, niece Claire and nephew Waylon. I especially thank my sister, Vicki Wright, for love, persever- ance, and wisdom and for the hard work of taking care of people. As well, her years of lawyering research helped me when she assisted in some late fact-checking. And she helped with the title! For you all, I hope you see good Moons. My dad was alive for part of the writing of this book so now I'll just say: I miss you very much.

To friends across years and miles—Natasha Kern, Paul Sutton, Jen Johnson, Ralph Harvey, Noemi Quagliati, and Joshua Marie Wilkinson—thank you for your care and your smarts. To the writers and artists who have supported and inspired me from afar, thank you: Richard Rhodes, Chet Raymo, Dava Sobel, Michael Soluri, and Kim Stanley Robinson.

Many thanks to Stan, Sarah Scoles, James R. Hansen, Earl Swift, Micheal Paterniti, and Meera Subramanian for their endorsements of this book. They mean a lot to me.

To those in Salt Lake City who been a part of journeys personal and profes- sional, thank you: Jason Yorgason and the Salt Lake Philosophy Apparatus, Steve Trimble, Kimberly Johnson, Bill Walsh and Lisa Gabbert, Paisley Rekdal and Pete Gomben.

And to friends on and off the trails in Logan: Paul Crumbley and Phebe Jensen, Ben Gunsberg and Andrea Melnick, Jen Peebles and Charles Waugh, and Jennifer Sinor and Michael Sowder. Thank you for listening, reading, supporting, for your love over the now-many years, for the help and good cheer for the homecoming.

In Tucson, Dustin Leavitt is a friend and fellow space nerd. The same for Julie Swarstad Johnson. You guys really should meet sometime. Gratitude to friends Kegan Tom and Jean-Luc Cuisinier, Larry Busbea and Michelle Strier, and Cory Gavito and Aileen Astorga Feng—thanks you two for hosting me during my Salt Lake City–Tucson commuting. It was good to show Luca the Moon.

At the University of Arizona's Department of English and elsewhere, I want to thank all the administrative staff who have helped me over the years with research accounts, travel support, and more; their labor is foundational: Sharonne Meyerson, Caitlin Conrad, Nick Smith, Reilly Rodriguez, Vicki Henry, Anne Shepherd, Stephanie Pearmain, Jeff Schlueter, and Marcia Simon. Department heads Larry Evers, Lee Medovoi, Cristina Ramirez, and Aurelie Sheehan were all stellar (or lunar, in this case). Lee, Aurelie (in memoriam), thank you for your collegial friendship. Several colleagues deserve mention for the same and for support down the line: Scott Selisker, Marcia Klotz, John Mellilo, Johanna Skibsrud, and Maritza Cardenas, among others.

In no particular order I want to thank some amazing students with whom it was a privilege to work and from whom I gleaned much, including friendships: Thomas Dai, Kend Morrison, Josh Riedel, Lucy Kirkman (in memoriam), Nick Greer, Nadia Moskop, Megan Kimble, Hea-Ream Lee, Hannah Hindley, Margo Steines, Matthew Morris, 'Suyi Okungbowa, Joi Massat, Kati Standefer, Katie Gougelet, Michelle Rubin, Natalie Cunningham, Nina Boutsikaris, Paulina Jenney, and Tony Colella. Other UA English and English-adjacent humans for whom I share appreciation: Simmons Buntin, Joela Jacobs, Sara Jean Deegan. There are others too numerous to name from whom I learned how to be a better person and teacher. (And got introduced to Frank Turner. Check out the second track on *Positive Songs for Negative People*.) I thank Ellen McMahon, Chris Impey, and Elliott Cheu for their science-communication chops and collaborations.

The Moon book also grew out of the most important teaching I've ever done, with the Carson Scholars Program in science communication at the University of Arizona. Appreciation to Kevin Bonine, Andrea Gerlak, Rachel

Gallery, Chris Scott, the late Rafe Sagarin, and Julie Cole. To Jonathan Overpeck, Diana Liverman, Betsy Woodhouse, Stephanie, Sklar, and Tina Gargus: those were heady days with the Institute of the Environment. Of late, Ariane Solwell kept the program running smoothly. The scholars themselves were and are incredible, and I thank them all. For your continued friendship, conversation, and amazing work, shout-outs to Eric Magrane, América N. Lutz Ley, Saleh Ahmed, Sonya Ziaja, Dervla Meegan Kumar, and, really, all of you. Kim Land at B2 always helped us.

For a 2019 Moon exhibit, I worked with Molly Stothert-Maurer at UA Special Collections, and for a talk on lunar folklore, Shipherd Reed at Flandrau Planetarium. Thanks to everyone at both facilities. At the Lunar and Planetary Laboratory, Jeff Hanna-Andrews, Jessica Barnes, Amanda Claire Stadermann, and Maria Schuchardt have my sincere appreciation. Tom Fleming in Astronomy has been an ongoing source of professional support.

Though the Imagination 1 simulated mission at B2's SAM—the Space Analog for Moon and Mars—was pushed back to spring 2024, planning for it was fundamental in my thinking about space ethics. I want to thank everyone with the wonderful Analog Astronaut Community and, especially, my crewmates Julie Swarstad-Johnson, Elizabeth George, and Ivy Wahome. Shoutout to Trent Tresch, he of the pressure suit, and Kai Staats, the almost literal driving force behind SAM. For a plant-growth experiment using powdered lunar meteorite, thanks to Mile High Meteorite's Matt Morgan and to Anna-Lisa Paul and Rob Ferl of the University of Florida Space Plant Laboratory.

The Planetary Society deserves a huge shout-out, especially Mat Kaplan and Casey Dreier.

The following publications have printed or posted my work related to the Moon: the *Los Angeles Times*, the *Space Review*, *Sky & Telescope*, *Astronomy*, *SkyNews*, *Discover.com*, and about a paragraph or two from my Apollo headnote in *Beyond Earth's Edge: The Poetry of Spaceflight* made its way into the von Braun material. I've been lucky to have great editors, especially Susan Brenneman at the *Times* and Alison Klesman, Mark Zastrow, and Jake Parks at *Astronomy*.

I also acknowledge quoting from "A Voyage to the Moon" from *Collected Poems 1917 To 1982* by Archibald MacLeish. Copyright © 1985 by The Estate of Archibald MacLeish. Used by permission of HarperCollins Publishers.

With the Moon Village Association, I thank the Cultural Considerations Working Group, especially Lisa Pettibone, Remo Rapetti, and Arthur Woods. Thanks to Madhu Thangavelu at UCLA for the chance to test-drive some ideas.

Thanks to David Blewett for discussions regarding Lunar Vertex and his help in getting me to a Moon map mosaic for this book.

This work owes much to writers and thinkers I have explored and absorbed, most especially Arthur Schopenhauer; two philosophers of contemporary nihilism, James Tartaglia and Tracy Llanera; along with Jonathan Crary, Russell West-Pavlov, David Nye, and Istvan Csicery-Ronay. Many of the poets in *Beyond Earth's Edge: The Poetry of Spaceflight* gave me pause for thought regarding how we can craft an equitable future on Earth and in space, in particular Nikki Giovanni. I have been moved and inspired by the writers Loren Eiseley, Ursula K. LeGuin, Brian Aldiss, and J. G. Ballard.

I have dedicated this book to the memory of my father, but I dedicate it here also to Kathe, with whom I traveled across each day and night for many Moons. There are more to come. And to our own, first small family of satellites, three dear cats—the boys Shackleton, Zinc, and Burchfield—who did not last long enough to rub their faces on the corner of this book but who, for years, orbited the telescope while I looked. You boys made us better. I hope they'd approve of letting Yoshi, the new boy, and Ursula, the new girl, pick up their traditions.

A BIBLIOGRAPHIC NOTE

Throughout this book, I have used informal, journalistic attribution to provide the motivated reader enough information to track down source material of interest. Such sources include oral histories, websites, photographs, broadcasts, podcasts, transcripts, popular news and features, scholarly articles, and more. Should you have a question about a source, or a correction, please contact me through my website at www.christophercokinos.com.

That said, certain books were especially important in my research—they are listed below—but I want first to offer some suggestions to those readers who wish to look at the Moon themselves.

If you are interested in looking at the Moon and don't have a telescope, you can borrow telescopes from many public libraries. Here is a resource for just that (which includes other helpful materials, including sky charts): librarytelescope.org.

One guide I found most helpful as a beginner—and I still use it—is Andrew Planck's self-published *What's Hot on the Moon Tonight? The Ultimate Guide to Lunar Observing* (2015), available through online sellers and AndrewPlanck.com. Planck's book is helpfully organized by days of lunar phases. Charles A. Wood's and Maurice J. S. Collins's *21st Century Atlas of the Moon* (Morgantown: West Virginia University Press, 2013) is an excellent guide with sharp photography, descriptions of specific features, and an introduction to lunar geology. It foregoes the usual and overly complicated material on lunar motions, orbits, and phases that begin other guidebooks. This atlas is organized by regions of the Moon. It is my go-to atlas.

The classic *Atlas of the Moon* by Antonín Rükl (Waukesha, WI: Kalmbach Publishing Co., 1996) is out of print but still available from used booksellers. For intermediate and advanced observers, its organization by region and its appealing drawings of features is indispensable. The book gives the biographical detail for feature names, of interest to history buffs.

Peter Grego's small-format *Moon Observer's Guide* (Buffalo, NY: Firefly Books, 2004) was the first I purchased and remains an important resource for me, especially on the nightstand. It is organized by phase and is a good, all-around introduction to all things lunar. You may need reading glasses, though.

You will need a sturdy, laminated Moon map that fits the image alignment of your telescope's view. I like *Sky & Telescope's Field Map of the Moon*, which is based on Rükl's maps.

Charles A. Wood's *The Modern Moon: A Personal View* (Cambridge, MA: Sky Publishing Corporation, 2003) is not an observing guide, but I include it here as an important introduction to lunar geology and science. Wood describes regions and features of the Moon with his usual clarity, combining science, history, and his perspectives as someone who helped pioneer lunar research in the Apollo and post-Apollo era.

Lunar science is, however, continually advancing. One can keep up via popular magazines like *Astronomy* and *Sky & Telescope* and through such organizations as the Lunar and Planetary Institute and the University of Arizona's Lunar and Planetary Laboratory. NASA's various lunar pages—from Apollo photo galleries to recent findings—are a treasure trove.

Websites you can turn to include the old but helpful https://www.shallow sky.com/moon/ and NASA's more sophisticated https://trek.nasa.gov/moon /index.html. You can find multiple apps for your phone that provide lunar observing information.

Intermediate and advanced observers can benefit from John Moore's self-published *Craters of the Near Side Moon* (2014) and *Features of the Near Side Moon* (2017). I also rely on Kwok C. Pau's *Photographic Lunar Atlas for Moon Observers* for hard-to-find observing targets. You can find it at various links. The Cloudy Nights online forum is the home to many sophisticated lunar observers with far more technical expertise than I possess, including in the area of lunar photography.

Finally, there many older lunar observing guides that the obsessed can seek out. These range from the exquisite *The Times Atlas of the Moon* to cheap paperback reprints of Dinsmore Alter's *Pictorial Guide to the Moon*.

Selected Works Consulted
Apollo 15 Preliminary Science Report. Washington, DC: NASA, 1972. NASA
    SP-289.

Baldwin, Ralph B. *The Face of the Moon*. Chicago: University of Chicago Press, 1949.

Beattie, Donald A. *Taking Science to the Moon: Lunar Experiments and the Apollo Program*. Baltimore, MD: Johns Hopkins University Press, 2001.

Bogard, Paul. *The End of Night: Searching for Darkness in an Age of Artificial Light*. New York: Little, Brown and Company, 2013.

Brenna, Virgilio. *The Moon*. New York: Golden Press, 1963.

Brunner, Bernd. *Moon: A Brief History*. New Haven, CT: Yale University Press, 2010.

Brush, Stephen G. *Fruitful Encounters: The Origin of the Solar System and of the Moon from Chamberlin to Apollo*. Cambridge: Cambridge University Press, 1996.

Burke, James. *The Day the Universe Changed*. Boston: Little, Brown and Company, Back Bay Books, 1995.

Cashford, Jules. *The Moon: Myth and Image*. New York: Four Walls Eight Windows, 2003.

Chaikin, Andrew. *A Man on the Moon: The Voyages of the Apollo Astronauts*. New York: Penguin, 1994.

Cortright, Edgar M. *Apollo Expeditions to the Moon*. Washington, DC: NASA, 1975. NASA SP-350.

Crim, Brian E. *Our Germans: Project Paperclip and the National Security State*. Baltimore, MD: Johns Hopkins University Press, 2018.

Crotts, Arlin. *The New Moon: Water, Exploration, and Future Habitation*. Cambridge: Cambridge University Press, 2014.

Crowe, Michael J., ed. *The Extraterrestrial Life Debate Antiquity to 1915: A Source Book*. Notre Dame, IN: University of Notre Dame Press, 2008.

David, Leonard. *Moon Rush: The New Space Race*. Washington, DC: National Geographic, 2019.

Dick, Steve J. *Plurality of Worlds: The Origins of the Extraterrestrial Life Debate from Democritus to Kant*. Cambridge: Cambridge University Press, 1982.

Dick, Steven J., ed. *Remembering the Space Age: Proceedings of the 50th Anniversary Conference*. Washington, DC: NASA, 2008. NASA SP-2008-4703.

Eiseley, Loren. *The Invisible Pyramid*. New York: Charles Scribner's Sons, 1970.

Ekirch, A. Roger. *At Day's Close: Night in Times Past*. New York: W. W. Norton, 2005.

Firsoff, V.A. *The Old Moon and the New*. London: Sidgwick & Jackson, 1969.

———. *Strange World of the Moon: Its Seas, Mountains, Atmosphere, Climate—and Its Possible Life*. New York: Basic Books, 1959.

de Fontenelle, Bernard Le Bovier. *Conversations on the Plurality of Worlds.* Translated by H. A. Hargreaves with an introduction by Nina Rattner Gelbart. Berkeley: University of California Press, 1990.

Galilei, Galileo. *The Starry Messenger.* Translated by Albert Van Helden. Chicago: University of Chicago Press, 1989.

Garfinkle, Robert A. *Luna Cognita: A Comprehensive Observer's Handbook of the Known Moon.* New York: Springer, 2020. 3 vols. One of four crucial titles for my work. An incredible compendium.

González, Carmen Pérez, ed. *Selene's Two Faces: From 17th Century Drawings to Spacecraft Imaging.* Boston: Brill, 2018.

Guthke, Karl S. *The Last Frontier: Imagining Other Worlds, from the Copernican Revolution to Modern Science Fiction.* Ithaca, NY: Cornell University Press, 1990.

Hartmann, William K., Andrei Sokolov, Ron Miller, and Vitaly Myagkov. *In the Stream of Stars: The Soviet/American Space Art Book.* New York: Workman Publishing, 1990.

Heilbron, J. L. *Galileo.* Oxford: Oxford University Press, 2010.

Hill, Harold. *A Portfolio of Lunar Drawings.* Cambridge: Cambridge University Press, 1991.

Irwin, James B., with William A. Emerson Jr. *To Rule the Night.* New York: Ballantine Books, 1973.

Kitt, Michael T. *The Moon: An Observing Guide for Backyard Telescopes.* Waukesha, WI: Kalmbach Books, 1991.

Laney, Monique. *German Rocketeers in the Heart of Dixie: Making Sense of the Nazi Past during the Civil Rights Era.* New Haven, CT: Yale University Press, 2015.

Launius, Roger D. *Apollo's Legacy: Perspectives on the Moon Landings.* Washington, DC: Smithsonian Books, 2019.

Mackenzie, Dana. *The Big Splat, or How Our Moon Came to Be.* Hoboken, NJ: John Wiley & Sons, 2003.

Maher, Neil M. *Apollo in the Age of Aquarius.* Cambridge, MA: Harvard University Press, 2017.

Marshack, Alexander. *The Roots of Civilization: The Cognitive Beginnings of Man's First Art, Symbol and Notation.* New York: McGraw-Hill Book Company, 1972.

Masursky, Harold, G. W. Colton, and Farouk El-Baz. *Apollo over the Moon: A View from Orbit.* Washington, DC: NASA, 1978. NASA SP-362.

Mee, Charles. *Life in the Renaissance.* NP: American Heritage, New Word City, 2016.

Miller, Ron, and Frederick C. Durant III. *The Art of Chesley Bonestell.* London: Paper Tiger, 2001.

Michigan Quarterly Review. The Moon Landing and Its Aftermath. Vol. 18, no. 2 (Spring 1979).

Montgomery, Scott L. *The Moon and the Western Imagination.* Tucson: University of Arizona Press, 1999. Alphabetically the second of my four most important references, this title is full of provocative and useful insights, especially regarding scientific and artistic visual representations of the Moon.

Moore, Patrick, and Peter Cattermole. *The Craters of the Moon.* New York: W. W. Norton & Company, 1967.

Nasmyth, James. *James Nasmyth: Engineer—An Autobiography.* Edited by Samuel Smiles. Milwaukee, WI: Lee Engineering Research Corporation, 1944. Reprint, 1883, Harper & Brothers.

Nasmyth, James, and James Carpenter. *The Moon Considered as a Planet, a World, and a Satellite.* London: John Murray, 1903.

Neison, Edmund. *The Moon and the Conditions and Configurations of Its Surface.* London: Longmans, Green, and Co., 1876.

Nesvold, Erika. *Off-Earth: Ethical Questions and Quandaries for Living in Outer Space.* Cambridge, MA: MIT Press, 2023.

Neufeld, Michael J. *The Rocket and the Reich: Peenemünde and the Coming of the Ballistic Missile Era.* New York: The Free Press, 1993.

———. *Von Braun: Dreamer of Space, Engineer of War.* New York: Knopf, 2007.

Nicolson, Marjorie Hope. *Voyages to the Moon.* New York: Macmillan, 1948.

North, Gerald. *Observing the Moon: The Modern Astronomer's Guide.* Cambridge: Cambridge University Press, 2007.

Panek, Richard. *Seeing and Believing: How the Telescope Opened Our Eyes and Minds to the Heavens.* New York: Viking, 1998.

Peterson, Michael B. *Missiles for the Fatherland: Peenemünde, National Socialism, and the V-2 Missile.* Cambridge: Cambridge University Press, 2009.

Phillips, Robert, ed. *Moonstruck: An Anthology of Lunar Poetry.* New York: Vanguard Press, 1974.

Phinney, William C. *Science Training History of the Apollo Astronauts.* Washington, DC: NASA. NASA/SP-2015-626.

Price, Fred W. *The Moon Observer's Handbook.* Cambridge: Cambridge University Press, 1988.

Rappaport, Margaret Boone, and Konrad Szocik, eds. *The Human Factor in the Settlement of the Moon: An Interdisciplinary Approach.* Cham, Switzerland: Springer, 2021.

Rowland, Wade. *Galileo's Mistake: A New Look at the Epic Confrontation between Galileo and the Church.* New York: Arcade, 2011.

Scott, David, and Alexei Leonov. *Two Sides of the Moon: Our Story of the Cold War Space Race*. New York: St. Martin's Press/Thomas Dunne Books, 2004.

Scott, David Meerman, and Richard Jurek. *Marketing the Moon: The Selling of the Apollo Program*. Cambridge, MA: MIT Press, 2014.

Sears, Derek W. G. *Gerard P. Kuiper and the Rise of Modern Planetary Science*. Tucson: University of Arizona Press, 2019.

Sevigny, Melissa L. *Under Desert Skies: How Tucson Mapped the Way to the Moon and Planets*. Tucson: University of Arizona Press, 2016.

Sheehan, William P., and Thomas A. Dobbins. *Epic Moon: A History of Lunar Exploration in the Age of the Telescope*. Richmond, VA: Willmann-Bell, Inc., 2001. One of the four indispensable books to those interested in the history of the Moon, this book is meticulously researched, profusely illustrated, and wonderfully written. It should be read by all lunar scientists and explorers. I could not have written this book without *Epic Moon*.

Simmons, Gene. *On the Moon with Apollo 15: A Guidebook to Hadley Rille and the Apennine Mountains*. Washington, DC: NASA, 1971.

Smith, Andrew. *Moondust: In Search of the Men Who Fell to Earth*. New York: HarperCollins, 2005.

Sprung, Jeffrey V., and William T. Farley Jr. *Apollo 11: Man's Greatest Adventure*. New York: American Broadcasting Company, 1969.

Spudis, Paul D. *The Value of the Moon: How to Explore, Live and Prosper in Space Using the Moon's Resources*. Washington, DC: Smithsonian Books, 2016.

Swanson, Glen E., ed. *"Before This Decade is Out . . .": Personal Reflections on the Apollo Program*. Gainesville: University Press of Florida, 2002.

Vookles, Laura, and Bartholomew F. Bland. *The Color of the Moon: Lunar Painting in American Art*. New York: Fordham University Press, 2019.

Whitaker, Ewen A. *Mapping and Naming the Moon: A History of Lunar Cartography and Nomenclature*. Cambridge: Cambridge University Press, 1999. The last of my most crucial books, this volume hides important and interesting scholarship beneath a dull title.

Wilhelms, Don E. *To a Rocky Moon: A Geologist's History of Lunar Exploration*. Tucson: University of Arizona Press, 1993.

Wilkins, H. Percy, and Patrick Moore. *The Moon: A Complete Description of the Surface of the Moon, Containing the 300-inch Wilkins Lunar Map*. London: Faber and Faber Limited, 1955.

Wlasuk, Peter T. *Observing the Moon*. London: Springer, 2000.

Worden, Al, with Francis French. *Falling to Earth*. Washington, DC: Smithsonian Books, 2012.

Worden, Alfred M. *Hello Earth: Greetings from Endeavor.* Los Angeles: Nash
Publishing, 1974.

Wright, Hamilton, and Helen and Samuel Rapport, eds. *To the Moon: A
Distillation of the Great Writings from Ancient Legend to Space Exploration.*
New York: Meredith Press, 1968.

ILLUSTRATION CREDITS

The Copernicus crater as seen from Lunar Orbiter 2 in 1966. *Credit: NASA and Lunar Orbiter Imagery Recovery Project.*

The Abri Blanchard bone. *Credit: Wikipedia.*

Fremont culture petroglyph. *Credit: Christopher Cokinos.*

Title page for Henry Cornelius Agrippa's 16th book of magic. *Credit: Wikipedia.*

Chang'e and her rabbit on the Moon. *Credit: Woodcut print by Tsukioka Yoshitoshi.*

Aztec depiction of a rabbit on the Moon. *Credit: Library of Congress/Florentine Codex.*

Jan Van Eyck's *The Crucifixion*. *Credit: Wikipedia/Metropolitan Museum of Art, New York.*

Portrait of Galileo by Justus Sustermans. *Credit: Wikipedia/National Maritime Museum, Greenwich, London.*

Replica of Galileo's telescope. *Credit: Christopher Cokinos.*

Drawing of the magnified lunar surface made by Galileo. *Credit: The Starry Messenger.*

Lunar phase map, *Selenographia* of Hevelius. *Credit: Christopher Cokinos/University of Arizona Special Collections.*

Scenes from a lively Moon. *Credit: Benjamin Henry Day*

William Henry Pickering. *Credit: Library of Congress.*

The gardens of Eratosthenes. *Credit: Christopher Cokinos/W. H. Pickering,* "Eratosthenes I, a study for the amateur," *Popular Astronomy.*

Plaster-of-Paris model of the Moon. *Credit: Christopher Cokinos photograph from Nasmyth and Carpenter.*

Ukert and Mare Vaporum. *Credit: Lunar Orbiter 3, NASA.*

Tycho. *Credit: NASA/ Lunar Reconnaissance Orbiter.*

The "fission" theory of the Moon's birth. *Credit: The Moon, Virgilio Brenna, New York: Golden Press, 1963.*

Astronomer Barbara Middlehurst. *Credit: American Institute of Physics/AIP Emilio Segrè Visual Archives, John Irwin Slide Collection.*

Astronomer Winifred Cameron. *Credit: Christopher Cokinos/original NASA photograph courtesy of Selene Green.*

Wernher von Braun. *Credit: Wikipedia.*

The Mittelwerk. *Credit: Wikimedia.*

Wernher von Braun's desk. *Credit: Christopher Cokinos.*

Arthur Rudolph's glasses. *Credit: Christopher Cokinos.*

Earthrise. Credit: NASA.

Chesley Bonestell-inspired Apollo art. *Credit: NASA.*

The Moon by Lucien Rudaux. *Credit: Sur les autres planètes, Conférence faite au Palais de la Découverte le 10 Avril 1943/University of Paris.*

Dave Scott and James Irwin at Hadley Rille. *Credit: NASA.*

The Lunar Module *Falcon. Credit: NASA.*

The Space Launch System. *Credit: NASA.*

Orion spacecraft. *Credit: NASA.*

Mount Wilson 60-inch telescope at night. *Credit: Christopher Cokinos.*

Author with the 60-inch telescope. *Credit: David Hasenauer.*

Region through the Alpine Valley through the 60-inch telescope. *Credit: David Hasenauer.*

Ptolemaeus, Alphonsus, and Arzachel. *Credit: David Hasenauer.*

Malapert Massif. *Credit: NASA.*

INDEX

ABOUT THE AUTHOR

Christopher Cokinos is the author or coeditor of several books, including *The Fallen Sky: An Intimate History of Shooting Stars, Hope Is the Things with Feathers: A Personal Chronicle of Vanished Birds,* and *Beyond Earth's Edge: The Poetry of Spaceflight.* He is the winner of awards and fellowships from, among others, New American Press, the Whiting Foundation, the Rachel Carson Center in Munich, and the National Science Foundation. His poems, articles, and essays have appeared in such venues as *Scientific American, High Country News, Astronomy, Discover.com,* and the *Los Angeles Times.* Having taught literature, writing, and science communication for more than three decades at three universities, he again lives and writes in Utah. His website is www.christophercokinos.com.

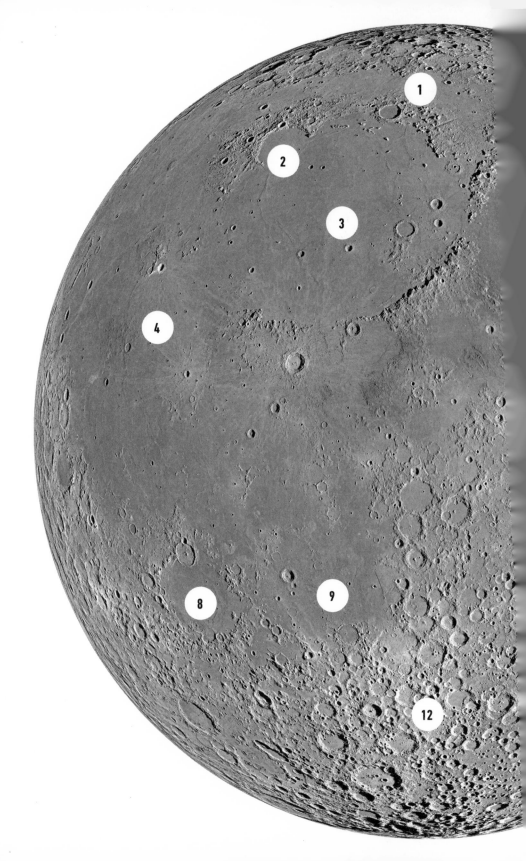